Functional Biochemistry of the Neuroglia

Functional Biochemistry of the Neuroglia

Leonid Z. Pevzner

I. P. Pavlov Institute of Physiology
Academy of Sciences of the U.S.S.R.
Leningrad, U.S.S.R.

Translation Editor
Brian Tiplady

Astra Clinical Research Unit
Edinburgh, United Kingdom

Consultants Bureau · New York and London

Library of Congress Cataloging in Publication Data

Pevzner, Leonid Zalmanovich.
    Functional biochemistry of the neuroglia.

    Translation of Funktsional'naia biokhimiia neïroglii.
    Bibliography: p.
    Includes index.
    1. Neuroglia. 2. Neurochemistry. I. Title. [DNLM: 1. Neuroglia — Metabolism. 2.
Neuroglia — Physiology. WL102.3 P514f]
    QP363.2.P4813   599'.01'88   78-26386
    ISBN-13: 978-1-4684-1637-4   e-ISBN-13: 978-1-4684-1635-0
    DOI: 10.1007/978-1-4684-1635-0

The original Russian text, published by Nauka in Leningrad in 1972, has been expanded and revised by the author for the present edition. This translation is published under an agreement with the Copyright Agency of the USSR (VAAP).

ФУНКЦИОНАЛЬНАЯ БИОХИМИЯ НЕЙРОГЛИИ

FUNKTSIONAL'NAYA BIOKHIMIYA NEIROGLII
Leonid Z. Pevzner

© 1979 Consultants Bureau, New York
Softcover reprint of the hardcover 1st edition 1979
A Division of Plenum Publishing Corporation
227 West 17th Street, New York, N.Y. 10011

TRANSLATOR'S NOTE

The Pavlov Institute is situated in Koltushi, a small village just outside Leningrad. A collection of timber buildings from Pavlov's original establishment, together with one or two larger and more modern buildings, some blocks of flats, and a grocery shop are irregularly distributed in a rather overgrown park.

One of those wooden buildings was intended by Pavlov to serve as accommodation for visiting scientists, and this now houses the Laboratory of Functional Neurochemistry. I worked in this laboratory with Leonid Pevzner, Vera Brumberg and Tanya Glushchenko for six months, on an exchange visit organized by the Royal Society and the Soviet Academy of Sciences.

I learned there to use the techniques of quantitative histochemistry, and to appreciate their usefulness in the range of techniques that have been brought to bear on the problems of neuronal-glial relationships, and of neurochemical effects of behavioral stimulation. Soviet workers have amassed a considerable body of evidence on these topics, and as Leonid Pevzner remarks, this is often not as accessible as it should be to workers in other countries. While some parts of this work have been published in English language journals, a comprehensive review has until now been lacking.

The present edition is not a straightforward translation of the work published in 1972, but contains additional material written in 1976. I have also made some corrections and suggestions, but always with the author's agreement. The abbreviations and conventions of the Journal of Neurochemistry have in general been used throughout. Russian names are transliterated according to the

British—American system used by Plenum Press. Where this leads to a discrepancy with the version preferred by the particular author (e.g., Demin, Doemin), I have used the standard system in the text and indexed references according to the spelling actually used in non-Russian publications. For non-Russian authors who have published in Russian, I have used the original form of the name in the text, but indexed Russian publications by the transliterated form.

I wish to express my thanks to N. N. Doemin, the Royal Society and the Soviet Academy of Sciences for making my visit to Leningrad possible, to Plenum Press for their interest in the work, and to JoAnn Frielink, for preparing the manuscript.

Brian Tiplady
Edinburgh, June 1978

FOREWORD TO THE AMERICAN EDITION

The dialectic of the development of modern science may be demonstrated as follows. On the one hand, science is constantly becoming more international. An ever greater number of countries is being drawn into the exponential growth of scientific research, and thus of publication. But, on the other hand, this growth and dispersal of scientific investigation over our multilingual planet leads to a situation where results obtained in one country do not always become known in another. This linguistic barrier, reflected in the legend of the tower of Babel, has long troubled mankind. The successful construction of the edifice of modern science is hindered when different research groups, often working in very similar directions, may not know each others works when they are written in different languages. There exist two main methods of surmounting this barrier - the maximum use of international journals and reviews; and widespread translation of the most interesting work from one language to another. The second is appropriate for larger works, particularly monographs.

Both as an author, and as a representative of national science, I am pleased and honored that my monograph has attracted the attention of a publisher as well known in the Soviet Union as Plenum Press. Naturally I do not attribute this to the author's own subjective qualities. The fact that my book was quickly sold out after being published in Leningrad, and that it has now been translated into English (now established as the international language of modern science) is a result of the very great interest aroused by the problem of the neuroglia. The glial cells have been known since the middle of the last century, but were long studied more in connection with pathomorphology than with physiology. Even fewer investigators studied the biochemistry of neuroglia until recently, even though glial cells make up a considerable proportion of any sample of nerve tissue taken for analysis.

It is becoming more and more clear that neither biochemical analysis of homogenates or slices, nor studies on the metabolism of individual neurones isolated from nerve tissue, can provide an understanding of the functional biochemistry of the nervous system. This can be obtained only by taking account of the roles of the neuroglia in neuronal - or more broadly of nervous system - function. The functional-biochemical aspect of this problem was formulated most clearly by Holger Hydén in Sweden in the late fifties. It is no accident that the work of Hydén's group receives a great deal of attention in the pages of this book. Subsequently, various groups have made great contributions to functional biochemistry, for example those of Ezio Giacobini, Otto Sellinger and William Norton (USA), Steven Rose (England), Anders Hamberger (Sweden), Paul Mandel (France), Yasuzo Tsukado (Japan), Antonio Giuditta (Italy), Karl Hemminki (Finland), Zdeněk Lodin (Czechoslovakia), Lubomir Venkov (Bulgaria), and Mieczislaw Wender (Poland).

One manifestation of this interest is the appearance of this book in the English language. Naturally in the years that have elapsed since the original publication in 1972, many new papers have appeared on the subject, and the original text is considerably out of date. Therefore I am very grateful to the publishers for giving me the opportunity to add a considerable amount to the text, which deals with the work that has appeared up to the middle of 1976. This has not necessitated any change in the basic conclusions drawn, indeed a number of suggestions have received added support from work done in the last two to three years.

I wish to express my thanks to the translator of this book, Dr Brian Tiplady. His work on the translation of the book is particularly valued in that Dr Tiplady is himself an established specialist in the biochemistry of the neuroglia, and his interest in the work is reflected in these pages.

My work on the functional biochemistry of the neuroglia began in the Laboratory of Functional Neurochemistry (part of the Pavlov Institute of Physiology of the Academy of Sciences of the USSR) in Leningrad in 1962. Since then a considerable body of experimental evidence has accumulated, much of which is presented in this book. However, it provided the occasion, rather than the cause, of the present monograph. The understanding of such important mechanisms as the participation of the neuroglia in neuronal function cannot be achieved by the work of a single group. Many different kinds of factual data must be obtained, and this involves various groups in different countries using different methodological approaches. The present book arose from a desire to bring together such diverse data, and to analyze them from my own perhaps somewhat subjective

point of view. It is not for the author to judge how successfully
he has dealt with his own problem. However if this book brings the
attention of an international scientific audience to the biochemical
study of the neuroglia, the author may say with a clear conscience
the words of the Roman Consuls at the expiry of their terms of
office:

        feci quod potui
        facient meliora potentes

                Leonid Z. Pevzner
                Leningrad, September 1976

CONTENTS

Translator's Note v

Foreword to the American Edition vii

Introduction 1

Chapter One. Prerequisites for a Functional-Biochemical Study
of the Neuroglia 5
    Morphological Data 6
    Electrophysiological Data 11

Chapter Two. Methods of Chemical Analysis of Neuroglia . . 15
    Topochemical Analysis 15
    Study of Pathological Material 16
    Enriched Fractions 18
    Micromanipulation 26
    Quantitative Cytochemistry In Situ 32

Chapter Three. Biochemical Properties of the Neuroglia . . 39
    Oxidative Processes 41
    Oxidative Phosphorylation and ATPase Activity 64
    Carbohydrate Metabolism 68
    Lipid Metabolism 70
    Mineral Composition 76
    Amino Acid Metabolism 79
    Neurotransmitter Metabolism 84
    Possible Enzyme Markers and Isoenzyme Ratios 90
    Protein Metabolism 93

Nucleic Acid Metabolism 106
    DNA content 106
    DNA turnover 109
    Changes in DNA metabolism correlated with changes
        in the functional state of the nervous system . 110
    RNA content 111
    RNA composition 113
    RNA turnover 115
RNA Metabolism as a Function of Changes in the Functional
    State of the Nervous System 116
Problems of Our Own Studies 120

Chapter Four.   Conditions of Cytospectrophotometric
Determination 125

    General Principles 125
    Object of Study 127
    Fixation and Histological Treatment 128
    Ultraviolet Cytospectrophotometry 128
    Visible Cytophotometry 130
    Calculation of Cell Volume 132

Chapter Five.   Investigation of Several Methodological
Questions 135

    Thickness of the Histological Sections 135
    Selective Extraction of RNA from Slices of Nerve
        Tissue 138
    Comparison of Cytophotometry in Visible Light and
        UV-cytophotometry 142
    Comparison of UV-cytospectrophotometry and
        Cytointerferometry 144

Chapter Six.   Nucleic Acids in the Neurone-Neuroglia Unit with
Differing Functional States of the Nervous System 147

    Circadian Variations of RNA and Protein Content . . . 147
    Motor Activity 152
    Electrostimulation 157
    Convulsive States 164
    Deprivation of Normal Sleep 170
    Forced Hypokinesia 174
    Acute and Chronic Hypoxia 176
    Cold Adaption 183
    Effects of Epinephrine 190
    Adrenalectomy and Substitution Therapy with
        Hydrocortisone 194
    Effects of Inhibitors of Nucleic Acid Biosynthesis . 198

Chapter Seven.   The Role of the Neuroglia in Neuronal Function:
A General Summary of the Experimental Evidence   201

    Scheme of Inter-relationships of Metabolism of Nucleic
      Acid and Protein in the Neurone-Neuroglia Unit . . .   205
    Neuroglia during Adequate Excitation of the Neurone . .   206
    Neuroglia during Prolonged or Sharp Excitation of
      the Neurone   208
    The Question of Transfer of RNA from Glia to Neurones .   209
    Neuroglia during Fatigue of the Neurones. Neuroglia as
      the Site of Trophic Influences on the Nervous System   213
    Neuroglia during Restorative Processes in the Nervous
      System   216
    Neuroglia during Secondary Changes in the Nervous
      System Metabolism   218
    The Neurone-Neuroglia Unit   219

Summary   227

References   231

INTRODUCTION

The first issue of the Sechenov Journal of Physiology began with an article by George E. Vladimirov entitled "Functional Biochemistry of the Brain - some conclusions and perspectives". Vladimirov wrote (p. 4):

> While there has been great success in the study of physiological aspects of higher nervous activity, we tend to ignore the physico-chemical nature of such fundamental physiological phenomena as the processes of excitation and inhibition in the nervous system. Elucidation of the chemical bases of these phenomena, and understanding of the spatio-temporal details of their functioning must be the aim of functional biochemistry of the brain, a subject of growing interest in our country.

And later, considering the difficulties of investigation of the brain by conventional biochemical methods, Vladimirov remarked, "In different samples of the brain taken for analysis, variation may occur in such factors as the structure of the cellular layers of the grey matter, the <u>ratio</u> <u>between</u> <u>neuronal</u> <u>and</u> <u>glial</u> <u>elements</u>, (emphasis mine, L.P.) and the proportion of intercellular material. In light of this, it is impossible to expect reproducibility of chemical analysis (p. 5)."

These lines were written at the end of 1952, that is several years before electron microscopic study of the central nervous system, and before the appearance of the first of the now classical papers of Hydén on the analysis of the interrelationships between neurones and glia.

1

In subsequent years, an enormous amount of material has been obtained, notably in the laboratories of A. V. Palladin, G. E. Vladimirov, and E. M. Kreps in the Soviet Union and by many neuro-chemists in other countries, laying the foundation of a true functional biochemistry of the brain. From the point of view of this science the unity of metabolism, structure and function in the living system is taken for granted. The recognition of this unity is the most important demand made of the functional neuro-chemist. Therefore, structureless tissue homogenates are more and more being replaced as the objects of such study by separated cells or by groups of cells forming morphological and functional units. Such preparations allow much greater preservation of the structure found in the living organism. In highly differentiated tissues, the basic cells which carry out the function characteristic of the tissue are surrounded by auxiliary cells. The role of these cell-satellites appears to be maintenance of the basic cells supplying them with necessary biochemical components, removing waste meta-bolites, and buffering the cells against harmful external influences. Examples of such associations are the myocyte-pericyte system in heart muscle, hepatocyte-Kupffer cell in the liver and oocyte-trophocyte in the ovary. We regard the neurone-neuroglia unit as being of this type.

It is no accident then, that the study of the neuroglia, which in the past had a primarily practical, medical character, has more and more become a general biological problem. Whereas previously the neuroglia were considered to play a primary role only in such pathological processes as edema, scar formation, and tumor growth, the data obtained in recent years indicate the participation of the glia also in the normal processes of neuronal function.

One of the difficulties in the study of neuroglial cells is that the glia do not consist of a single homogeneous population. First of all there is the distinction between macro- and microglia. The latter seem to have a mesenchymal origin, and participate mainly in the phagocytic reaction of nervous tissue. The macroglia, on the other hand, originate, like the neurones, from the neural tube, hence the name "neuroglia". There are two kinds of macroglia, astrocytes, and oligodendrocytes. Doubtless there are many bio-chemical differences between these two cell types (and possibly within each cell type also).

A functional biochemical study of the neurone-neuroglia unit must include a comparison of changes occurring in neurones and glia under different functional conditions of the nervous system. Such comparisons, both under physiological conditions (such as the transition of neurones from rest to excitation, from excitation to fatigue, from fatigue to rest, from excitation to inhibition and vice versa), and also under various extreme or pathological conditions, may give a basis for conclusions about the biochemical characteristics of glial cells compared to the neurones, and about the role of the neuroglia in neuronal function.

As will be clear from the subsequent account, such studies, comparing metabolic shifts in the neurone-neuroglia unit resulting from well-defined dynamic functional changes in the nervous system are few in number. Therefore we consider it possible to review all the data throwing light on the biochemical characteristics of the glial cells, constantly comparing them with neuronal characteristics, and not to limit ourselves to a consideration of functional biochemical changes in the neurone-neuroglia unit.

At the same time the emphasis will of course be on such data concerning the biochemistry of the neuroglia as seems relevant to functionally conditioned changes of metabolism in the neurone-neuroglia unit. In particular, our own studies on nucleic acid and protein metabolism in neuronal and glial cells of various parts of the nervous system have taken this approach.

Considering both our own data and other data in the literature we are able to present a general picture of the participation of the neuroglia in the functioning of the neurone. This is described in the concluding part of the present book. This picture is based on very early results of the biochemical and cytochemical analysis of neurones and glia. Subsequent work will show whether our representation of the role of the neuroglia in the functioning of the neurone is broadly correct or not. In any case our attempt at systematization of contemporary data on functional biochemistry of the neuroglia should help future investigators to formulate the problem more clearly. The elucidation of neurone-neuroglia relationships is vital for our understanding of specific functions of the nervous system.

CHAPTER 1

PREREQUISITES FOR A

FUNCTIONAL-BIOCHEMICAL STUDY

OF THE NEUROGLIA

One of the characteristic features of nervous tissue is that its principal functional structure - the neurone - occupies only a small proportion of its total volume (see reviews by Kuhlenbeck, 1970; Peters et al., 1970; Palay and Chan-Palay, 1973). Therefore, with the development of more sensitive methods of biochemical analysis, separated neurones are tending to be used as well as whole tissue homogenates both for straightforward biochemical analysis and for studies of metabolic changes resulting from changes in the functional state of the nervous system.

The pioneers in this field were Oliver Lowry in the USA (Lowry, 1941; Lowry and Bessey, 1946; Lowry et al., 1951) and Holger Hydén in Sweden (Hydén, 1943, 1947, 1955), whose works are both numerous and well known. Using microchemical and quantitative cytochemical techniques, they and their co-workers and, later on, a number of other neurochemists obtained data which made it possible to draw preliminary conclusions concerning the basic functional states of the nerve cell - excitation and inhibition. (For reviews, see: Hydén, 1960, 1962, 1964, 1972; Pevzner, 1963a, 1966a, 1966b; Brodskii, 1966; Gaito, 1969; Jakoubek and Semiginovský, 1970; Geinisman, 1974; Jakoubek, 1974).

However, it soon became clear that the study of the biochemical properties of a single cell type was insufficient for an understanding of nervous system function. Hydén (1959a,b) first made the fruitful suggestion that definite biochemical links exist between the neurones and the surrounding glial cell-satellites, and that the neurone-neuroglia system forms a functional unit. Morphological, pathological, and electrophysiological data also point to the importance of the neuroglia in nervous system function. These

data, which will be described briefly below, are essential to any
functional-biochemical study of the neuroglia.

Morphological Data

    Glial cells originate in a similar way to neurones. During
embryonic development both neuroblasts and glioblasts develop from
the medulloblasts (or matrix cells), the precursor cells of nervous
tissue (Fig. 1). The former develop into neurones, the latter into
glia - astrocytes and oligodendrocytes (Wechsler and Kleihues, 1968;
Pyl'dvere, 1971).

    During the early stages of phylogenesis, glial development
parallels that of the neurones. Glia are clearly absent only in
coelenterates, being present in the ciliary worms (<u>Turbellaria</u>)
(Oksche, 1968; Kuhlenbeck, 1970).

    At the same time, a number of criteria indicate that glial cells
are less differentiated than neurones. They divide actively until
the birth of the animal, and some continue to proliferate well after
birth. In the cat, for example, the number of glial cells reaches
a maximum at the beginning of the second month of life (Fleischauer,
1968). Experiments with labelled thymidine showed that in the rat
glial cells continue to proliferate until at least 20 weeks of age
(Haas et al., 1970), and in specific regions of the brain this continues even in old animals (Dalton et al., 1968).

    As would be expected, glial cells show considerably more
activity than neurones in tissue culture, surviving better <u>in vitro</u>,
and showing directed movements and energetic pulsations (Pomerat
et al., 1967; Lodin et al., 1970; Svanidze et al., 1973). The
character and rate of this pulsation is considerably influenced by

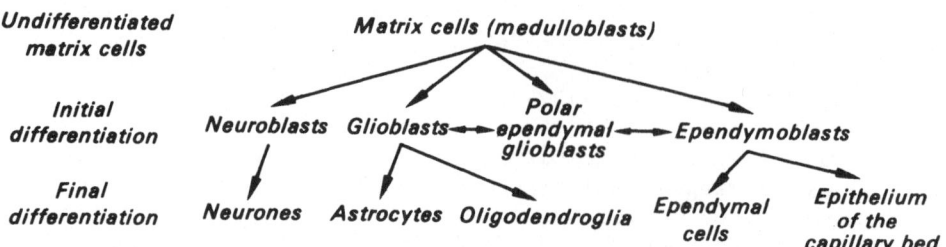

Fig. 1. Scheme for development of new cell types in the
nervous system.

such agents as serotonin, adrenalin, and eserine (Geiger, 1963). The motor activity of glial cells in tissue culture is also increased by a raised concentration of K$^+$ ions in the medium (Svanidze et al., 1973).

The sensitivity of glial cells to toxins, infection, trauma, inflammation, tumor growth, and other pathological states of the nervous system is well established (Penfield, 1932; Snesarev, 1959; Nikulesku, 1963; Colmant, 1968).

An important role is also ascribed to the neuroglia in the development of such pathological states as atherosclerosis (Gorizontov, 1940), and psychosis (Aleksandrovskaya, 1950).

Morphological and cytochemical studies show that a glial response also occurs during recovery from mechanical damage to the nervous system (Sjöstrand, 1965; Reznikov, 1968; Guth and Watson, 1968; G"l"bov, 1969).

The oligodendroglia are responsible for myelin formation, the outer membrane and cytoplasm forming the myelin sheath of the axon (Geren, 1954; Lusen, 1956a; DeRobertis et al., 1958; Korey, 1960; Caley and Butler, 1974; Dermietzel, 1974. For reviews, see: Adams and Davison, 1965; Johnston and Roots, 1970; Knobler and Stempak, 1972; Peters et al., 1970).

The possible participation of the neuroglia in the blood-brain barrier has been widely discussed. The classical morphological picture of neurones surrounded by glial cells led to the idea that the neuroglia mediate between neurones and the capillary bed of brain tissue (Glees, 1955; Cammermeyer, 1960a; Polak, 1965). Cammermeyer (1960a), moreover, established a direct correlation between the degree of vascularity and the density of glial cells in different brain regions.

Electron microscope studies have shown that neurones are almost nowhere contiguous with the basement membrane of brain capillaries. Neuroglial elements always intervene between these structures (Kuhlenbeck, 1970; Peters et al., 1970; Manina, 1971; Palay and Chan-Palay, 1973). Electron microscopy has refuted the idea of a structureless neuropil and shown that the space between neurones is filled with glial cell bodies and processes (Luse, 1956b, 1960; Schultz et al., 1957; Hesse, 1968). The true extracellular space of brain tissue consists of a narrow cleft between cell membranes which does not exceed 10-15 nm (Nakai, 1963; Kaplan, 1965; Sarkisov and Bogolepov, 1967; Kuhlenbeck, 1970; Peters et al., 1970; Manina, 1971; Glees, 1973). On this basis it has been calculated that the volume of the extracellular space is not more than 5% of the volume of the brain tissue (Horstmann and Meves, 1959).

   On the other hand, calculations of the size of the extracellular space from data obtained using substances which are
generally considered not to penetrate cells (e.g. sucrose, inulin,
chloride, aminohippuric acid, and thiocyanate) give values varying
from 2-4% to 30-32%, while in edema, the thiocyanate or chloride
space can double (Barlow et al., 1961; Streicher et al., 1964).
A realistic value, taking into account the definition of extracellular space in the earlier and later work, is probably 12-15%
(Dawson and Spaziani,1959; Dawson and Bradbury, 1965; Woodward
et al., 1967; Goodman et al., 1973). This is three times greater
than the volume accounted for by the intercellular clefts, and the
question thus arises whether the neuroglial cells are, in part at
least, analogous to the extracellular space of other tissues. A
number of investigators are inclined to this view. For example, the
electrolyte composition of glial cells (in contrast to that of
neurones) is more similar to extracellular than to intracellular
fluid (Gerschenfeld et al., 1959; Torak et al., 1959; Katzman, 1961;
Koch et al., 1962; Patel et al., 1971). Gerschenfeld et al. (1959)
induced edema of nervous tissue in rats and rabbits. Electron
microscopy, however, revealed neither swelling of the neuronal cell
bodies, nor enlargement of the surrounding extracellular space. In
all cases the swelling was confined to the astrocyte cell bodies.
Analogous selective swelling of the neuroglial cells during acute
edema has been repeatedly confirmed by other authors. (For reviews,
see: DeRobertis and Gerschenfeld, 1961; Ule, 1968; Rechardt, 1969;
Ruscakova, 1969; Torack et al., 1970; Baethmann and Van Harreveld,
1973).

   Much interest has been aroused by the observation that almost
the entire surface of the brain capillaries is closely contiguous
with a special kind of astrocytic process, the vascular foot. This
observation has been confirmed by electron microscopy (for reviews,
see: Lierse, 1968; Friede, 1970). The neuroglial cells may be
thought of as forming a natural barrier between the neurones and the
vascular bed; not simply passing various substances from the
capillary to the neurone by passive transport, but actively regulating this intercellular transport (DeRobertis et al., 1960; Glees,
1963; Zadunaisky et al., 1965). On the basis of morphological and
histochemical data, many workers support the idea of DeRobertis
(DeRobertis et al., 1960; DeRobertis and Gerschenfeld, 1961) that
the neuroglia in many respects represent the site of the blood-brain
barrier (Quastel and Quastel, 1961; Coxon, 1964; Aleksandrovskaya,
1965; Vernadakis and Woodbury, 1965; Dimova et al., 1966; Lord
et al., 1967; Vernadakis, 1975).

   All these results and the conclusions drawn from them must be
viewed critically, since the work cited above depends for the most
part on visual observations, which are not always sufficiently
reproducible, and which are not strictly quantitative. There are
also data which are not in agreement with these conclusions, for

example work which suggests that apart from the neuroglia, the myelin sheath or the dendrites could fulfill the functions of extracellular space in nervous tissue (Lumsden, 1955; Lajtha, 1957). In her study of edema of brain tissue in vivo and in vitro, Pappius (1965) came to the conclusion that edema could be only partly explained by acute swelling of neuroglial cells. Lasansky (1965) studied the uptake of ferricyanide by toad retina in vitro, and showed that the ferricyanide penetrated to the neurones through the intercellular clefts, not through the glial cytoplasm. In contrast to Hertz (1968), who considered that the high rate of diffusion of ions and amino acids which he obtained in brain slices could not be accounted for by diffusion through the narrow intercellular clefts, Nicholls and Kuffler (Nicholls and Kuffler, 1964, 1965; Kuffler and Nicholls, 1965, 1966; Kuffler, 1967) have defended the view that the basic flow of ions (Na^+, K^+) into the nerve cells is through the intercellular canals, not through the glial cell bodies.

The thorny question of the interrelationships between neuroglial cells and the extracellular space of brain tissue is discussed in the reviews of Robin (1965), Soëtens (1968), Appel (1968), and Katzman (1970).

However, this discussion concerns details of the present problem, and by no means challenges the general conception that the neuroglia are actively involved in specific neuronal functions. This conception, moreover, is supported by morphological studies which show marked structural changes in glial cell satellites as a result of changes in the functional state of neurones. Thus Kuntz and Sulkin (1947) first noted that prolonged electrical stimulation of the superior cervical sympathetic ganglion in cat and dog resulted in increased numbers of both capsular and interstitial oligodendroglia; some of the glial cells underwent amitotic division. Aleksandrovskaya et al. (1964) found that daily administration of chlorpromazine for 7-14 days caused a marked increase in the number of neuroglial cells in rat cerebral cortex. A similar increase was also observed in the anterior horn of rat spinal cord after swimming. This effect was already detectable by 20 min (Aleksandrovskaya et al., 1965). As the experimental period was short, and cell division was not observed, a functionally dependent migration of glial cells toward the neurones is postulated. The same authors later showed that various kinds of stimulation (e.g. light, mechanical, electrical) increased the number of perineuronal glia without changing the overall number of glial cells (Aleksandrovskaya et al., 1968; Aleksandrovskaya and Chizhenkova, 1973).

Lodin et al. (1968) stimulated the vestibular apparatus by infusing water at 50°C into the ear for one hour, and showed an accumulation of perineuronal glia in the Purkyně cell layer of the cerebellum.

Brazovskaya (1968) demonstrated an increase in the mean number of glial cells surrounding the neurones of layer V of rat motor cortex following two months in the dark. The published data lead her to conclude that prolonged visual deafferentation results in a compensatory activation of motor neurones. It is interesting to note that under these conditions an even greater accummulation of pericapillary glia occurred.

In rats reared in total darkness from birth to 8 weeks, an increase in the number of perineuronal glial cells in visual cortex occurred (Busnyuk and Pigareva, 1974). For a review and critical discussion of the literature on functionally conditional changes in glial cell number, the reader is referred to the monograph of Geinisman (1974).

We now turn our attention to the work carried out over many years by Rosenzweig and his co-workers (Krech et al., 1960; Diamond et al., 1964, 1966; for reviews, see: Rosenzweig, 1970; Rosenzweig et al., 1972). Young rats were raised for 80 days in an enriched environment with stairs, wheels, swings, shelves etc. in a group cage. Control rats of the same age were raised for the same period in individual cages with a minimum of external stimulation. In the experimental rats, the cerebral cortex was markedly thicker than in the controls, the effect occurring chiefly in layers II and III, and there was also a general increase in the number of glial cells. It is, of course, possible to interpret this experiment in the opposite sense, and regard the experimental animals as the normals, and the control animals as being subjected to partial sensory deprivation, which is equivalent to deafferentation. The effect of this deafferentation would then appear as a reduction in glial cell number, and a reduction in the thickness of the cerebral cortex. In either case, however, the data demonstrate the dependence of the glial cell population on the specific function of the nervous system. More detailed morphological analysis, carried out by an Australian group (Walsh et al., 1969) showed an even greater increase in thickness of the hippocampus (5.7% as compared to 4.9% for visual cortex). The number of neurones and microglia of hippocampus was unchanged, while the number of oligodendroglia was increased by 39%. The number of astrocytes, on the other hand, declined slightly (by 9.8%). A sharp increase (more than 3.5 times) was noted in the number of cells classified by the authors as intermediate between astrocytes and oligodendroglia.

In other work, marked changes were observed in the sizes of glial cells of the spinal motor region as a result of strenuous motor exercise (Kuhlenkampff and Wüstenfeld, 1954; Mats et al., 1970). During electrical skin stimulation in rats (leading first to a sharp excitation, and then to fatigue of the animals) and subsequent recovery, Khaidarliu (1967a) showed a corresponding series of changes in the volume of the glial cell-sattelites of the sensory spinal ganglia, and to a lesser extent of the anterior horn of the spinal cord.

Our own studies, using automatic high speed microscopic analysis (Ivanitskii et al., 1967) have shown that daily administration of adrenalin for 2 weeks results in an increase in the volume of those glial cells adjacent to the neuronal cell membrane (Pevzner and Litinskaya, 1968).

Differences in the volume changes of glial cells in spinal cord, dependent on the type and function of the adjacent neurone have also been shown by Brumberg (1969) in hyper- and hypokinetic mice, and by Piven (1973) for hypothermic rats. Finally, even without the interference of the investigator, there are marked changes in glial cell volume in various regions of rat nervous system associated with the circadian variation of its functional state. Such variations may coincide with the variations of volume of the corresponding neurones, or be in the opposite direction (Pevzner et al., 1973, 1974).

It is interesting that only at certain stages of electro-stimulation do reductions occur simultaneously in the dimensions of neurones and their glial cell-satellites in the anterior horn of the spinal cord (Mats et al., 1970).

Hypertrophy of the neurone following axonal trauma (Pannese, 1964), hypothermia and circulatory hypoxia (Collewijn and Schade, 1965), inhibition of neuronal activity by a high frequency electro-magnetic field (Aleksandrovskaya, 1969), and cardiazole (pentylene-tetrazole) convulsions (DeRobertis et al., 1969) all produce swelling, an increase in intensity of staining, and other morpho-logical signs of an increased activity of the neuroglial cells. Similar changes have been shown to occur in the neuroglia of cerebral neurosecretory nuclei during intense neurosecretory activity (Polenov, 1968; Yurisova, 1970).

Electrophysiological Data

The first electrophysiological studies of the neuroglia were carried out by Tasaki and co-workers on cultured nervous tissue (Hild et al., 1958; Chang and Hild, 1959; Hild and Tasaki, 1962). A resting potential similar to that of neurones was demonstrated in the neuroglia. However, the frequency of glial biopotentials was two orders of magnitude less. Single electrical pulses (1.5 msec in duration and 15-20 μA in intensity) produced a reduction of the glial resting potential only after an interval of at least 5 seconds. No increase in the frequency of the glial response could be obtained. It is important to note that the electrical resistance of the glial membrane is in the range 3-10 ohm/cm^2, compared to a value for neurones of 400-2000 ohm/cm^2, so it can be assumed that the current arising from the neuronal action potential passes through the glial cell membrane and not through the intercellular space.

This electrical activity of the glia does not occur only in tissue culture. In frog, 4-7 days after motor nerve transection there was disappearance of axonal terminals. Nevertheless the formation and release of acetylcholine quanta continued due to Schwann cells being in contact with muscle fibers; the resulting membrane potential was characterized by a constancy with respect to a number of external factors (Birks et al., 1960). Studies of the resting rhythms of the cerebrum (Sokolov, 1962) and of medulla (Aladzhalova and Kol'tsova, 1964) show slow (3-6 Hz) and ultra-slow (2-3 and 5 oscillations/minute) rhythmic changes in the electrical activity of the brain localized in regions where neuro-glial cells predominate (these authors show that in the cerebral cortex, for example, the frequency of these oscillations at rest is on average 14 Hz).

The membrane potentials of the astrocytes and oligodendroglia of visual nerve are approximately equal (77-85 mV). Action potentials are not observed in these cells. The size of the membrane potential depends on the concentration gradient of K$^+$, but not on that of Na$^+$ or Cl$^-$ (Dennis and Gerschenfeld, 1969).

In the experiments of Grossman et al. (1969), depolarization of cortical neuroglia of cat began within 10 msec of the beginning of cortical activity induced by thalamic stimulation. The depolari-zation reached its maximum value of 0.5 mV after 30-40 msec, and lasted in all for 90-150 msec. The timing of the glial changes coincides with the potential changes in the whole cortex leading the authors to conclude that the glia are involved in neuronal function.

The time constant of glial cell membranes of the sensorimotor and association areas of cat cerebral cortex has been measured (Trachtenberg and Pollen, 1970). The value of 385 msec is more than 20 times slower than that of the giant pyramidal neurones of layer V of the motor cortex (Betz cells). The specific resistance of the glial cell membrane was 200-500 ohm/cm^2. This is rather higher than that obtained by Tasaki from culture cells, but still much lower than that of neurones. These authors consider the neuroglia to be an ideal buffer system, preventing excessive increases of K$^+$ in the extracellular space surrounding the areas of synaptic contact during synaptic spike activity, and protecting the neurones from the effects of such activity in neighboring regions.

Laborit and Laborit (1965) show that the slow waves which are seen on the electroencephalogram (EEG) during deep sleep have a glial origin, and are replaced by a strengthening of bioelectrical

activity (the paradoxical phase of sleep) when the neurones are
activated by Na$^+$ ions diffusing from the neuroglia. Orkand et al.
(1966) consider that both under in vivo and in vitro conditions
electrostimulation of the optic nerve of frog leads to a depolarization of the glia, as indicated by the release of K$^+$ from the
axons and its accumulation in the intercellular space, which should
lead to a lowering of the glial cell membrane potential. Higashida
et al. (1971) have described changes in glial membrane potentials
in rabbit cerebral cortex during spreading depression by local
application of KCl to the surface of the cortex.

Roitbak's studies (1963, 1965, 1968, 1970) have substantiated
the glial origin of the slow, surface-negative potentials of the
cortex, which have a duration measured in hundreds of milliseconds.
He considers the glia to be the site of action of a number of
neuropharmacological agents, in particular analgesics. It is
interesting that the glial depolarization in cerebral cortex
clearly increases both with the frequency and the voltage of tetanic
stimulation (Roitbak and Fanardzhyan, 1973). Roitbak (1969, 1970,
1973) has suggested a mechanism for the formation of these temporary
connections. According to this scheme, a synapse changes from a
"potential" to an "actual" one due to the release of a specific
neuromediator which is triggered by the electrotonic spreading of
the action current. This spreading is in turn connected with the
myelination of the presynaptic terminal. The role of the neuroglia
in myelination is well established, as indicated above. In this
mechanism the formation of a temporary connection is closely
connected with the neuroglia. The action potential spike is the
most obvious and easily demonstrated result of unconditional
stimulation. But another result is the prolonged depolarization of
the glial cell membrane. According to the model, this depolarization
triggers myelin formation, which then provides for more effective
function of the axon terminal and thus for greater release of
neurotransmitter. In this way, a circuit is completed, and the
formation of such circuits is considered an important part of neuroglial function.

This and other neurophysiological data (for detailed reviews,
see: Kuffler and Nicholls, 1966; Roitbak, 1968, 1973; Watson, 1974;
Somjen, 1975; Shelikhov et al., 1975) indicate that glial cells
exert an influence on the most characteristic function of neurones -
bioelectrical activity. It is not surprising that a number of
authoritative neurophysiologists have expressed the view that the
neuroglia play an important role in neuronal function (Galambos,
1961, 1965; Roitbak, 1963, 1964, 1965; Hertz, 1965, 1968;
Svaetichin et al., 1965; Tasaki, 1965; Chernigovskii, 1971).

CHAPTER 2

METHODS OF CHEMICAL ANALYSIS

OF NEUROGLIA

The morphological and physiological data presented in Chapter One indicate the important role played by the neuroglia in specific functions of the nervous system. The main task of functional neurochemistry must be a careful analysis of the molecular mechanisms of neuroglial function. However the possibility of such an analysis is limited by the methods available for biochemical analysis of glial cell composition.

Topochemical Analysis

This method is based on the comparative analysis of tissue samples taken from regions of the CNS which differ in their morphology and in the proportion of neuronal and glial cells they contain.

For example one might compare cerebral cortex with corpus callosum, or gray and white matter from cerebrum or spinal cord (Abood et al., 1952; Palladin, 1959; Friede, 1966). But this is a very indirect way of determining the composition of glial cells. In any region of the CNS there are many structural elements present besides glia and neuronal cell bodies, in particular axons. With the development of methods of microchemical analysis, much data has accumulated which indicates that complex biochemical processes occur in the axons.

Transport of a wide variety of chemical components along the axon has been demonstrated, e.g. amino acids, neurotransmitters, RNA and lipids. Particular attention has been paid to the axonal flow of protein (for reviews, see: Lubinskaya, 1971; Dahlström, 1971; Ochs, 1972, 1974, 1975; Barondes, 1973; Jeffreys and Austin, 1973;

Kristensson and Olsson, 1973; Dahlström et al., 1973; Droz, 1973; Weiss, 1974; Kerkut, 1975; Heslop, 1975).

The proteins migrating along the axons consist of several electrophoretically distinct components. The various components differ in their speed of migration, in their association with particular subcellular fractions, in their sensitivity to inhibitors, and in their time of appearance during postnatal development. Besides this, in situ protein synthesis has been demonstrated both in axons (Koenig, 1972; Ingoglia et al., 1974) and in nerve endings (Appel, 1972; Deanin and Gordon, 1973; Carton and Appel, 1973; Kruglikov et al., 1975). Discussion of the problems of axonal flow and in situ protein synthesis may be found in the reviews by Mezentsev and Messinova (1971), Droz (1973), Glebov (1974) and Heslop (1975).

Gerebtzoff (1966) has obtained interesting data on the iso-enzymes of lactate dehydrogenase in spinal ganglia. In the neuronal cytoplasm the predominant isoenzyme is that most active under aerobic conditions, while the other isoenzyme, which is more active under anaerobic conditions, was found in the axons and dendrites. Wachsmuth et al. (1975) showed by immunochemistry that aldolase A is present in the neuronal cell body, while aldolase B is localized in the neuronal processes as well as in peripheral nerve axons. As Rutter et al. (1963, 1968) showed, isoenzyme A of aldolase mainly breaks down fructose-1,6-diphosphate into two phosphotriose molecules, while the other isoenzyme is involved in gluconeogenesis and fructose biosynthesis. Finally the classical microchemical work of Lowry (Lowry et al., 1954; Lowry, 1955, 1957, 1967) confirms the high rate of metabolic activity in dendrites. Arbogast and Arsenis (1974) calculate that 80-90% of cerebral cortex metabolism occurs in processes, mainly dendrites.

Thus the presence of a considerable proportion of axons and dendrites in nervous tissue complicates the interpretation of data from whole tissue homogenates. The vascularization, and hence the fraction of blood in a tissue sample may also vary significantly between brain regions.

Study of Pathological Material

This approach most commonly involves the study of gliomas, which are tumor growths consisting almost entirely of glial cells. They have been extensively studied in the Soviet Union by Promyslov and his co-workers (Promyslov, 1963, 1966; Promyslov and Andreeva, 1967; Promyslov et al., 1968, 1970).

An analysis of the voluminous biochemical and cytochemical literature on brain tumors leads to the conclusion that the

interpretation of results from glial tumors in terms of normal
glial function is very difficult (Pevzner and Tomina, 1967; Haglid,
1973; Wollemann, 1974).

As is clear from Table I, the DNA content of cell nuclei from
an astrocyte line increases with the degree of malignancy. (In
agreement with Promyslov, we consider astrocytomas and glioblastomas
to be successive stages in the dedifferentiation of astrocytes).
At the same time, the increase in DNA content in medulloblastomas
was considerably less than in either astrocytomas or glioblastomas.
The protein content in the cell nucleus of astrocytomas was markedly
higher than in normal astrocytes. However the increase in the
glioblastomas was less marked than in the astrocytomas.

Table I

DNA and protein content of normal astrocytes
and tumor cells in human brain
(Pevzner, Tomina and Chaika, 1964;
Tomina and Pevzner, 1965)

Type of cell	DNA content (pg per cell)	Protein content (in arbitrary units)	
		per nucleus	per pg DNA
Normal astrocytes	6.2+0.1	57+2	9.2
Astrocytoma	11.4+0.3	81+6	7.1
Multiform glioblastoma	14.5+0.6	69+5	5.6
Medulloblastoma	9.5+0.3	-	-

There is no doubt that the process of tumor development leads
to major changes in the biochemical organization of the proliferating
cells. As tumor growth proceeds, increasing dedifferentiation of
the cells need not be accompanied by further changes in chemical
composition. Therefore it is far from reliable to use the stability
of a given biochemical characteristic in cells at different degrees
of malignancy as a guarantee that one is dealing with a property of
the normal neuroglia.

The study of glial scar tissue (Dimova, 1966; Ruščak et al.,
1968) also demands similar caution as does the use of regions of
the brain from which specific neurones have degenerated, for example
the thalamus (Koch et al., 1962; Cicero et al., 1970) or the lateral

or medial geniculate body (Uttley, 1963; Margolis et al., 1968).
The medial geniculate body has been extensively studied by Uttley
(1963, 1964). Defined regions of the cerebral cortex were removed
in cats, resulting in selective retrograde degeneration of the
neurones in the medial geniculate body. The result, according to
these authors, was a region containing only neuroglial cells.
However, the neuroglia remaining after such degeneration can hardly
be considered normal. In fact, Margolis et al. (1968), using
Uttley's method found a reactive gliosis in the lateral geniculate
body after enucleation and removal of visual cortex in rabbits.

Enriched Fractions

A series of attempts have been made to separate brain tissue
into neuronal and neuroglial fractions on the basis of differences
in size, mass, density, and other physical characteristics. The
methods used have been filtration and centrifugation (by analogy
with the differential centrifugation of subcellular fractions).
From the start it was clear that to obtain pure fractions of
particular cell types would be even more difficult than in the case
of subcellular fractionation. Hence the phrase "enriched fractions"
was coined, indicating fractions with a marked predominance or
increase of concentration of one or other of the cell types under
study.

The first attempts to separate enriched fractions seem to have
been those of Korey and his co-workers (Korey, 1957, 1958; Korey
et al., 1958; Korey and Orchen, 1959). White matter from centrum
ovale of lamb brain was homogenized in 0.25 M sucrose containing
3 mM K_2HPO_4, and filtered through a silk mesh of 400 μm pore size,
then through one of 250 μm. The final filtrate was centrifuged
on a discontinuous sucrose density gradient at 500 g_{AV} for 20 min
and then at 18,400 g_{AV} for 15 min. A tight layer formed between
1.0 and 1.75 M sucrose which contained glial cells, red blood cells
and fragments of the myelin sheath. The yield of glial material
after a 5 h procedure was 18 mg protein from ten lambs.

The authors themselves underline that theirs is the first,
very imperfect, attempt to solve the problem of a separate analysis
of neurones and glia. The main drawback of Korey's method seems to
be that the enriched glial fraction is in fact more a fraction
depleted in neuronal cell bodies; it contains all kinds of glial
cells (including microglia), blood cells, connective tissue, and
myelin fragments. The last, obviously, have their own characteristic,
complex biochemical processes, which have not been adequately studied.
The proportion of glial cells and axonal material in the fraction,
often described by Korey as the "non-neuronal elements" remains
unknown. Korey and his colleagues have been mainly concerned with
glial cells, and their work suffers from a lack of attention to the

nerve cells, since the basic problems in the biochemistry of the glia are those of the interactions between neurones and glial cells.

Subsequent attempts to perfect methods for obtaining enriched fractions were made by Roots and Johnston (1964, 1965) in England and by Satake and Abe (1966) in Japan.

Roots and Johnston used a filtration method, which did not involve centrifugation. Brain stem of ox was taken into chilled 0.25 M sucrose, homogenized, and filtered through nylon meshes first of pore area 300-350 μm^2, and then of 108-110 μm^2. The filtrate was placed in a depression in a special microscope slide filled with 0.25 M sucrose, and individual neuronal cell bodies taken with a nylon loop. This method results in relatively undamaged neurones but the yield is so small (300-600 neurones from a single animal) that conventional biochemical determinations are subject to considerable error.

Satake and Abe (1966) suspended chopped cerebral cortex in a chilled mixture of acetone-glycerol-water (1:1:1); the tissue was then transferred to a mixture of glycerol-0.25 M sucrose (3:1) and subjected to 13 gentle strokes in a teflon homogenizer. The homogenate was then filtered twice under vacuum through flannel cloth. The filtrate was suspended in 0.5 M sucrose, layered onto 1.75 M sucrose and centrifuged for 30 min at 14,500 rev/min. The pellet contained the nerve cell bodies, the yield being $1-3 \times 10^6$ neurones per rat. The quantity of RNA and DNA per cell calculated by the authors is in good agreement with a number of previous studies, including our measurements on cortical neurones using UV-cytospectrophotometry (Pevzner, 1960, 1962a).

Both these methods have been subsequently modified by the authors themselves, and other methods of obtaining enriched fractions of a monocell type have been suggested (for reviews, see: Johnston and Roots, 1970, 1972; Pomazanskaya, 1970, 1974; Rose and Sinha, 1974). However, a drawback of all these studies is the inability to make parallel preparations of both neuronal and glial cells. Rose (1965, 1967) was the first to develop a sufficiently rapid and relatively simple method of obtaining enriched fractions of neuronal cell bodies and of glial cells from the same piece of nervous tissue. In this method cleaned and chopped tissue was teased through a nylon gauze of 30 μm pore diameter into a chilled solution of 10% Ficoll (a sucrose polymer of molecular weight about 400,000) containing KCl, filtered under gentle vacuum through a steel mesh of 40 μm pore diameter; and centrifuged on a discontinuous gradient with 30% Ficoll above 1.45 M sucrose at 0°C for 45 min at 39,000 rev/min (125,000 g). Four fractions are obtained, one being enriched in neurones, another in glia. The chemical composition of the fractions showed good agreement with data obtained by other methods. Rose and Sinha (1969) showed that uptake of oxygen, accumulation of CO_2 and lactate, and incorporation of ^{3}H-lysine by the cells was considerably greater than those prepared by the method of Satake and Abe.

However, an examination of the morphological and biochemical properties of the enriched fractions prepared by Rose's method (Volpe and Giuditta, 1967; Cremer et al., 1968) shows that the neuronal fractions contain fragments of capillaries and endothelial cells, while the glial fraction includes various processes, nerve endings, and synaptic endings.

Subsequently Mandel and his co-workers (Freysz et al., 1968) suggested a modification offering some advantages. Cerebral cortex of rat was disaggregated in a mixture of glycerol-acetone-water (1:1:1) and filtered through a nylon mesh of 82 μm pore size in a mixture of glycerol-0.25 M sucrose (3:1). The filtrate was transferred into a physiological solution, filtered through a 58 μm mesh, centrifuged for 15 min at 2,500 rev/min, the pellet homogenized in 0.25 M sucrose, and the homogenate centrifuged 30 min at 20,000 rev/min in a sucrose concentration gradient (0.25-1.00-1.75 M). The pellet formed the neuronal fraction. The material at the 1.00-1.75 M interface was drawn off, filtered in glycerine-0.25 M sucrose (3:1), transferred to physiological solution, centrifuged at 2,500 rev/min and the pellet resuspended in 10% Ficoll. This suspension was centrifuged for 90 min at 25,000 rev/min on a concentration gradient (10% Ficoll, 30% Ficoll, 1.5 M sucrose). The interface between 30% Ficoll and 1.5 M sucrose formed the glial fraction. Light microscopy indicated that the purity of the neuronal fraction was 80%, and that of the glial fraction was 90%. There was satisfactory preservation of the structure of both cell types.

Zonal centrifugation has also been used for preparation of enriched fractions (Flangas and Bowman, 1968). In this method cerebral cortex of rat was pushed first through a 37 μm mesh, and then after rapidly weighing a suspension was prepared in a Ficoll solution of final concentration 10% containing 5 mM $CaCl_2$ and 50 mM Tris buffer (pH 7.8). This suspension was further dispersed by drawing it into a syringe and expelling it 5-10 times, then diluted four-fold by the addition of 10% Ficoll, and filtered through two 37 μm meshes using a microsyringe. The filtrate (usually around 40 ml) was centrifuged in a B XIV zonal rotor at 3,500 rev/min on a discontinuous gradient of 30% Ficoll (density 1.102) containing 5 mM Tris buffer (pH 7.8), 58% sucrose (density 1.282) containing 5 mM $CaCl_2$ and 50 mM Tris buffer (pH 7.8) and potassium citrate (density 1.49). Centrifugation time was 45 min. The contents of the rotor were displaced by pumping in potassium citrate and 20 ml fractions collected. Each of these was divided into two 10 ml portions, diluted with 0.25 M sucrose, and centrifuged for 30 min at 30,000 rev/min in the angle head rotor of the Spinco L ultracentrifuge. It was shown that fractions 16-21 contained principally glial cells, fractions 18-20 being astrocytes, the remainder

oligodendroglia and microglia.   Fractions 24-25 contained the nerve
cells.

     It is very significant that the preparation of enriched
fraction has since found a place in Hydén's laboratories.   Blomstrand
and Hamberger (1969, 1970) used the same principle as Rose, but
introduced several modifications   (Fig. 2).   After mechanical
disruption of rabbit cerebral cortex (1) and incubation in a buffered
salt solution (2), the material was forced through a syringe closed
with a nylon mesh (3), and then through a mesh with 50 μm pore
diameter (4).   The filtered material was centrifuged for 5 min at
150 g in a salt solution containing sucrose (5).   The authors
consider that this intermediate centrifugation at low speed helps
to obtain an improved resolution of cells from the various contaminants.   The precipitate was resuspended in Ficoll of final concentration 20%, and this suspension formed the middle layer of a discontinuous Ficoll concentration gradient (6).   In the opinion of
the authors this position is useful principally because during the
subsequent high speed centrifugation the majority of the glial cells
migrate upwards, while the neuronal cell bodies descend (7), thus
reducing the possibility of mutual contamination.   In the electron
microscope, well preserved neuronal cell bodies were seen, but
there was some damage to the plasma membrane.   Oligodendroglia were
attached to some of the neurones.   More heterogeneity was found in
the glial fraction (on the criterion of fragments of neuronal and
glial processes).   Microglia were present along with astrocytes and
oligodendroglia, however the plasma membranes of the glial cells
seemed less damaged than with the neurones (Blomstrand, 1971).

     An interesting modification of Rose's method has been introduced
in Korey's laboratory.   Norton and Poduslo (1970), besides introducing several stages of low speed centrifugations, suggested an
initial treatment of the cells with a 1% solution of crystalline
trypsin.   This helps to reduce damage to the cells, although it
considerably reduces the yield by comparison with the other modifications of Rose's method: of the order of 20%-30% of the neurones
and 5%-10% of the glial cells of cerebral cortex.   Morphological
examination of the fractions shows that in the neuronal fraction,
neuronal cell bodies account for over 90% of the total fraction,
while in the glial fraction, glial cell bodies account for only
50% of the total.   Later a detailed study was published by these
authors (Norton and Poduslo, 1971) in which they showed that their
glial fraction consisted mainly of astrocytes.   In the discussion,
the authors point out that the divergent data obtained from various
separation methods (for enriched fractions) may be mainly due to
the fact that small changes in the separation scheme may significantly
alter the proportions of oligodendroglia, astrocytes, and microglia
in the glial fraction.   These three cell types are far from
identical in their metabolic properties.

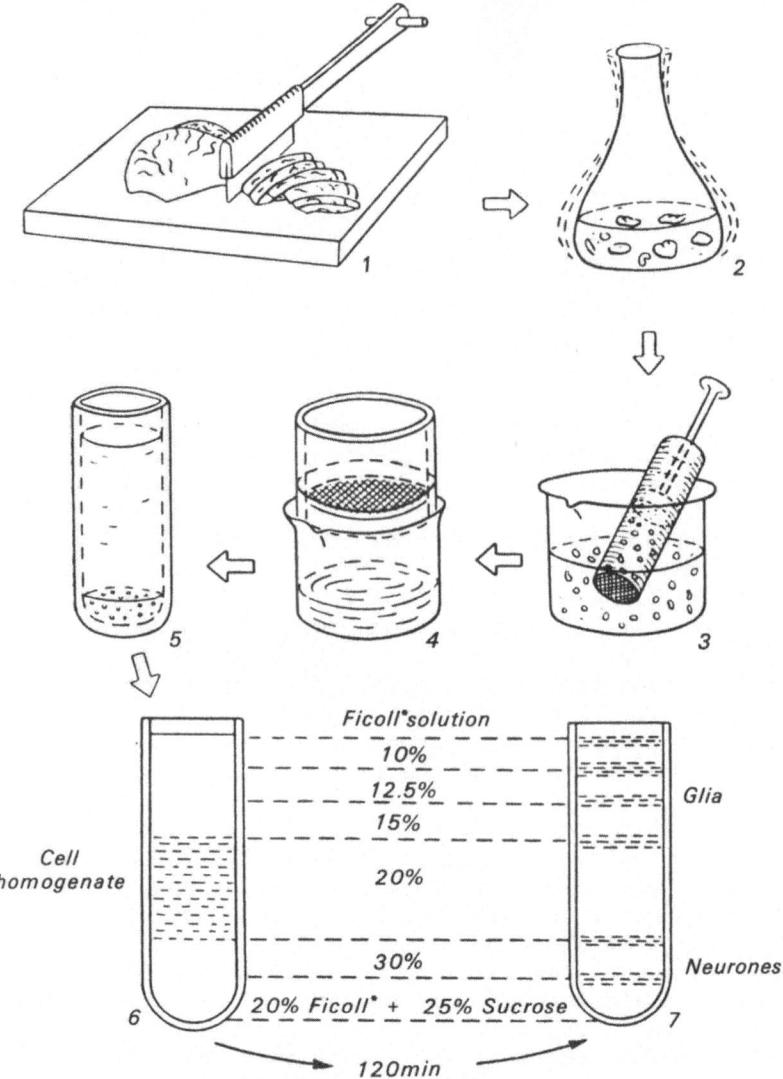

Ficoll is a trade name (Pharmacia, Sweden)

Fig. 2. Scheme for preparation of enriched fractions from
brain tissue (Blomstrand, 1971). 1- mechanical disruption
of brain tissue; 2- incubation in buffered saline solution;
3- passage through nylon cloth; 4- filtration to obtain the
initial cell suspension; 5- low-; and 6- high-speed centrifugation in a discontinuous Ficoll gradient; 7- final
separation of neuronal and glial enriched fractions.

Subsequently, these authors reported a new modification of the method (Poduslo and Norton, 1972) which permits the separation of fractions enriched specifically in oligodendroglia. Calf white matter (centrum ovale and corpus callosum) was finely chopped, and incubated in 1% trypsin at 37°C for 90 min, followed by cold filtration through a steel gauze (200 mesh) and then layered onto a discontinuous concentration gradient of sucrose (0.9 M-1.40 M-1.55 M). After 10 min centrifugation at 3,300 g, three layers and a pellet were obtained. The interface between 1.40 and 1.55 M sucrose contained predominantly oligodendroglia. The contamination with other cell structures did not exceed 10%. The overall yield of oligodendroglia was 1.1×10^7 cells from 1 g protein.

A study of the fractions obtained by this method using both light and electron microscopy shows good preservation of structure of both neurones and oligodendroglia, while the condition of the astrocytes was rather poorer (Raine et al., 1971).

A promising modification has been suggested by Giorgi (1970). A combination of treatment with trypsin, filtration through four different meshes, and zonal ultracentrifugation enabled the author to separate six different enriched fractions from rat cerebrum: large neurones (containing 8% of the total DNA of the cerebrum), small neurones (4% of the DNA), neurones and capillary cells (5% of the DNA), capillary cells and oligodendroglia (2%), oligodendroglia (2%), and astrocytes (3%). The overall yield was 25% of the brain DNA.

Hemminki's group have produced another modification, also envolving enzyme treatment, but using 0.2% collagenase and 0.1% hyaluronidase in place of trypsin (Hemminki et al., 1970; Hemminki, 1970, 1972). After treatment of the starting material with these enzymes for 1 h at 37°C under 100% oxygen, sequential filtration was carried out through gauzes of pore diameter 500, 200, 135 and finally 65 μm, followed by a low speed centrifugation (300-800 g) in the cold. In some cases, sonication at 10 kHz was used after the first filtration (through 350 μm gauze)(Hemminki et al., 1970).

A widely used method for preparing enriched fractions is that of Sellinger et al. (1971) which they have used successfully with rats of various ages, for adult rabbits, and for dogs. As can be seen from Fig. 3 this scheme involves three stages of filtration through nylon gauze (333, 111, and 73 μm mesh respectively) and three centrifugation steps on sucrose density gradients. This method has the disadvantage that the preparation time for the glial fractions is greater than that for the neurones. Thus the possibility arises that metabolic differences might appear during the two additional centrifugations that the glial fractions undergo.

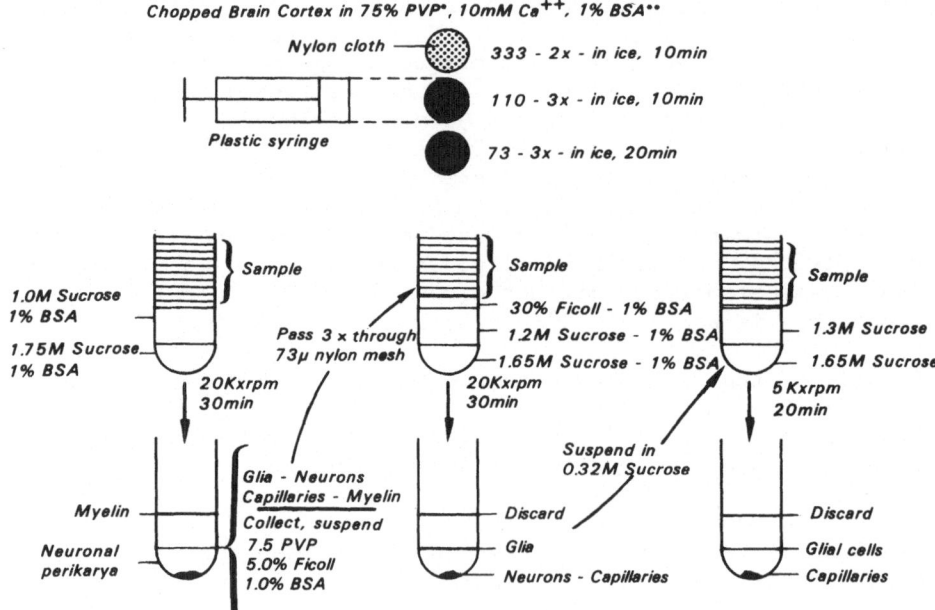

Fig. 3. Preparation of enriched fractions of neurones and glia according to Sellinger et al. (1971). PVP = Polyvinylpyrollidine; BSA = bovine serum albumin.

To counter this possibility, Tsukada's group (Nagata et al., 1974) have suggested the scheme illustrated in Fig. 4 in which both fractions are obtained from a single centrifugation step. Chemical treatments such as acetone (Freysz et al., 1968) collegenase and hyaluronidase (Hemminki and Holmila, 1971) trypsin (Norton and Poduslo, 1972) are completely excluded. In place of these, as can be seen from Fig. 4, the authors have introduced a series of four filtration steps under pressure. Nylon gauzes of 435-55 µm pore diameters are used. The authors remark that this may cause substantial mechanical damage to the cells, but consider that the biochemical integrity will be better preserved.

The method of Iqbal and Telléz-Nagel (1972) is less widely used. These authors avoid all enzyme treatment of the minced tissue, as well as organic solvent, but incubate for 1 h at 37°C in a phosphate buffer containing 1% Ficoll, 10% glucose and 10% fructose. For human brain this period is shortened. The tissue is next filtered through nylon gauze of 149 µm pore diameter (100 mesh) and then five times through a steel gauze of 74 µm pore diameter (200 mesh).

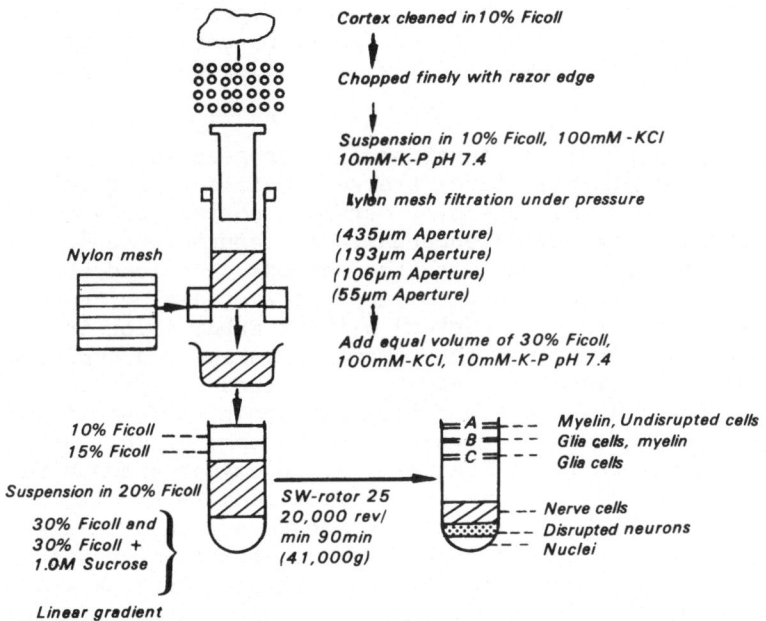

Fig. 4. Preparation of enriched fractions of neurones and glia according to Nagata et al. (1974).

The filtrate was diluted twice with 50% sucrose, and layered onto a sucrose concentration gradient (50%, 45%, 40% and 35%). After 10 min centrifugation at 4,500 g five layers were obtained, two of which (40-45% and 45-50% sucrose) contained large and small neurones respectively, while the layer between 35-40% sucrose was an impure neuroglial fraction. A further centrifugation on a 35-40% sucrose concentration gradient was carried out at 4,500 g for 15 min. The 35-40% interface contained the glial cells.

A scheme for the separation of isolated cell nuclei, which may find limited application has been suggested by Lovtrup-Rein and McEwen (1966). Cerebral tissue was homogenized by hand in a sucrose solution, and centrifuged for 30 min at 30,000 rev/min on a sucrose concentration gradient. Five layers were obtained, of which one contained primarily neuronal nuclei, a second astrocyte nuclei, and a third a mixture of nuclei, primarily of oligodendroglia and microglia. A similar scheme was suggested by Kurokawa and his co-workers (Kurokawa et al., 1966; Kato and Kurokawa, 1967) using

a combination of sucrose and Ficoll density gradient centrifugation.
After several centrifugations, two fractions were obtained, in one
of which the ratio of neuronal to glial nuclei was 51:49, in the
other 4:96.

The morphological and functional heterogeneity of the brain is
well known. For example, there are well-defined topochemical
differences between the layers of cerebral cortex, between the cortex
and sub-cortical regions, diencephalic structures, basal nuclei, and
so on. One of the problems with the methods described above for the
preparation of enriched fractions is that they deal with whole brain
(or at least with cerebral cortex) without subdivision into the
various regions. Therefore the scheme of Hazama and Uchimura (1973)
represents a significant advance. This method permits the separation
of neuronal and glial fractions from ten brain regions of a single
rat, each piece of tissue weighing no more than 10 mg. Slices of
tissue are incubated for 1 h at 37°C in an oxygen atmosphere in a
medium containing 10% glucose and fructose and 2% Ficoll in 0.01 M
phosphate buffer pH 7.2. The slices are then placed on a wax block,
and the required areas removed under a dissecting microscope, cut
into pieces about 0.4 mm diameter, and filtered through a polythene
gauze of 286 µm pore diameter. They are then filtered through nylon
gauzes of 111 and 40 µm pore diameter, the latter filtration being
repeated 10 times. The final filtrate is introduced into a capillary
(length 73 mm, internal diameter 1 mm) containing a sucrose concentration gradient - 0.9 M; 1.2 M; 1.55 M and 2.0 M. The capillary is
centrifuged for 10 min at 2°C in a microhematrocrit centrifuge at
10,000 g. The required fractions are drawn off with a micropipette.
The purity of the glial fraction is less than that of the neurones,
and so the layer containing the glia is diluted to 0.5 M sucrose
and introduced onto a gradient of 0.9 M sucrose over 1.35 M in the
capillary. The capillary is again centrifuged at 2°C for 10 min at
10,000 g. The purified glia are collected from the 0.9 M-1.35 M
interface.

A detailed analysis of the advantages and disadvantages of the
various methods for preparation of enriched fractions may be found
in a number of special methodological reviews (Sinha and Rose, 1971;
Johnston and Roots, 1970, 1972; Pomazanskaya, 1974; Sellinger and
Azcurra, 1974; Rose and Sinha, 1974).

Micromanipulation

This method is sometimes called microsurgery or microdissection,
as it depends on cutting out individual cells from the surrounding
tissue under the microscope. According to the originator of the
method and those trained by him, the separation procedure, especially
for large cells, is not very difficult. The main obstacle to the
wider use of this technique is the necessity of carrying out chemical

analyses on tiny samples weighing fractions of a milligram or even
of a micnogram. Such ultrasensitive microchemical determinations
of the composition of neurones and glia have been developed over
the years in the laboratories of Lowry in the USA and of Hydén and
Edström in Sweden.

While the studies of Lowry (1953, 1955, 1957, 1962, 1967) were
mainly concerned with analysis of nerve cells, a broad series of
works by Hydén and his co-workers (Hydén, 1959a, b, 1960, 1961, 1963,
1964, 1967a) was devoted to the comparative study of the biochemical
characteristics of neurones and glia. The Swedish neurochemists
prepare slices of the required brain area (≃ 2 mm thick) in the
cold and place them in chilled 0.25 M sucrose containing 1-2 mM
methylene blue, and under the stereo microscope dissect out large
neurones with a steel needle of 30 μm diameter. The glial capsule
surrounding the neurone is then removed (Fig. 5). Using the ocular
micrometer they select a clump of glial cells approximating in
volume to the neuronal cell body. In some work (Hamberger, 1961,
1963; Hamberger and Hydén, 1963) a similar clump of glial cells was
removed from around the small blood vessels.

Subsequent analysis of the separated material necessitates the
use of ultramicrochemical methods which were either developed
directly in Hydén's laboratory (Edström, 1953, 1958, 1960; Edström
and Hydén, 1954; Hydén, 1955) or adapted from the work of the
Carlsberg laboratory in Denmark, using the Cartesian diver
(Linderstrøm-Lang, 1937; Holter and Linderstrøm-Lang, 1951; Zajicek
and Zeuthen, 1956; Holter, 1961; Zeuthen, 1961).

In particular, the RNA determination involves placing the nerve
cell body or clump of glia in a layer of liquid paraffin under a
cover slip and introducing a ribonuclease solution for digestion of
RNA. A homogeneous drop of hydrolysate is photographed under UV-
light and compared with a scale of optical density. Knowing the
volume of the drop enables the quantity of RNA in the nerve cell or
glial clump to be calculated (Fig. 6). For determination of the
nucleotide composition of the RNA, an acid hydrolysate is subjected
to microelectrophoresis on specially prepared microfibrillar
cellulose. The fibers are photographed in UV-light to reveal the
separated spots corresponding to the individual nitrogenous bases.
Automatic analysis gives the ratio of the bases from the optical
density scale (Fig. 7).

The uptake or release of oxygen in the separated cells is
measured by the Cartesian diver method (Fig. 8). A detailed
description of the use of this micromanometric method for neuroglia
may be found in the work of Giacobini (1961, 1962, 1964) and
Hamberger (Hamberger, 1961, 1963; Hamberger and Hydén, 1963). For
a general review, see Krasnov, (1967).

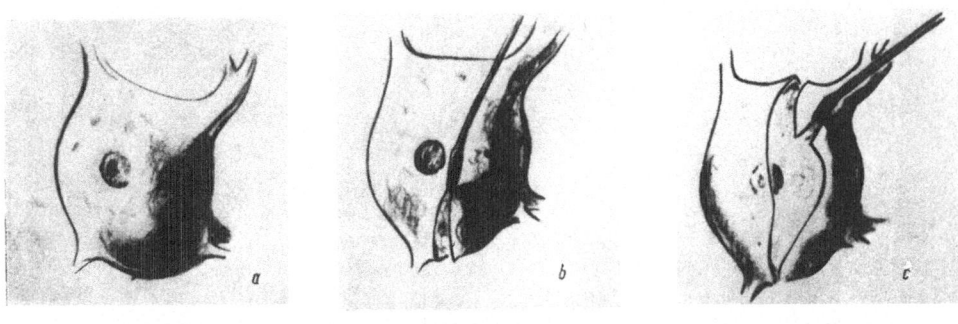

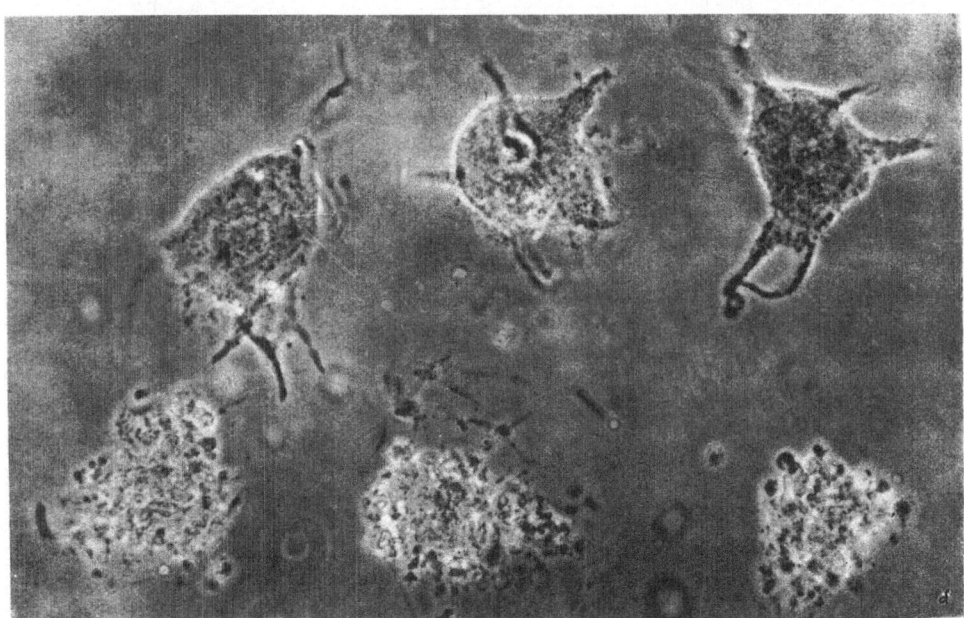

Fig. 5. Scheme for separation of neuronal cell bodies and
their glial capsules by micromanipulation   (Hydén, 1961).
Above – successive stages of microdissection of neuronal
cell body and its separation from the surrounding capsule.
Below – isolated neuronal cell bodies, and below them matched
volumes of glia.

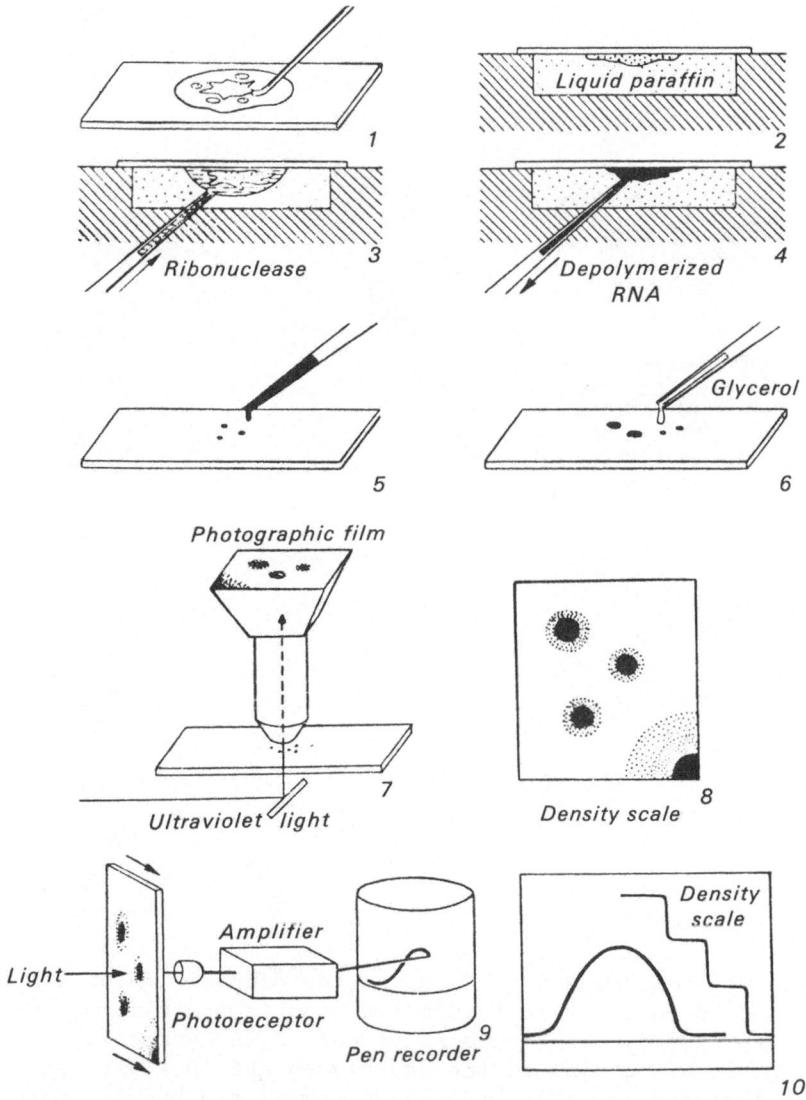

Fig. 6. Scheme for quantitative determination of RNA in isolated cells (Hydén, 1961). 1- microdissected neuronal cell bodies (or their surrounding glia) are placed on the slide; 2- immersed in liquid paraffin; 3- treated with ribonuclease, and 4- the depolymerized RNA removed; 5- the latter is placed on a slide; 6- covered with glycerol; 7- the glycerol droplet, containing RNA is photographed in monochromatic UV-light, together with a density scale; 8- the negative is developed and fixed; 9- then subjected to scanning densitometry; 10- comparison of the optical density of the glycerol droplet with the standard density scale is used to calculate the RNA content.

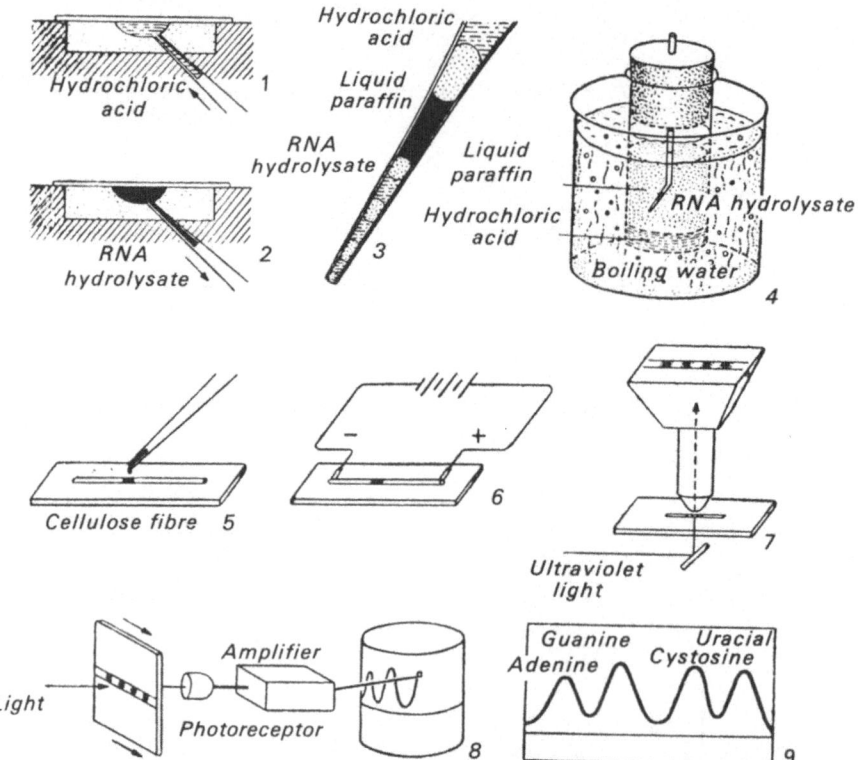

Fig. 7. Determination of nucleotide composition of RNA of separated cells by Edström's microelectrophoretic method (Hydén, 1961). 1- treatment of cell with hydrochloric acid; 2- withdrawal of the hydrolyzed RNA in a micropipette; 3- pipette containing hydrolyzed sample and hydrochloric acid; 4- completion of hydrolysis in hot hydrochloric acid; 5- application of the mixture of nucleotides to cellulose microfiber; 6- electrophoresis; 7- photography of the microfiber in UV-light; 8- scanning densitometry of the resulting negative; 9- densitogram of the separated nucleotides.

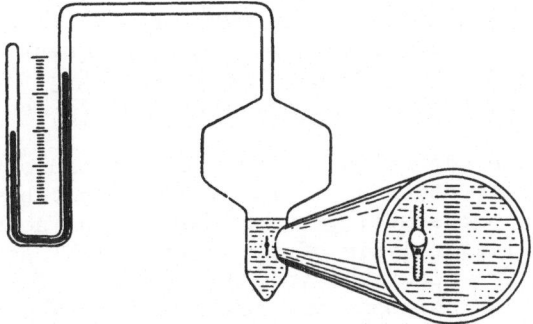

Fig. 8. Cartesian diver for micromanometric determination
of tissue respiration in isolated cells   (Hyden, 1961).

The development of these methods has been a major achievement
of modern neurochemistry and has resulted in the accumulation of a
large body of facts substantiating the idea of the unity of the
neurone and neuroglia as an integral metabolic system. However, in
spite of the value of the clear separation of neurones and glia
under visual control, and their subsequent chemical analysis, this
method also has its drawbacks.

Firstly, the very procedure of removal of the neurones and
stripping of the glial capsule is highly traumatic. Although Hydén
considers that the viability of the neuroglia after such a procedure
is demonstrated by the ability of the separated glial cells to grow
in tissue culture in vitro, this hardly demonstrates that their
physiological condition (as opposed to their general viability) has
not been sharply altered at the moment of separation. Moreover, the
question of the effect of the separation procedure on neurones
including the breaking of the axon and damage to the dendrites was
not investigated. Meanwhile, the extensive work of McIlwain
(McIlwain, 1956, 1962; McIlwain and Bachelard, 1971) has made it
clear that stimulation of surviving nerve cells in vitro can lead
to considerable changes in metabolism in particular of oxidative
processes and ionic composition.

Secondly, it must not be forgotten that the surface membrane
of the neurone and that of the surrounding glia are separated by
no more than 15 nm; during mechanical separation of the neurones

from the glia it can scarcely be possible to ensure corresponding separation of the outer membranes. Rather one would expect part of the neuronal cell membrane to be torn off from the cell along with the outer glial cell membrane. Thus subsequent analysis of the biochemical differences between neurones and glia (in particular difference in transport of ions) may give a distorted picture.

Thirdly, in the separation of the glial capsule to obtain the clump of glial cells, other cell elements are undoubtedly taken - neuronal and glial processes, parts of dendrites, neuronal cell membrane etc. - while analysis of the neurone will sample only the nerve cell body. The dendritic tree differs from the nerve cell body both in chemical composition, and in its metabolic processes. It cannot be ruled out, therefore, that the metabolic differences between neurones and glia reported by Hydén's group are related to the different purities of the samples under study.

Fourthly, the clump of glial cells analyzed only approximates in volume that of the corresponding nerve cell body. As neither the mass of glial cells making up the capsule of a single neurone, nor the number of cells in the sample taken for analysis are known, it is difficult to assess the data on the basis of a single glial cell, or on the whole neurone-neuroglia unit.

More detailed critical evaluation of the micromanipulative method as used to study neuroglia may be found in reviews by Rose (1968a, 1969), Pevzner (1969a, 1972b), and Aleksidze (1974).

Quantitative Cytochemistry In Situ

In recent years the development of numerous histochemical methods has permitted the visual estimation of chemical substances in individual cells directly from histological preparations. This method preserves the integrity of the tissue structure. However a major disadvantage of many such methods is the lack of a quantitative approach. The most significant methods, then, are those which permit quantitative estimation of chemical components in histological sections. The principal of these is cytospectrophotometry.

The principles of this method, as well as its practical basis were laid down by Caspersson (1936, 1940, 1950, 1955). He used the basic law of spectrophotometry, the Bouguer-Lambert-Beer law:

$$I = I_o e^{-Xc\ell}$$

where
I_o = intensity of incident light beam,
I = intensity of light beam after passing through object,

χ = extinction coefficient, a constant for each substance at a defined wavelength,

c = concentration of the absorbing substance,

ℓ = thickness of the absorbing layer (in this case of the microscopic section).

Caspersson showed that if certain conditions were observed, this law could be applied to the measurement of concentrations of substances within the cells of a histological preparation. He suggested also the first plans for a cytospectrophotometer, that is a combination of microscope and spectrophotometer (see Caspersson, 1955; Argoskin, 1958, 1964, 1967; Wied, 1966). Either the natural absorption of the substance at a particular wavelength was used, or stained preparations were made with specific histochemical dyes reacting stoichiometrically with the component being studied.

Fig. 9 shows the schematic layout of a cytospectrophotometer. Light from a stabilized source (1) is passed through a monochromator (2) and then to the microscope (3) in which is placed the object (4). The light transmission is measured either directly by a photomultiplier (5) displaying the reading on a meter (6), or a pen recorder (7); or by first photographing the object together with a standard density scale (8) and then scanning the film (9).

The history of the cytospectrophotometric method, its basic technical variations, and the accuracy of the method are discussed in detail in a series of monographs and reviews (Caspersson, 1955; Mellors, 1955; Brodskii, 1966; Wied, 1966).

Following Caspersson's account of the theory of cytospectrophotometry, a number of papers appeared casting doubt on Caspersson's method, and demonstrating the possibility of considerable error in cytospectrophotometric determinations (Commoner and Lipkin, 1949; Danielli, 1950; Naora, 1951; Gomori, 1952; Glick, 1953). The most important sources of error are the heterogeneous distribution of the substances within the cell and non-specific light losses (absorption or dispersion of light by cell structures). Cytophotometric errors are discussed from all angles in a monograph by Brodskii (1966) and in a special methodological review by Wied (1966). In response to the above criticisms both Caspersson himself, and a number of other workers have carried out extensive methodological research resulting in the development of Caspersson's initial theory, and suggesting a number of technical modifications of the apparatus and methods of measurement (for reviews, see: Pevzner, 1963a, 1966a; Brodskii, 1965, 1966). This work has removed or considerably reduced the possibility of such errors occurring. Comparison of determinations of nucleic acid by cytospectrophotometry and by various chemical methods (for instance direct measurement using orcinol or diphenylamine, followed by calculation of the nucleic acid content per cell) shows that the divergence between these methods does not exceed the

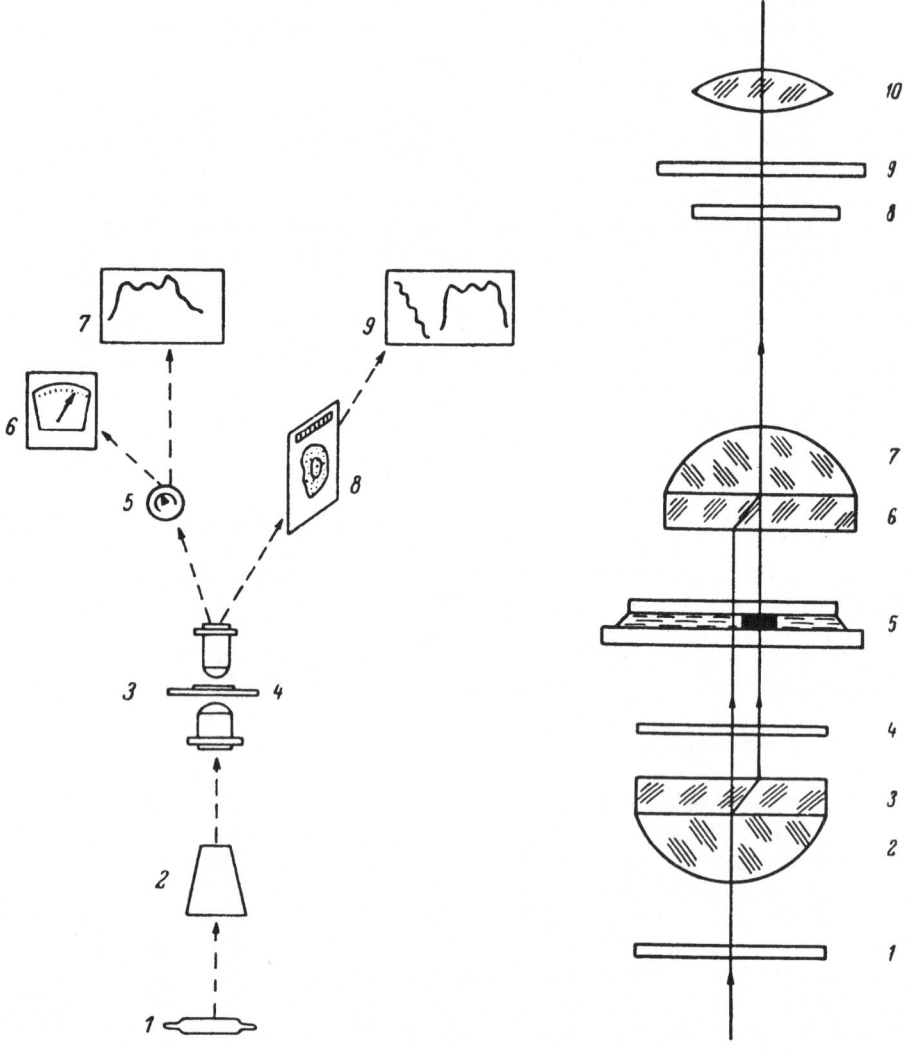

Fig. 9. Basic scheme for cytospectrophotometry. 1-stabilized light source; 2-monochromator; 3-microscope; 4-sample; 5-photomultiplier; 6-galvanometer; 7-pen recorder; 8-photographic negative; 9-densitometer.

Fig. 10. Optical scheme of the interference microscope MBIN-4. 1-polarizer; 2-condenser; 3- anisotropic spar plate; 4-half-wave retardation plate; 5-sample; 6- compensating spar plate; 7-objective; 8-quarter-wave retardation plate; 9-rotating analyzer; 10-eyepiece.

5-10% error inherent in cytophotometry (Caspersson, 1950, 1955; Brodskii, 1966; Sandritter, 1966).

Cytointerferometry has now been added to cytospectrophotometry (Barer, 1956; Davies, 1958; Zakhar'evskii and Kuznetsova, 1961a, b). This method permits the determination of dry mass of cells in the histological section. Under the conditions of fixation used over 90% of such optically dense material consists of protein, so this represents a method for the determination of total cell protein. This makes the analysis of a number of functionally conditioned changes in protein metabolism of nerve cells possible (Brodskii and Kuznetsova, 1961; Pevzner et al., 1964; Pevzner, 1965a; Brodskii, 1966, 1970).

In interference microscopy, the unstained section is mounted in distilled water on the microscope stage (Fig. 10). The incident light first passes through a polarizer made of either calcite or polaroid (1), then through a condensor lens (2), then through a double-refracting anisotropic spar crystal (3) which makes it birefringent. The two unipolarized light beams then pass through a half-wave retardation plate (4), and then pass through the section, one through the cell structure to be measured, the other through a neighboring "blank" area. The greater the dry mass of the cell, the greater is the retardation of the light beam (i.e. the phase difference between the two beams). To measure this difference a compensating spar lens (6) is combined with the objective lens (7), and then a compensator, which in the Soviet Union is generally of Sénarmont type. It consists of a quarter-wave retardation plate (8) and a rotatable analyzer (9) with a graduated scale. The scale shows the retardation due to the cell, as compared to the blank area near the cell.

The dry mass (P) of the cell is given (in g) by the formula

$$P = \frac{\delta S}{100\alpha}$$

where
  δ = optical retardation or phase difference (in cm),
  S = area of the structure (in cm^2),
  α = specific refractive increment of the material.

For protein, α is constant over a wide range of concentration, being 0.0018 in an aqueous medium (Hale, 1958; Davies, 1958).

There seem to be three main drawbacks to the cytophotometric and cytointerferometric methods.

1. At present it is possible to determine only a limited number of chemical components of the cell, in practise nucleic acids and protein. Although in the early years of the method there

was optimism that in the near future stains would become available which would allow the determination of many constituents of the cell, up to now this prognosis has not been fulfilled. Reliable quantitative estimations can still be carried out only on DNA, total nucleic acids, total protein, on individual groups in the protein molecule, e.g. sulphydryl, and amino groups, and on a few enzyme activities.

2. Even these few components may be determined only in toto and not as particular fractions. It is true that the method of successive extraction (Brodskii and Suetina, 1960) suggests possibilities in this direction. However these first successes appear modest in comparison with the achievements of conventional biochemistry in recent years for fractionation of nucleic acids and proteins. In particular, cytophotometric determination of RNA measures only ribosomal RNA, the messenger RNA and transfer RNA being washed away in the fixation process (Watson 1965a; Hogenhuis and Spaulding, 1967; Hogenhuis et al., 1967).

3. As was stated above, the majority of the theoretical sources of error in cytophotometry are either not significant under the actual conditions used, or there exist techniques for reducing them to a defined minimum. Nevertheless, the accuracy of cytophotometric analysis is still considerably less than that of conventional biochemical determinations, and a large number of repeated observations is necessary to reduce the error in the mean to around 5-10%.

* * * *

Thus of the five methods currently used for investigation of the neuroglia, two - topochemical analysis and the study of pathological material - may be considered indirect. Their popularity may be explained by their simplicity, and accessibility to any biochemical laboratory. However the undoubted disadvantage of the first method is the absence of reliable information concerning the proportion of glial cells in the total mass of nervous tissue under study. The second method involves the use of pathological material in which the neuroglia themselves may show (and in the great majority of cases undoubtedly do show) chemical change. Therefore the usefulness of these methods is strictly limited.

The three direct methods for analysis of neuroglia - enriched fractions, micromanipulation, and cytophotometry - are based on entirely different principles. The great advantage of the first method is the comparatively large yield of material for study, which is sufficient to permit the use of routine biochemical techniques. It is possible that improvements in the homogeneity and purity of the fractions will make this the principal method for the study of the neuroglia. At present, however, the basic inadequacy of the

method is the low purity of the fractions obtained. Besides this
it is necessary to consider the degree of damage caused to the cells
by the procedure, and its influence on different aspects of the
metabolism of the neurones and glia.

The advantages of the micromanipulation method are the abilities
to select the desired cells under visual control, and to analyze
different types of neurones and of neuroglia (e.g. perineuronal and
pericapillary glia). However the yield of material is so small
that chemical analysis is possible only in a laboratory set up for
specialized microchemical techniques. Besides that, the disadvantage of the method (as with enriched fractions) is the mechanical
separation of the glial cells from the neurones, and the disruption
of the integrity of nervous tissue during separation of the cellular
elements. The use of the method has enabled Hydén's group to make
a number of important generalizations about the role of the neuroglia
in neuronal function, so the value of the method is not in doubt.
The difficulties in making a critical appraisal of the work arises
from the fact that so far the results have not been confirmed in
other laboratories.

The methods of cytospectrophotometry and cytointerferometry
do not involve the mechanical disruption of nervous tissue. There
is also the possibility of visual selection of the cell type to be
studied. The disadvantages of the method are the limited range of
chemical components that can be analyzed, the comparatively low
accuracy of individual determinations, and the necessity of histological preparation, which may wash a portion of the substance to
be analyzed from the cells. Therefore quantitative cytochemistry
is most suitable for comparative estimations of changes in total
RNA or protein in individual cells.

It should be clear from this short survey that all the methods
discussed have both advantages and disadvantages. It is necessary
to keep this in view as in many cases contradictory data have been
obtained by various authors due mainly to the use of different
separation methods.

CHAPTER 3

BIOCHEMICAL PROPERTIES OF THE NEUROGLIA

In this chapter we have collected data from the literature on the metabolic properties of the neuroglia as compared to neurones. We made an attempt of this sort several years ago (Pevzner, 1969b). At that time, there was so little quantitative data that we also included results from qualitative histochemical studies of glial cells.

Histochemical investigations have several important advantages. The analysis of tissue elements without disruption of their spatial relationships, a parallel study of the structural and chemical features of cells and organelles, and the determination of very diverse substances within the cell - e.g. protein, carbohydrate, nucleic acids, lipids, and enzymes - are possible.

However, comparison of biochemical components of different structures, or following dynamic changes of a particular component by qualitative histochemistry can only give a first approximation. And the reason for this is not only that visual appraisal ("more - less" or "two star - four star", etc.) is not sufficiently objective. The human eye is in general a sensitive organ, capable of detecting very small changes in brightness. But it has one particular drawback. Its perception of a given intensity depends on that of the background against which it is viewed.

Apart from this, in analyzing colored histochemical reactions, the eye mainly assesses the intensity of color, which depends on the dye concentration. The concentration is a function of the amount of dye present, and of the volume of the structure containing it. If identical quantities of DNA are present, for example in the large neuronal nuclei and in the small glial nuclei, the concentration of DNA, and therefore of the corresponding dye, will be greater in the glial nuclei than in those of neurones. It would probably not be necessary to make such a trivial methodological point were it not for the fact that the histochemical literature is full of statements of the type "these cells are rich in glycogen (DNA, RNA, protein, phosphate, etc.)" or "these cells are poor in glycogen (DNA... etc.)". The microphotographs presented in these works usually demonstrate the different sizes of the cells under comparison, and as a rule the "impoverished" cells turn out to be the large ones and the "enriched" cells the small ones.

The evaluation of functional changes in cell metabolism is particularly difficult with quantitative histochemistry. As was pointed out in the preceeding chapter, changes in functional state may be accompanied by changes in cell volume. In this case the intensity of staining will depend on two variables, the quantity of substance in question, and the cell volume. These variables may differ not only in magnitude but in sign, and so determination only of the intensity of staining (whether by visual or photometric estimation) in general has no significance.

Finally, it is necessary to mention briefly the question of specificity and stoichiometry. The dye may not be specific. That is in addition to the component we are interested in, it may stain other compounds within the cell.

Thus Nissl staining of neurones with toluidine blue is sometimes used for estimation of cytoplasmic RNA. However this dye also binds to all the basic components of the Nissl substance - RNA, protein, and polysaccharide. Their ratios might be changed in different directions by factors affecting the nervous system. But even if the dye is specific, it may not be strictly stoichiometric, i.e. it may not bind in a fixed proportion to the corresponding compound. In practice, only a very few histochemical stains are stoichiometric, the great majority reflect only the trend of a chemical change in the cell, rather than giving a quantitative measure. For example, pyronine binds to RNA, but by adsorption and not by a covalent linkage as is the case for gallocyanin. Thus there is a possibility that conformational changes in the RNA molecule will cause changes in the amount of pyronine bound by the cell. More detailed discussion of the methodological questions involved in the evaluation of histochemical color reactions will be found in the reviews by Pevzner (1965b), Sandritter (1966), and Govardovskii et al. (1969).

All these factors probably contribute to the considerable amount of disagreement on the relative composition of neuronal and glial cells in the histochemical literature. This applies particularly to histochemical studies of enzymes in which the possibility of artefact is particularly high (Pevzner, 1972b) In the present chapter, we have considered it possible to exclude the qualitative histochemical data almost completely and to confine ourselves to quantitative studies on neuroglial biochemistry, since the number of such studies has greatly increased during the last few years.

Oxidative Processes

Topochemical analysis of oxygen uptake by different layers of the cerebral cortex of rat was carried out by Epstein (1970) using the Cartesian diver. It was shown that in the absence of glucose a sharp reduction in the rate of respiration occurs in those layers, II and VI, of the cortex which are particularly rich in neuronal cell bodies. In layer I, in those regions of layer V free of neuronal cell bodies, and also in samples of glial cells from tissue culture, oxygen uptake in the presence of glucose was less than that of the regions enriched in neurones, but this difference was less when the incubation was carried out in a glucose-free medium.

In the experiments of Ruščak et al. (1968) mechanical damage was caused to the cortex of rats, resulting in a reduction in the number of neurones, and a reactive proliferation of glia. In these regions, a reduction in glucose uptake was observed. On the basis of their own and other published data on the number of neurones in the cerebral cortex, the authors calculate the oxygen uptake by a single nerve cell body to be on average 62×10^{-6} µl/h, and in non-neuronal cells to be of the order of 7×10^{-6} µl/h. From this, the conclusion was drawn that the observed respiration of cerebral cortex consists mainly of neuronal respiration. This conclusion can hardly be considered convincing, however, because of the possibility of differential effects of damage as discussed in the previous chapter.

In the early work of Korey and his co-workers (Korey, 1958; Korey and Orchen, 1959) it was also shown that the respiration in the enriched nerve cell fraction was ten times higher than in glial fractions. However, the latter were undoubtedly highly contaminated with myelin structures, which have the lowest oxygen uptake of any fraction. Rose (1965, 1967), using a much improved scheme for separation of enriched neuronal and glial cells showed that for endogenous respiration the value for glia was 1.5 times higher than for neurones. On the addition of substrate, however, respiration in both fractions was increased, but more so in the neurones. With glutamate as substrate, for example, the respiration rates were about equal, while for glucose or pyruvate neuronal respiration slightly exceeded that in the glia (Table II).

TABLE II

Tissue respiration of neuronal and glial enriched fractions from rat cerebral cortex (Rose, 1967)

Incubation time and substrate	Whole Cell Suspension		Neurones		Neuroglia		Ratio of neuronal to neuroglial respiration
	Uptake of O_2 (in nmol/mg dry weight)	Stimulation with added substrate (%)	Uptake of O_2 (in nmol/mg dry weight)	Stimulation with added substrate (%)	Uptake of O_2 (in nmol/mg dry weight)	Stimulation with added substrate (%)	
60 min							
No added substrate (endogenous respiration)	234 ± 15	-	117 ± 6	-	173 ± 13	-	0.68
Glucose	430 ± 37	+88	394 ± 14	+235	378 ± 33	+119	1.02
Pyruvate	330 ± 25	+41	368 ± 37	+215	345 ± 26	+100	1.03
Glutamate	250 ± 17	+7	213 ± 18	+82	216 ± 12	+25	0.99
120 min							
No added substrate (endogenous respiration)	329 ± 23	-	174 ± 8	-	249 ± 22	-	0.70
Glucose	642 ± 45	+95	823 ± 75	+375	860 ± 70	+245	0.96
Pyruvate	555 ± 26	+69	690 ± 83	+295	596 ± 30	+138	1.16
Glutamate	342 ± 23	+8	344 ± 25	+99	345 ± 24	+38	1.00

The data of Rose are substantially confirmed by Hamberger
(1961) using the micromanipulation method. Although the endogenous
respiration in the Deiters' cells was only one-seventh (13%) of
that in the surrounding glial capsule, on addition of substrate the
neuronal oxygen uptake was stimulated to a much greater degree than
that of the glia. Thus with glutamate the respiration in the
neurones became greater than the glia, and with α-ketoglutarate it
was about the same. Only with a mixture of pyruvate and malate as
substrate did neuronal respiration remain below that of the glia.
(Table III). Hamberger (1963) later showed that chemical stimulation
of Deiters' nucleus with tricyanoaminopropene had virtually no
effect upon neuronal respiration (with α-ketoglutarate as substrate)
but somewhat activated the respiration of the glial capsule.

According to Hertz (1966) the respiration of glial cells
separated by Hydén's method, using glucose as substrate, was 6-7
times lower than that of isolated neurones. In a medium containing
a high concentration of potassium (up to 100 mM) the glial respiration was stimulated, but in a medium without Na^+ it was depressed.
In this work however, the isolated neurones were obtained in the
main from cats and the glia from rats. It is interesting that in
slices of rat cerebrum incubated in vitro in a sodium-free medium
the respiration soon falls to 10% of the initial value, and then
remains stable. Hertz suggested that the value of 10% represents
the neuronal contribution to normal respiration. The same author
together with Mandel's group later carried out micromanometric
determinations of oxygen uptake in cells in tissue culture (Hertz
et al., 1973). In chick embryo, a value of 6.6 pl O_2/cell/h was
found for small neurones, while that of the large glial cells was
7.5 pl O_2/cell/h. When the same data were calculated per gram wet
weight, the tissue respiration in the neuronal cell bodies was found
to be 570 μmol O_2/g/h, while that in the glia was 130 μmol O_2/g/h.
However, it is difficult to say to what extent these data represent
the situation in vivo.

Hemminki and Holmila (1971) showed that the respiration of their
glial enriched fraction from rat cerebral cortex, using glucose as a
substrate, was markedly lower than that of the neuronal cell body
fraction. These results are closer to those of Hertz for isolated
single cells than to Rose's data obtained from enriched fractions.

Hinzen et al. (1972) and Murai (1973) prepared enriched
fractions from dog and rabbit cerebral cortex, and found no significant differences in rate of oxygen uptake between neurones and
glia.

Nagata et al. (1976) found a somewhat higher respiration in
glial enriched fractions than in neuronal fractions from rat brain.
However this difference was only 20%, and the authors unfortunately
do not give the standard errors of these values, making it

impossible to assess the statistical significance of their figures.
It is interesting though, that the rate of respiration in the
initial unseparated cell suspension was intermediate between that
of neurones and glia, though rather closer to the glial value.

Bradford and Rose (1967) compared the effect of potassium ions
on oxygen uptake of enriched fractions incubated in an oxygenated
medium containing glucose. An excess of K^+ led to approximately
equal stimulation of tissue respiration ($\frac{1}{4}-\frac{1}{3}$) in both neurones and
glia.

The effect of K^+ on tissue respiration has also been studied
in Hydén's laboratory. Aleksidze and Blomstrand (1969) studied
neurones and perineuronal glia isolated from the lateral vestibular
nucleus of rabbit by Hyden's method (Fig. 11). The respiration of
neurones and glia in the presence of glucose was almost identical.
Addition of K^+ produced a somewhat different effect from that
reported by Bradford and Rose. K^+ ions sharply increased the
oxygen uptake in the glial capsule, but respiration in the neuronal
cell bodies was practically unaffected.

This relationship was later confirmed for enriched fractions
in the same laboratory (Haljamäe, Hamberger, 1971). Increasing the
concentration of KCl in the incubation medium to 50 mM led to an
increase in oxygen consumption of 85% in the glial enriched fractions,
and only of 15% in the neuronal fraction.

A preferential activation of tissue respiration by potassium
ions in glia, but not in neurones, has also been shown by Hertz et
al. (1973) in tissue culture. However, use of the original highly
sensitive micromethod for spectrophotometric determination of oxygen
uptake (by reduction of HbO_2 to Hb) allowed Hultborn and Hyden
(1974) to show an almost two-fold K^+-activation of microdissected

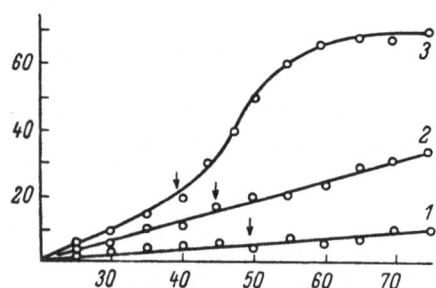

Fig. 11. Effect of K^+ ions on rate of tissue respiration of isolated cell bodies of Deiters' cells and the surrounding glial cell
capsule of lateral vestibular nucleus of rabbit (Aleksidze and
Blomstrand). Ordinate- O_2 saturation (in arbitrary units); Abs-
cissa- incubation time (in minutes); 1- control; 2- neuronal cell
bodies; 3- glial cell capsules; Arrows- times of addition of KCl to
the medium.

TABLE III

Tissue respiration (in 10^{-4} μl O_2 per h) of isolated Deiters' cell bodies and their glial capsule of lateral vestibular nucleus of rabbit (Hamberger, 1961)

Substrate	Neurones		Glia		Ratio of neuronal to neuroglial respiration
	Uptake of oxygen	Stimulation with added substrate*	Uptake of oxygen	Stimulation with added substrate*	
No added substrate (endogenous respiration)	0.04 ± 0.03	-	0.3 ± 0.06	-	0.13
Pyruvate + malate	0.8 ± 0.10	20	1.9 ± 0.47	6.3	0.4
α-ketoglutarate	1.1 ± 0.15	27.5	0.9 ± 0.19	3	1.2
Glutamate	2.2 ± 0.34	55	0.6 ± 0.14	2	3.7

*Ratio of respiration in the presence of added substrate to endogenous respiration.

Deiters' neurones. However in this case, no comparison with glial
cells was made.

As tissue respiration is mainly localized in the mitochondria,
and their concentration might differ between neurones and glia (see,
for example, Pysh and Khan, 1972), the separate study of neuronal
and glial mitochondria is an important step forward in the investigation of oxidative processes in these types of cell. Hamberger et
al. (1970) prepared enriched fractions from rabbit and beef brain,
and separated mitochondria from these fractions. No substantial
differences were found between neuronal and glial mitochondria in
the oxidation of various substrates (succinate, α-glycerophosphate,
pyruvate, and malate).

As long ago as 1956, Lowry et al. compared the activity of a
series of oxidative enzymes in neuronal cell bodies and the
surrounding glial capsule in the spinal cord using their microchemical methods. The activity of glutamate dehydrogenase in the
neuronal cell bodies of spinal ganglia, and anterior horn of spinal
cord was 1.5-2 times higher than in the glial capsule.

In enriched fractions from cerebral cortex of rat (Rose, 1968b)
the relationship was similar. In the neuronal fraction the activity
of glutamate dehydrogenase was higher than in the glial fraction by
2.5 or 4 times, depending on the presence or absence of detergent
in the incubation medium (0.1% triton X-100).

A relatively high activity of other oxidative enzymes in the
neurones of spinal cord has also been noted, e.g. lactate and
malate dehydrogenases, while the activities of glucose-6-phosphate
and 6-phosphogluconate dehydrogenases were higher in the glia
(Lowry, 1957). When the lactate dehydrogenase activity was investigated in enriched fractions, either no difference was found
between neurones and glia (Arbogast and Arsenis, 1974; Nagata et
al., 1974) or on the other hand a four-fold excess of activity was
found in the glia as compared to neurones (Packman et al., 1974).
Comparable (four-fold) differences were also found in determinations
of malate dehydrogenase in mitochondria separated from enriched
fractions (Venkov and Rusanov, 1976). The activity of glucose-6-
phosphate dehydrogenase in enriched fractions from adult rats did
not differ markedly between neurones and glia, except that in the
end of the first month after birth there was a significant but
transient excess of this enzyme in glia, which was probably connected
with the phase of intensive myelination (Arbogast and Arsenis, 1974).

Catalase activity was slightly higher in the glial fraction of
10-60 day old rats; citrate synthetase, however, was much higher in
the glia than in the neuronal fraction (Arbogast and Arsenis, 1974).

The activity of α-glycerophosphate dehydrogenase has been shown
to be 1.5 times higher in glial enriched fractions of mouse brain

than in the corresponding neurones (Svoboda and Lodin, 1972); for rat cerebral cortex this enzyme activity was three times higher in the glial fraction (Sellinger et al., 1973).

Aleksidze and Blomstrand (1968) measured the activity of the succinoxidase system in separated Deiter's neurones and their glial capsule of rabbits one hour after intraperitoneal injection of γ-aminobutyric acid (GABA) 20 mg/kg. They found a marked reduction in enzyme activity in the neurones, and an increase in the neuroglia. Intravenous injection of hydroxylamine (4.8 mg/kg) produced the opposite effect, while thiosemicarbazide (8 mg/kg) led to an activation of oxidation of succinate in the neurones without any changes occurring in the glia (Table IV).

Table V shows the activity of the succinoxidase system in neurones and glia of various brain regions. Deiters' cells of the lateral vestibular nucleus (Hamberger and Hyden, 1963) and neurones of the spinal ganglia (Hydén et al., 1958) showed a lower activity than their glial capsules, while the opposite is the case for hypoglossal nucleus (Hamberger and Sjöstrand, 1966), for the gigantocellular and oral nuclei of the reticular formation (Hydén and Lange, 1965) and for the nucleus of the trigeminal nerve (Hamberger et al., 1966). This enzyme has been studied in some detail by Hamberger. In particular he showed that intravenous injection of urea leads to a rapid increase in activity in the Deiters' neurones, but to a slow increase in the surrounding glia (Hamberger and Løvtrup, 1964). A sudden increase in intracranial pressure led to a considerable activation of oxidation of succinate in the neurones, while the glia showed practically no change (Hamberger and Rinder, 1966). During regeneration however, the succinoxidase activity of neurones of the nucleus of the hypoglossal nerve showed only a slight increase, while that of the glial capsule was greatly increased. This effect was particularly marked six days after section of the nerve (Hamberger and Sjöstrand, 1966). During natural sleep, succinate oxidation was found to be markedly increased in neurones of the gigantocellular and oral nuclei of the reticular formation, but was unchanged during barbiturate sleep; in the neuroglia, the activity was reduced in both types of sleep. In artificial insomnia induced by worrying a rabbit 5 h/day for 12 days, the activity of this enzyme system was reduced in the neurones and increased in the perineuronal glia (Hydén and Lange, 1965; Hamberger et al., 1966; Hydén, 1967a).

In enriched fractions from rat cerebral cortex the succinate dehydrogenase activity was practically the same in both neurones and glia (Rose, 1967; Sellinger et al., 1973). Similar results were obtained by Svoboda and Lodin (1972) for fractions from mouse brain. However Venkov and Rusanov (1976) separated mitochondria from each enriched fraction of rat brain, and showed that the succinate dehydrogenase activity per mg protein in glial mitochondria was four times higher than in the neuronal mitochondria.

TABLE IV

Effect of various substances on the oxidation of succinate by
isolated Deiters' cell bodies and their glial capsule from
lateral vestibular nucleus of rabbit
(Aleksidze and Blomstrand, 1968)

Substance and dose	Oxygen uptake (in 10^{-4} µl O_2/h)	
	Neurones	Glia
Control	3.8 ± 0.2	2.0 ± 0.3
GABA (20 mg/kg)	2.8 ± 0.2	3.0 ± 0.3
Hydroxylamine (4.8 mg/kg)	8.1 ± 0.6	1.2 ± 0.2
Thiosemicarbazide (8 mg/kg)	5.6 ± 0.7	2.0 ± 0.4
Sodium nitrite (30 mg/kg)	2.9 ± 0.5	1.3 ± 0.3

Animals were killed 1 h after administration of the substance; GABA was injected
intraperitoneally, the other substances intravenously. Control animals were injected
with physiological saline. The effects of sodium nitrite on neurones and glia and of
thiosemicarbazide on glia were not statistically significant. All other differences
from control were significant at the 5% level.

TABLE V

Oxidation of succinate by various types of isolated neuronal
cell bodies and their glial capsules

Region of the nervous system	Oxygen uptake (in 10^{-4} μl O_2 per hour)		Ratio neurones/glia	Authors
	Neurones	Glia		
Spinal ganglion	3.4	5.1	0.67	Hydén et al., 1958
Lateral vestibular nucleus	2.2	4.2	0.52	Hamberger, 1963.
Nucleus of the hypoglossal nerve	1.4	1.1	1.30	Hamberger and Sjöstrand, 1966
Gigantocellular nucleus of the reticular formation	2.74	2.16	1.28	Hydén and Lange, 1965; Hamberger et al., 1966
Oral nucleus of the reticular formation	5.41	3.72	1.43	Hamberger et al.,1966
Nucleus of the trigeminal nerve	3.21	1.15	2.79	" " "

Micromanometric determination of oxygen uptake using the
Cartesian diver has shown that cytochrome oxidase is 2-3 times
higher in the perineuronal glia than in neuronal cell bodies of
spinal cord anterior horn (Hydén et al, 1958) or Deiters' cells of
the vestibular nucleus (Hydén and Pigon, 1960). Hypoxic hypoxia
(subjecting the animal to an atmosphere containing 8% oxygen for
12 h) and vestibular stimulation (rotation for 25 min per day for
7 days) led to an increase in cytochrome oxidase activity in the
Deiters' cells, but did not influence the activity in the perineuronal glia (Hydén and Lange, 1961; Hamberger and Hydén, 1963).
Chemical stimulation (intravenous injection of rabbit with tricyanoaminopropene at a dose of 20 mg/kg) produced an activation of
neuronal cytochrome oxidase after one hour while no change occurred
in glial activity (Hamberger, 1963). Inhibition of monoamineoxidase
with tranylcypromine led to an activation of cytochrome oxidase in
the Deiters' cells one hour after administration; at that time no
changes in the glia were observed. During the following few days
the enzyme activity returned to normal in the neurones, while
progressively decreasing in the neuroglia (Hydén and Egyhazi, 1963).

According to Rose (1967) the activity of cytochrome oxidase in
enriched fractions from cerebral cortex of rat was 54.3 units/mg
protein in the neurones while in the glia it was 29.6 units/mg.
Subsequently Hamberger et al. (1970) working with mitochondria
separated from enriched fractions of beef brain, showed that the
activity of cytochrome oxidase in neuronal mitochondria was 0.865
μatoms O_2/min/mg protein, while in the glial fraction it was
1.066 μatoms O_2/min/mg protein. Arbogast and Arsenis (1974)
measured cytochrome oxidase activity during postnatal development,
and showed no differences between the values for neurones and glia
over the period from 10-60 days of age. However the activity of
NADPH-cytochrome-c reductase in the neurones tended to remain at a
consistently low level during development while in the glial fraction
it steadily increased, so that at the end of the first month the
level was an order of magnitude higher than in the neurones.

Venkov and Rusanov (1976) also separated mitochondria from
enriched fractions of rat brain (prepared by a modification of
Sellinger's method) and confirmed the observation of Hamberger that
cytochrome oxidase is mainly present in glia rather than in neurones.
Moreover, malate and succinate dehydrogenases also predominated
(four-fold) in glial mitochondria. The authors conclude that the
majority of the mitochondrial oxidative enzymes in nerve tissues are
localized in the glia.

The early biochemical work on separated neurones and glia
agreed with much of the histochemical data in suggesting that
oxidative metabolism was more active in neurones than in glia.
However the direct quantitative determinations outlined above show
that this conclusion can no longer be supported (see Table VI).

TABLE VI

Enzyme activities in neurones and glia

Enzyme	E.C. No.	Unit of activity	Cell type or organelle studied	Activity in neurones	Activity in glia	Reference
GROUP 1. OXIDOREDUCTASES						
Lactate dehydrogenase	1.1.1.27	pmol substrate/h/mg dry weight	Large neuronal perikarya of rabbit spinal ganglion; surrounding glial capsule	50±5	28±1.6	Lowry et al., '56; Lowry, '57
Malate dehydrogenase	1.1.1.37	"	"	231±11	90±14	"
"	"	nmol NADH/min/mg protein	Mitochondria from enriched fractions of rat brain	187±15	800±30	Venkov & Rusanov, '76
Isocitrate dehydrogenase	1.1.1.42	pmol substrate/h/mg dry weight	Large neuronal perikarya of rabbit spinal ganglion; surrounding glial capsule	4.5±0.1	6.2±0.3	Lowry et al., '56; Lowry, '57
6-phospho-gluconate dehydrogenase	1.1.1.44	"	"	1.29±0.06	2.2±0.16	"
Glucose-6-phosphate dehydrogenase	1.1.1.49	"	"	2.15±0.35	4.8±0.2	"

TABLE VI (contd.)

Enzyme	EC no.	Units	Source			Reference
α-glycero-phosphate dehydrogenase	1.1.99.5	Arbitrary units/min/mg protein	Enriched fractions of neurones and glia from mouse brain	1.26±0.07	2.4±0.64	Svoboda & Lodin, '72
Succinate dehydrogenase	1.3.99.5	" "	"	1.97±0.25	2.58±0.38	"
"	"	nmol 2,6-di-chlorophenol-indophenol/min/mg protein	Mitochondria separated from enriched fractions of rat brain	6.4±0.2	28.6±3.4	Venkov & Rusanov, '76
Glutamate dehydrogenase	1.4.1.2	pmol substrate/h/mg dry weight	Large neuronal perikarya of rabbit spinal ganglion; surrounding glial capsule.	2.5±0.2	1.6±0.2	Lowry et al., '56; Lowry. '57
"	"	μmol substrate/h/mg protein	Enriched fractions of neurones and glia from rat cerebral cortex	1.99	0.69	Rose, '68b
Monoamine oxidase	1.4.3.4	Arbitrary units/h/mg protein	"	4017±67	2188±115	Sinha & Rose, '72a
"	"	pmol substrate/h/mg protein	"	470±60	364±41	Hemminki et al., '73
"	"	μmol substrate/h/g protein	Enriched fractions of neurones and glia from rat cerebrum	33.9±1.3	24.1±2.6	Hazama et al., '76

TABLE VI (contd.)

Enzyme	EC no.	Units	Description			Reference
Monoamine oxidase	1.4.3.4	μmol substrate/h/mg protein	Enriched fractions of neurones and glia from rat hypothalamus	72.2±4.0	33.3±5.9	Hazama et al., '76
"	"	"	Enriched fractions of neurones and glia from rat hippocampus	57.2±2.3	20.1±6.9	"
"	"	"	Enriched fractions of neurones and glia from rat amygdala	62.2±1.2	27.0±3.2	"
"	"	"	Enriched fractions of neurones and glia from rat caudate nucleus	26.5±2.3	24.2±5.4	"
"	"	"	Enriched fractions of Purkyně cells, granule cells, and glia from rat cerebellum	49.2±7.3 (Purkyně cells) 31.1±1.4 (granule cells)	18.3±7.0	"
Cytochrome oxidase	1.9.3.1	$10^{-4}\mu$l O_2/h/ neuronal cell body (or equivalent volume of surrounding glial capsule)	Deiters' neuronal cell bodies from rabbit lateral vestibular nucleus; surrounding glial capsule	4.2±0.6	11.5±0.8	Hamberger & Hydén, '63
"	"	μmol substrate/min/mg protein	Enriched fractions of neurones and glia from rabbit brain	0.21±0.05	0.27±0.03	Hamberger et al., '70

TABLE VI (contd.)

	EC No.	Units	Description	Neurones	Glia	Reference
Cytochrome oxidase	1.9.3.1	µmol substrate/min/mg protein	Mitochondria separated from enriched fractions of neurones and glia from rabbit brain	2.78±0.29	2.90±0.45	Hamberger et al., '70
"	"	natom O_2/min/mg protein	Mitochondria separated from enriched fractions of neurones and glia from rat brain	122±12	322±20	Venkov & Rusanov, '76
Succinate: cytochrome c reductase	"	"	"	7.9±0.3	32.0±3.1	"
Succinate oxidase		10^{-4} µl O_2/h/neuronal cell body (or equivalent volume of glial capsule)	Deiters' neuronal cell bodies from rabbit lateral vestibular nucleus; surrounding glial capsule	2.2±0.3	4.2±0.5	Hamberger, '63
"		"	"	3.8±0.2	2.0±0.3	Aleksidze & Blomstrand, '68
"		"	Neuronal cell bodies from the gigantocellular nucleus of rabbit reticular formation; surrounding glial capsule	1.3±0.25	3.06±0.24	Hamberger et al., '66; Hyden, '67

TABLE VI (contd.)

Enzyme	EC no.	Units	Material	Neuronal	Glial	Reference
Succinate oxidase		$10^{-4} \mu l\ O_2/h/$ neuronal cell body (or equivalent volume of glial capsule)	Neuronal cell bodies from the oral nucleus of rabbit reticular formation; surrounding glial capsule	4.0±0.5	2.9±0.2	Hamberger et al., '66; Hydén, '67
"		"	Neuronal cell bodies from the trigeminal nerve nucleus of rabbit brain; surrounding glial capsule	3.1±0.8	1.1±0.1	"

GROUP 2. TRANSFERASES

Enzyme	EC no.	Units	Material	Neuronal	Glial	Reference
Choline acetyl-transferase	2.3.1.6	nmol substrate/ h/mg protein	Enriched fractions of neurones and glia from rat cerebral cortex	2.2±0.3	2.0±0.4	Hemminki et al., '73
Galactosyl transferase	2.4.1.23	nmol substrate/ 90 min/mg dry weight	Enriched fractions of neurones and glia from 16-17 day old rat brain	0.66	0.54 (astrocytes) 1.01 (oligodendroglia)	Radin et al., '72
Sialyl transferase	2.4.99.1	c.p.m./nmol N-acetylneuraminic acid	Enriched fractions from temporal lobe of rat brain	72000	30000	Gielen & Hinzen, '74
Aspartate amino-transferase	2.6.1.1	pmol substrate/ h/mg dry weight	Large neuronal perikarya from rabbit spinal ganglion; surrounding glial capsule	49±5	24±5	Lowry et al., '56 Lowry, '57

TABLE VI (contd.)

Aspartate amino-transferase	2.6.1.1	μmol substrate/h/mg protein	Enriched fractions of neurones and glia from rat cerebral cortex	5.73	5.75	Rose '68b
Hexokinase	2.7.1.1	pmol substrate/h/mg dry weight	Large neuronal perikarya from rabbit spinal ganglion; surrounding glial capsule	7.2 ± 0.2	2.2 ± 0.2	Lowry et al., '56; Lowry, '57
DNA polymerase	2.7.7.7	Arbitrary units/μg DNA	Enriched fractions of nuclei of neurones and glia from rat brain	50.9 ± 1.3	30.5 ± 0.1	Suzuki & Kato, '73
"	"	dis/30 min/mg protein	"	100	400	Szijan & Burdman, '76
Acetylneur-aminate cytidyl-transferase	2.7.7.43	c.p.m./nmol N-acetylneur-aminic acid	Enriched fractions of neurones and glia from temporal lobe of rat brain	3400	2700	Gielen & Hinzen, '74
Cerebroside sulpho-transferase	2.8.2.11	pmol sulphatide/15 min/mg protein	Enriched fractions of neurones and oligo-dendroglia from calf brain	0.54-0.55	7.86-7.95	Benjamins et al., '74
GROUP 3. HYDROLASES						
Phospho-lipase A_2	3.1.1.4	nmol monoacyl-glycero-3-phos-phorylcholine/h/mg protein	Enriched fractions of neurones and glia from rabbit brain	50.5	9.2	Woelk et al., '73

TABLE VI (contd.)

Enzyme	EC number	Units	Description			Reference
Butyryl-cholinesterase	3.1.1.8	μmol substrate/ h/neuronal cell body (or equivalent volume of glial capsule)	Neuronal cell bodies from the gigantocellular nucleus of human reticular formation; surrounding glial capsule	19.0±3.0	17.0±0.4	Pavlin, '71
	"	"	Neuronal cell bodies from human putamen; surrounding glial capsule	23.0±3.5	9.0±4.2	"
	"	Arbitrary units/ h/mg protein	Enriched fractions of neurones and glia from rat cerebral cortex	29.1±2.3	8.0±1.1	Sinha & Rose, '72a
Acetyl-cholinesterase	3.1.1.17	μmol substrate/ h/neuronal cell body (or equivalent volume of glial capsule)	Neuronal cell bodies from the caudal nucleus of guinea pig reticular formation; surrounding glial capsule	1.56	0.94	Pavlin, '71
	"	"	Neuronal cell bodies from the gigantocellular nucleus of human reticular formation; surrounding glial capsule	62.0±3.8	20.0±4.6	"
	"	"	Neuronal cell bodies from human putamen; surrounding glial capsule	55.0±5.7	25.0±5.9	"

TABLE VI (contd.)

Enzyme	EC no.	Units	Source	Values		Reference
Acetyl-cholinesterase	3.1.1.17	pmol substrate/h/mg protein	Enriched fractions of neurones and glia from rat cerebral cortex	2040 ± 280	2010 ± 220	Hemminki et al., '73
Phospho-lipase A_1	3.1.1.32	nmol monoacyl-glycero-3-phosphorylcholine/h/mg protein	Enriched fractions of neurones and glia from rabbit brain	97.7	12.0	Woelk et al., '73
Alkaline phosphatase	3.1.3.1	µmol substrate/h/g protein	Enriched fractions of neurones and glia from rat brain	19.3 ± 0.2	3.2 ± 0.1	Sinha & Rose, '72b
Acid phosphatase	3.1.3.2	"	"	267.0 ± 34.3	188.8 ± 37.3	"
DNase: acid alkaline	3.1.4.5	" "	" "	383.1 ± 17.9 55.4 ± 2.7	62.9 ± 8.4 13.0 ± 1.1	" "
2',3'-cyclic nucleotide 3'-phospho-hydrolase	3.1.4.166	µmol inorganic phosphate/min/mg dry weight	Enriched fractions of neurones and astrocytes from rat brain; and of oligodendroglia from calf white matter	0.42	0.41 (astrocytes) 3.23 (oligodendroglia)	Poduslo & Norton, '72
"		µmol substrate/min/mg protein	Enriched fractions of neurones and astrocytes from cerebral cortex, and of oligodendroglia from white matter of rat brain	0.31	1.57 (astrocytes) 4.64 (oligodendroglia)	Desmukh et al., '74

TABLE VI (contd.)

Enzyme	EC number	Units	Substrate/fraction			Reference
2',3'-cyclic nucleotide 3'-phosphohydrolase	3.1.4.166	Arbitrary units/mg protein	Enriched fractions of neurones and glia from rat cerebral cortex	0.18	3.04	Nagata et al., '74
Aryl sulphatase	3.1.6.1	nmol substrate/h/mg protein	Enriched fractions of neurones and glia from rat brain	4.1	4.1	Raghavan et al., '72
β-glucosidase	3.2.1.21	"	"	17.2	15.3	"
"		μmol substrate/h/mg protein	Enriched fractions of neurones and glia from rat cerebral cortex	29.2 ± 2.8	4.3 ± 1.8	Sinha & Rose, '72b
α-galactosidase	3.2.1.22	nmol substrate/h/mg protein	Enriched fractions of neurones and glia from rat brain	20.4	21.6	Raghavan et al., '72
β-galactosidase	3.2.1.23	"	"	96.5	110.3	"
"		μmol substrate/h/mg protein	Enriched fractions of neurones and glia from rat cerebral cortex	199.3 ± 12.6	20.1 ± 3.6	Sinha & Rose, '72b
α-mannosidase	3.2.1.24	nmol substrate/h/mg protein	Enriched fractions of neurones and glia from rat brain	36.0	42.5	Raghavan et al., '72

TABLE VI (contd.)

Enzyme	EC number	Units	Description	Values	Reference
α-manno-sidase	3.2.1.24	μmol substrate/h/mg protein	Enriched fractions of neurones and glia from rat cerebral cortex	2.3 ± 0.9 1.7 ± 0.4	Idoyaga-Vargas et al., 1973
β-glucos-aminidase	3.2.1.30	nmol substrate/h/mg protein	Enriched fractions of neurones and glia from rat brain	3334 3807	Raghavan et al., 1972.
"		μmol/substrate/h/mg protein	Enriched fractions of neurones and glia from rat cerebral cortex	55.6±11.7 45.1±7.7	Idoyaga-Vargas et al., 1972.
"		"	"	414.8±98.2 151.3±29.8	Sinha and Rose,1972b
"		μmol substrate/h/mg dry weight	Enriched fractions of neurones and astroglia from rat cerebral cortex and oligodendroglia from white matter of bovine brain	0.42 1.57 (astrocytes) 4.64 (oligodendroglia)	Desmukh et al., 1972
β-glucu-ronidase	3.2.1.31	nmol substrate/h/mg protein	Enriched fractions of neurones and glia from rat brain	5.4 3.5	Raghavan et al., 1972.
"		μmol substrate/h/mg protein	Enriched fractions of neurones and glia from rat cerebral cortex	22.3±1.4 6.6±2.1	Sinha and Rose,1973.
Glucocerebro-sidase	3.2.1.45	nmol substrate/h/mg protein	Enriched fractions of neurones and glia from rat brain	29.3 20.2	Raghavan et al., 1972.

TABLE VI (contd.)

Enzyme	EC no.	Units	Description	Values		Reference
Galacto-cerebrosidase	3.2.1.46	nmol substrate/h/mg protein	Enriched fractions of neurones and glia from rat brain	0.08	0.07	Raghavan et al., '72
Galacto-cerebroside-cerebrosidase	3.2.1.47	nmol cerebroside hydrolized during 3 h incubation of sample (2 mg dry weight)	Enriched fractions of neurones and glia from 16–17 day old rat brain	6.06	5.73 (astrocytes) 6.36 (oligodendroglia)	Radin et al., '72
N-acetyl-β-galactos-aminidase	3.2.1.53	nmol substrate/h/mg protein	Enriched fractions of neurones and glia from rat brain	185	180	Raghavan et al., '72
"	"	μmol substrate/h/mg protein	Enriched fractions of neurones and glia from rat cerebral cortex	5.5±0.9	5.7±0.3	Idoyaga-Vargas et al., '72.
"	"	arbitrary units h/g protein	"	40.0±2.4	64.9±6.3	Sinha and Rose, '73
Arabino-sidase	3.2.1.55	μmol substrate/h/g protein	Enriched fractions of neurones and glia from rat brain	31.1±2.6	9.0±1.9	Sinha and Rose, '72b
Cathepsin	3.4.4.9	μmol tyrosine/h/g protein	"	270±8	39.7±2.6	"

TABLE VI (contd.)

Enzyme	EC no.	Units	Tissue			Reference
ATPase Mg-activated	3.6.1.3	pmol substrate/h/neuronal cell body (or equivalent volume of surrounding glial capsule	Deiters' neuronal cell bodies from rabbit lateral vestibular nucleus; surrounding glial capsule.	0–0.3	1.7+0.3	Cummins and Hydén, '62.
K, Na activated	"	"	"	0–0.3	2.0+0.4	"
"	"	pmol substrate/h/mg protein	Enriched fractions of rabbit cerebral cortex	0.4	1.04	Henn et al., '72.
"	"	μmol substrate/h/mg protein	Enriched fractions of neurones and glia from rat cerebral cortex	1.5+0.5	7.0+2.0	Medzihradsky et al., '72.
Ouabain-insensitive	"	"	"	4.0+0.5	16.0+2.0	"
GROUP 4. LIASES						
Glutamate decarboxylase	4.1.1.15	μmol substrate/h/mg protein	Enriched fractions of neurones and glia from rat cerebral cortex	0.24	0.15	Rose, '68b
DOPA decarboxylase	4.1.1.26	pmol substrate/h/mg protein	"	4.1+0.4	4.3+0.3	Hemminki et al., '73.

TABLE VI (contd.)

Enzyme	EC No.	Units	Material	Neurones	Glia	Reference
Carbonic anhydrase	4.2.1.1	nl CO_2/h/neuronal cell body (or equivalent volume of the surrounding glial capsule)	Deiters' neuronal cell bodies from rabbit lateral vestibular nucleus; surrounding glial capsule.	0.4	4.5	Giacobini, '64.
"		arbitrary units/mg protein	Enriched fractions of neurones and glia from rabbit cerebral cortex	0.19	2.14	Murai, '73.
"		arbitrary units	Enriched fractions of neurones and glia from rat cerebral cortex	1.51	8.42	Nagata et al., '74.
Guanylate cyclase	4.6.1.1	pmol substrate/min/mg protein	7-day cultures of chick embryo brain cells containing principally neurones or corresponding glial cells	15±1	7±1	Goridis et al., '74.
GROUP 5. ISOMERASES						
Glucose-phosphate isomerase	5.3.1.9	pmol substrate/h/mg dry weight	Large neuronal perikarya of rabbit spinal ganglion; surrounding glial capsule	49±5	24±5	Lowry et al., '56; Lowry, '57.
GROUP 6. LIGASES						
Glutamine synthetase	6.3.1.2	μmol substrate/h/mg protein	Enriched fractions of neurones and glia from rat cerebral cortex	1.40	0.84	Rose, '68b.

Firstly, data from different authors are often in direct contradiction. This is even the case for data from rabbit Deiters'
neurones and their surrounding glia obtained by two different
authors in the same laboratory: according to Hamberger (1963)
oxidation of succinate occurs at half the rate in neurones than
that found in the glia, while Aleksidze and Blomstrand (1968) found
the rate in neurones to be twice that of glia. Moreover, the
differences that have been shown are not particularly large, and
not sufficient to convince one that oxidative processes are in
general more intensive in glia. Finally, the data summarized in
Table VI show a great deal of variety: while one particular oxidative step may be higher in the neurones, another is found to predominate in the glia. It seems to us, then, that we should not
consider the neurones as showing a more rapid oxidative metabolism
than glia, but that the oxidative process, in particular its
regulatory steps, is organized rather differently in the two cell
types. The functional significance of these differences is a
problem for future study.

    According to histochemical data, the activity of practically
every oxidative enzyme studied is considerably increased in glia
during the period of intensive myelination, in vivo, and in cultured
brain tissue in vitro both with edema, and in particular with
various types of glial proliferation (for reviews, see: Friede,
1965, 1966; G"l"bov, 1969, 1971; Pevzner, 1969b; Adams, 1969;
Gerebtzoff, 1970; Ruščak and Ruščakova, 1971; Szydlowska and Kaluza,
1972).

                Oxidative Phosphorylation and ATPase Activity

    In their studies of isolated Deiters' cells and the surrounding
glial capsule, Cummins and Hyden (1962) found the ATP content of a
neuronal cell body to be 123 pg, while a corresponding volume of glia
contained only 5 pg. The activity of ATPase however was somewhat
higher in the glia, and showed a sharp pH optimum at 8.0, while the
neuronal pH optimum was a broad range about pH 7.4.

    In neurones separated from cultured spinal ganglia of 12-14
day chick embryos by means of a Leitz micromanipulator, the content
of ATP was found to be 79×10^{-15} M per cell, while in astrocytes
microdissected from spinal cord of rat the value was 11.9×10^{-15} M
ATP. It is interesting that incubation of the cells in a high
potassium medium (54 mM instead of 4 mM) led to a reduction in ATP
content in the astrocytes to 6.4×10^{-15} M per cell, while in the
neurones practically no change occurred (Schousboe et al., 1970).
The value of this work (as with the work of Hertz et al. cited above)
is limited by the fact that the neurones and glia were obtained from
different types of tissue. It is obvious that neurones or glia
obtained from different regions of the nervous system are likely to

differ in their properties. Thus Schousboe et al. (1970) were
unable to detect ATP in neurones or astrocytes from cultured
cerebral cortex of newborn rats, although they used the same highly
sensitive luciferase assays for ATP.

The comparison of enriched fractions with mitochondria isolated
from them (Table VII) showed an almost identical ATPase content in
both fractions, and in their mitochondria. Addition of K^+ and Na^+
activated the ATPase in the whole fractions to the same degree,
while in the mitochondria, the activation was higher in the glial
fraction (Hamberger et al., 1970).

Sellinger et al. (1971) studied enriched fractions from rat
cerebral cortex. The activity of both ouabain-insensitive ATPase
and of Na^+-K^+-stimulated ATPase was an order of magnitude higher in
the glial fraction than in the neurones. Comparison of activity in
rats of different ages (5-32 days) showed that in the glial fraction
the activity of both forms of ATPase sharply increased (almost 12-
fold) between days 6-12 after birth. In the neurones however, no
substantial changes were observed during postnatal development
(Medzihradsky et al., 1971, 1972). The authors in discussing the
discrepancy between their own results and those of Hamberger et al.
(1970) are careful to take into account the results of their previous
work (Johnson and Sellinger, 1971). This work showed that neuronal
cell bodies prepared by Sellinger's method contain practically no
synaptic membranes; and it is in these structures that the greater
part of the neuronal ATPase activity is apparently localized.

In Japan, Tsukada's group has also studied ATPase during
development, using their own scheme for separation of enriched
fractions (Nagata et al., 1974). On the whole, they confirmed
Sellinger's observations, showing that both general and Na^+-K^+-
activated ATPase show stable levels in neurones during postnatal
development, in contrast to the glial fraction in which both types
of ATPase showed considerable increases. In the adult rat ATPase
activity was only 1.5 times higher in the glial fraction than in
the neuronal fraction. It is interesting that the activity of, for
example, Mg^{2+}-ATPase in the neuronal enriched fraction is equal to
that in cerebral white matter, while that in the glial fraction was
closer to the value found in gray matter (cerebral cortex). This
emphasizes again the care which must be taken in interpreting data
from topochemical analysis, and in particular warns against the
tendency to assume that the properties of neurones resemble those
of gray matter, while those of the glia resemble the white matter.

Nagata et al. (1976) showed that the activity of pyruvate
phosphokinase in glial enriched fractions of rat cerebral cortex
only slightly exceeded that in the neuronal fraction. However, the
pyruvate kinase activity in the glia was markedly stimulated by an
increased concentration of K^+ in the medium, while this effect was

TABLE VII

ATPase activity (in μmoles inorganic phosphate/10 min/mg protein) in neuronal
and glial enriched fractions of cerebral cortex of rabbit
(Hamberger et al., 1970)

Fraction	Activity in the absence of Na$^+$ and K$^+$ (control)	Increase in activity on addition of K$^+$ and Na$^+$	Percentage activation
Whole cells:			
Neurones	0.69 ± 0.03	0.65 ± 0.04	+94
Glia	0.58 ± 0.05	0.62 ± 0.06	+103
Ratio: neurones/glia	1.19	1.11	
Mitochondria:			
Neurones	0.86 ± 0.10	0.24 ± 0.07	+28
Glia	0.86 ± 0.03	0.53 ± 0.06	+62
Ratio: neurones/glia	1.00	0.79	

TABLE VIII

Half-lives (in days) of various phospholipid fractions
of neuronal and glial enriched fractions from rat
cerebral cortex (Freysz et al., 1969)

Phospholipid	Neurones	Glia	Whole cerebrum
Phosphatidylcholine	11.0	14.0	18.5
Phosphatidylethanolamine	11.5	14.5	19.5
Phosphatidylserine	13.0	16.5	-
Phosphoinositide	7.0	9.5	12.5
Sphingomyelin	11.0	20.0	40.0

much weaker in the neuronal fraction. An interesting observation
was that ·in the newborn animal, the K⁺-activation in the glial
fraction was almost as weak as in the neurones, but this activation
gradually increased in the glia during postnatal development.

Kurokawa et al. (1966) have suggested a method of separating
enriched fractions of neuronal and glial nuclei. The activity of
NAD pyrophosphorylase (E.C. 2.7.7.1) in the neuronal nuclei was
shown to be 10 times higher than in the glial nuclei.

Using another method of obtaining neuronal and glial nuclei
(Løvtrup-Rein and McEwen, 1966), Rapava et al. (1973) found that
the rate of oxidative phosphorylation was higher in neuronal than
in glial nuclei. Creatine kinase activity was also higher in
neuronal nuclei than in isolated glial cells (Murone and Ogata,
1973).

Carbohydrate Metabolism

Hamberger and Hydén (1963) found that the rate of anaerobic
glycolysis in Deiters' neurones and the surrounding glia was almost
identical. Hypoxic hypoxia increased neuronal glycolysis by 57%
but the glial rate increased by only 16%.

After a period of vestibular stimulation (moderate rotation of
the animal) glycolysis was reduced in the neurones and increased in
the glia; since the authors have also shown an activation of
oxidative enzymes as a result of vestibular stimulation (Hamberger
and Hydén, 1963; Hydén, 1963) they must conclude that stimulation of
the neurone-neuroglia system of the lateral vestibular nucleus
leads to the Pasteur effect in neurones and the Crabtree effect in
glial cells.

Lowry's group has shown that the hexokinase reaction – the first
step in glycolysis – is more active in neuronal cell bodies of
rabbit spinal ganglion than in the glial capsules surrounding these
neurones (Lowry et al., 1956; Lowry, 1957).

However, in enriched fractions from cerebral cortex of three
week old rats, hexokinase turned out to be much more active in the
glia than in the neuronal fraction. In two month old rats, the
activities were equal in the two fractions (Arbogast and Arsenis,
1974).

Glucosephosphate isomerase activity in glial capsules microdissected from spinal ganglion was shown to be 1.5-2 times higher
than in the corresponding neurones, or than in neurones of spinal
cord anterior horn (Lowry et al., 1956). Comparison of the layers
of Ammon's horn, which differ in the proportions of neurones and

glia, showed no significant differences in aldolase activity (Lowry et al., 1954).

According to Lowry (1957), in the spinal cord the neuroglia show higher activities of the pentose cycle dehydrogenases than do the corresponding neurones. Pope and Hess (1957) and Robbins et al. (1957) also conclude that the hexosemonophosphate pathway is more important in the glia than in the neurones on the basis of a topochemical analysis of the layers of cerebral cortex, and of different brain regions. Subsequently, in a number of histochemical papers, the activities of glucose-6-phosphate dehydrogenase, 6-phosphogluconate dehydrogenase and transketolase were consistently found to be higher in glia than in neurones (for reviews, see: Friede, 1965; G"l"bov, 1969; Pevzner, 1969b). Taken in conjunction with histochemical data on the predominance of Krebs cycle enzymes in the neurones, this suggests a spatial separation of oxidative processes (Laborit, 1964, 1965; Friede, 1965). Whereas in the neurones glucose is mainly broken down via the Embden-Meyerhof pathway, the products of which are oxidized by the citric acid cycle, in the glia the hexosemonophosphate shunt (pentose cycle) predominates. How far this picture corresponds to reality can only be determined after confirmation of the histochemical data by direct quantitative determination. The predominance of glucose-6-phosphate dehydrogenase in glial cells has also been confirmed by Arbogast and Arsenis (1974) in enriched fractions from young rats (up to 1.5 months of age). In adult rats, however, the activity of this enzyme (per mg protein) turned out to be almost equal in the two cell types.

Margolis and Margolis (1974) assayed mucopolysaccharides in enriched fractions of neurones, astrocytes, and oligodendroglia from calf brain, and showed that there was more heparan and chondroitin sulphate (per mg lipid-free dry weight) in neurones, while hyaluronic acid predominated in the astrocytes; low concentrations of all three compounds were found in oligodendroglia. However the concentrations of N-acetylglucosamine, N-acetylgalactosamine, N-acetylneuraminic acid, galactose and mannose separated from glycoproteins were much higher in the oligodendroglia than in the neurones. Only fucose was an exception, being highest in the astrocytes, and equal in the other two cell types. In the same work, three week old rats were given intraperitoneal injections of ^{3}H-1-glucosamine or Na^{35}SO$_4$. 16 h later, separated fractions were prepared according to Norton and Poduslo. The turnover of hyaluronic acid was shown to be much lower in the astrocytes than in the neuronal fraction, while that of glycoproteins was only slightly less. The turnover of heparan sulphate, by contrast, was higher in the astrocytes than in the neuronal fraction, while chondroitin sulphate showed an approximately equal turnover rate in the two fractions (Margolis and Margolis, 1974).

Lipid Metabolism

Topochemical analysis of human autopsy material has led to the
conclusion that the principal lipid components of the neurones are
gangliosides, and those of the glia, cerebrosides (Levin and Hess,
1965). Subsequently, Hinzen and Gielen (1973) came to a similar
conclusion from a comparison of enriched fractions prepared by
Rose's method. They found that gangliosides predominated in the
neurones, glactocerebrosides in the glia.

Promyslov (1963, 1966) showed that in glial tumors the total
lipid content, and the composition of various lipid fractions (free
and bound, cerebrosides, phospholipids, cholesterol, and mucolipids)
were closely similar to the values for normal brain tissue. Gliomas
differing in their degree of malignancy were also found to have
similar lipid compositions. If we accept this author's suggestion
that gliomas may be used as a model of normal glial cells, the
conclusion follows that the lipid composition of neurones and
neuroglia is rather similar, if not identical.

However, Davison et al. (1967) have carried out experiments on
enriched fractions prepared by Rose's method and concluded that
cerebrosides were completely absent from rat brain neurones, that
the concentration of cholesterol was almost the same in the two cell
types, and that the amounts of sphingomyelin and phospholipids were
1.5-2 times higher in the neurones than in the glia. This is in
good agreement with the previous observation of Hydén and Pigon
(1960), that treatment of isolated Deiters' cells with chloroform
reduced their dry weight only by 20%, while in the surrounding glial
capsule the reduction was 80%.

Davison and his co-workers also compared the lipid composition
of the glial fraction of the brain with that of glioma tissue. It
is interesting that unlike Promyslov, they found a much lower concentration of phospholipid and cerebroside in the glioma cells.

Studies carried out in Mandel's laboratory using enriched
fractions prepared by a modification of Rose's method, showed that
in the glial fraction from rat brain total lipid and cholesterol
were about 1.5 times higher than in the neuronal fraction, while
phospholipids were 2.5 times higher (Freysz et al., 1967). Thus
the proportion of cholesterol to total lipid was similar in the two
fractions, while the proportion of phospholipid in glial lipids was
markedly higher than in neuronal lipid. In other work from the
same laboratory (Freysz et al., 1968) the number of cells in a known
mass of the fraction was measured in a hemocytometer, from which it
could be shown that the content of total lipid and cholesterol in a
single glial cell was 5.6-9 times higher than in a single neurone.
The ratio between the different types of phospholipid was practically
identical in neurones and glia. However the turnover of all the

TABLE IX

Lipid composition of neuronal and glial enriched fractions
from rat cerebrum (Norton and Poduslo, 1971)
(Figures are percentages of total lipid content)

Lipid	Neurones of 10–30 day rats**	Glia of 10–30 day rats**	Whole Brain		
			10 day	20 day	30 day
Cholesterol*	10.6 + 0.6	14.0 + 1.9	18.8	21.1	21.5
Total phospholipid	72.3 + 2.9	70.9 + 2.0	76.1	67.5	66.5
Phosphatidylcholine	39.9 + 2.5	36.3 + 2.1	37.8	28.3	26.3
Phosphatidylethanolamine	18.2 + 0.6	20.1 + 1.8	21.1	20.0	20.5
Phosphatidylserine*	3.9 + 0.3	5.2 + 0.4	9.8	7.3	7.8
Phosphoinositide*	4.9 + 0.4	3.5 + 0.5	2.7	2.2	2.5
Sphingomyelin	3.2 + 0.3	3.7 + 0.6	2.9	4.1	4.2
Galactolipids	2.1 + 0.4	1.8 + 0.4	1.5	6.6	11.2

*Statistically significant differences found between neurones and glia.

**Since there were no statistically significant differences between 10, 20 and 30 day-old
rats, the figures for these three groups of animals were pooled.

types of phospholipid studied was markedly higher in the neuronal
than in the glial fraction (Freysz et al., 1969). It is interesting
that the metabolism of these phospholipids in whole homogenate was
even slower than that in the enriched glial fraction (Table VIII),
indicating the presence of other structural elements containing
metabolically inert phospholipids. These structural elements
presumably consist of the neuronal processes and their myelin
sheaths. However, as was suggested above, it is exactly these
components which form the major contaminant of the glial fraction,
and this could be responsible for the observed metabolic difference
between the two fractions. To resolve this, data are necessary
from glial fractions which are as free of myelin fragments as the
neuronal cell body fractions.

Norton and Poduslo (1971) using the procedure for preparing
enriched fractions of neurones and astrocytes described above,
reported a considerably higher content of total lipid in glial
cells. Unlike Freysz et al., they found no predominance of phospholipids in the glia; however they demonstrated a difference in the
proportions of different types of phospholipid. In the neurones
there was a somewhat higher proportion of phosphatidylinositol,
and in the astroglia - phosphatidylserine (Table IX).

In contrast to the conclusions of Lewin and Hess (1965) and of
Hinzen and Gielen (1973) that gangliosides predominate in the
neurones and cerebrosides in the glia, Norton and Poduslo report a
striking excess of ganglioside N-acetylneuraminic in the glial
fraction (1.6 pg/cell) by comparison with the neuronal fraction
(0.123 pg/cell) while galactolipid content was rather similar in
the two fractions (Table IX).

It is possible that these differences could be due to the fact
that the glial fraction of the latter authors contains largely
astrocytes, which might differ in lipid composition from other cell
types. This same consideration might explain some of the other
observations; surprisingly they did not find any difference in lipid
composition of neuroglia in a comparison of 10, 20, and 30-day old
rats, that is during the period of most rapid myelination. Clearly,
the biochemical reorganization of lipid metabolism leading to
myelination is localized mainly in the oligodendroglia, not in the
astrocytes.

Subsequently, Poduslo and Norton (1972) separated an enriched
oligodendroglial fraction. In these cells the mean lipid content
was 8-9 times lower than in cells of the enriched neuronal fraction.

Thus, three different methodological approaches to the biochemical study of neuroglia (topochemical analysis, investigation
of pathological material, and enriched fractions) have given
completely contradictory results. At the present time, it is hardly

possible to decide which data merit the greatest confidence in
relation to conclusions about the composition of neuroglia, since
none of the three approaches are free of drawbacks. As is clear
from the work that has been described, even a slight modification
of Rose's method leads to considerable differences in the types of
lipid found in the enriched fractions by different authors.

Pomazanskaya et al. (1969) compared the fatty acid composition
of lipids from neuronal and glial fractions of rat cerebral cortex.
Lipid analysis (phosphatidylserine, monophosphoinositide, sphingomyelin) showed that the fatty acid characteristics of neuronal and
glial phospholipids differed considerably for a number of fatty
acids. Glial phospholipids in this study are very similar to those
for whole gray matter of cerebral cortex (Table X). The authors
suggest that this is due to the much greater number of glial cells
than neurones, and the high phospholipid content per glial cell.
Thus in their opinion, the major part of cerebral lipid is of glial
origin.

A marked similarity of fatty acid composition in whole brain
and neuroglia has also been shown for sphingolipids (Abe and Norton,
1974).

Hamberger and Svennerholm (1971), analyzing enriched fractions
of cerebral cortex in rabbit did not find such a marked difference
in fatty acid composition between lipids from neuronal and glial
enriched fractions from rabbit cerebral cortex. Only in the case
of arachidonic acid was some predominance in the neurones confirmed.
The authors agree with Norton and Poduslo (1971) that gangliosides
predominate in the neuroglia, although the proportions of the various
types of gangliosides did not vary between neurones, glia, mitochondria, synaptosomes, and the whole brain tissue of cortex. They
suggest that the much higher percentage of gangliosides in the
glial fraction is not a real property of glial cells, but rather a
result of the preferential contamination of the enriched glial
fraction by synaptosomes which are rich in gangliosides (Hamberger
and Svennerholm, 1971).

Hinzen and Gielen (1973) determined the fatty acid composition
of galactocerebrosides extracted from enriched fractions of cerebral
cortex. In the glial cerebrosides 22% of the fatty acids were
hydroxy fatty acids, while this type of fatty acid was not found in
neuronal cerebrosides. The fatty acid composition of neuronal and
glial gangliosides was identical. Hamberger and Svennerholm (1971)
also found no differences between neuronal and glial gangliosides.

Comparison of the triphosphoinositide phosphomonoesterase
activity in enriched fractions prepared from rat brain by the Rose
method showed a slightly higher enzyme activity in the glial
fraction than in the neuronal fraction (Salway et al., 1967).

TABLE X

Content of individual fatty acids (as % of total fatty acids) in separated phospholipids of neuronal and glial enriched fractions from rat cerebral cortex (Pomazanskaya et al., 1969)

Fatty acid	Serinephosphatide			Monophosphoinositide			Sphingomyelin		
	Neurones	Glia	Whole cortex	Neurones	Glia	Whole cortex	Neurones	Glia	Whole cortex
Palmitic (16:0)	11.8	3.8	3.3	19.6	10.6	10.5	18.0	9.0	7.0
Palmitoleic (16:1)	3.4	0.9	traces	3.5	2.3	2.9	5.5	2.7	1.0
Stearic (18:0)	36.6	46.4	42.0	28.3	35.4	31.4	23.1	61.7	58.0
Vaccenic (18:1)	15.2	14.5	15.1	9.9	9.7	10.5	11.1	6.5	5.0
Linoleic (18:2)	1.0	traces	-	2.3	1.8	3.7	2.3	0.8	1.4
Arachidic (20:0)	1.2	0.7	traces	1.8	1.3	0.6	5.0	2.4	2.1
Gadoleic (20:1)	1.7	0.6	traces	-	-	-	-	traces	2.4
Arachidonic (20:4)	4.9	2.7	5.2	20.4	29.8	30.8	-	-	-
Behenic (22:0)	-	-	-	1.4	0.9	0.9	2.8	2.1	1.5
Lignoceric (24:0)	-	-	-	-	-	-	5.3	1.3	1.2
Nervonic (24:1)	-	-	-	-	-	-	7.9	2.8	2.8

In brackets - number of carbon atoms, and number of double bonds.

According to Ansell and Spanner (1968) the activity of plasmalogenase is found mainly in the glial cells (data from topochemical
analysis), and rises considerably in the processes during
demyelination. Comparison of isolated myelin fractions with other
brain fractions led O'Brien and Sampson (1965) and Horrocks (1968)
to conclude that the lipid composition of glial cells (principally
oligodendroglia) is very close to that of myelin. Analogous data
has also been obtained by direct lipid determination in the glial
fraction, obtained by the Rose method (Davison et al., 1968).
This biochemical data is in good agreement with the well established
morphological data indicating that the myelin sheath of the axon
is formed directly from the outer membrane and cytoplasm of the
oligodendroglia (for review, see: Adams and Davison, 1965; Peters
et al., 1970; Johnston and Roots, 1972). Norton and Poduslo (1973)
have calculated that a single oligodendroglial cell, with a dry
weight around 50 pg, synthesized on average about 175 pg myelin per
day.

In vitro measurements of the rate of myelin synthesis from
labelled precursors by enriched fractions showed that the ratio
between the incorporation by neurones and glia depended both on
the type of lipid concerned, and on the particular labelled
precursor used. For example, incorporation of ^{32}P-orthophosphate
into phospholipids (Freysz et al., 1969; Woelk et al., 1974) ^{14}C-
mevalonic acid into neutral lipids (Jones et al., 1971), ^{14}C-linoleic
and linolenic acids into phospholipids (Cohen and Bernsohn, 1973),
^{14}C-galactose into galactosides (Desmukh et al., 1974), and ^{14}C-
acetate into sterol (Jones et al., 1975) occurred at a higher
rate in the neuronal fraction. On the other hand, incorporation of
^{14}C-labelled serine and ethanolamine (Goracci et al., 1973) and
choline (Binaglia et al., 1973) into phospholipids, mevalonic acid
into cholesterol esters (Jones et al., 1971) and N-acetylneuraminic
acid into gangliosides (Jones et al., 1972) occurred more rapidly
in the glial enriched fraction. In vivo, however, incorporation of
^{14}C-ethanolamine into ethanolamineglycerophosphatides after intraperitoneal injection in rabbits was higher in the neuronal than the
glial fraction (Goracci et al., 1975).

Even when similar experimental conditions are used, different
authors do not always agree. Thus Abdel-Latif et al. (1974) found
a more intensive incorporation of ^{14}C-choline into phospholipids of
the neuronal fraction in vivo, in contrast to the results of Binaglia
et al. (1973). One cause of this discrepancy may be that the
former authors used the Norton and Poduslo method, while the latter
authors used Hamberger's scheme. In the first case, the glial
fraction consists mainly of astrocytes, while in the second case,
a mixture of glial cell types is obtained.

The data on the ratios between various enzyme activities in
neurones and glia are also rather contradictory. Thus the activity

of CDP-choline-1,2-diglycerophosphocholinetransferase was twice as
high in the neuronal fraction from chick cerebral cortex as in the
corresponding glial fraction (Freysz and Mandel, 1974). Ethanolamine phosphotransferase also predominated (1.7-2.8 times) in
neuronal enriched fractions from rabbit cerebral cortex (Roberti et
al., 1975). Glucosyl transferase activity was shown only in the
neuronal fraction of rat cerebral cortex, being absent from the glia,
while galactosyl transferase and β-galactosidase activities were
practically identical in the two fractions (Radin et al., 1972).
Cerebroside sulphotransferase activity turned out to be eight
times higher in an enriched oligodendroglial fraction from calf
cerebral cortex than in the neuronal fraction, while in the astrocyte and neuronal fractions of 10 day old rat cerebral cortex the
activity of this enzyme was very similar; by 16 days, however, the
neuronal activity had increased several times, while that in the
astrocytes remained almost unchanged (Benjamins et al., 1974).
These results are summarized in Table VI, together with many other
studies in enzymology.

        Data from comparison of lipid metabolism in enriched fractions
must be interpreted with even more care than other kinds of
metabolic data. For example, the age of the animals used (and thus
the stage of myelination) will profoundly affect the quantitative
and qualitative lipid composition in oligodendroglia (Norton and
Poduslo, 1971) while such parameters are relatively constant in the
neurones during development. This is underlined by the data of
Haglid and Hamberger (1973) showing that the yield of the glial
enriched fraction from their separation scheme increased 10 times
during development (from 10 days old to adult) in rats, while that
of the neuronal fraction did not change. More work is required on
this question.

                              Mineral Composition

        Data on the mineral composition of the neuroglia are not
numerous, and are rather contradictory. Studying a homogenate of
brain tissue, Katzman (1961) showed that the potassium content of
brain homogenate was only 2 times higher than that of sodium. In
neurones the main intracellular ion is K^+, while the extracellular
ion is Na^+. However the true extracellular space (the cleft
between the neuronal and glial membranes) is very small. Hence the
author suggests that the ratio of Na/K obtained in the homogenate
indicates a high sodium content in the glial cell body. Two lines
of evidence are presented for this suggestion. 1) Acidic lipids
extracted from brain tissue contain Na^+ ions bound to them in
higher concentration than K^+, and 2) glioma tissue contains almost
twice as much sodium as potassium.

Koch et al. (1962) used a different approach: comparison of
ion concentrations in thalamus before and after selective degeneration of the neurones. On completion of such retrograde degeneration, the concentration of Na^+ and Cl^- in the thalamus increased,
while K^+ and water remained unchanged. On this basis, the authors
draw the conclusion that the glial content of Na^+ and Cl^- is higher
than that of the neurones while the content of K^+ and water is the
same. The suggestion of a high content of sodium in glial cells is
supported by the data of Villegas et al. (1966) who compared intact
squid axons with cleaned axons, stripped of the layer of Schwann
glial cells. In the cleaned axon the concentration of Na^+ was
found to be 6 times lower, while the concentration of Cl^- remained
almost the same as in the intact axon. The concentration of K^+ was
increased by 1.5 times.

The opposite point of view has been consistently argued by
Kuffler and Nicholls (1964, 1965, 1966; Kuffler and Potter, 1964;
Nicholls and Kuffler, 1964, 1965; Kuffler et al., 1966; Kuffler,
1967). The authors calculate the concentration of Na^+ and K^+ both
on the basis of electrophysiological data on the changes of membrane
potentials in neurones and glia, and on direct determination using
flame photometry. The data obtained indicate no significant
differences in the ion composition of neuronal and glial cells.
These results undoubtedly deserve attention, but they were obtained
mainly from nerve cells of leech, where the ratio of neurones and
glia is unusual - each glial cell is so large that it is surrounded
by a group of 60 or more neurones. To what extent the properties
of such an arrangement resemble those of the more familiar system
consisting of large neurones surrounded by small glial cell-
satellites is not clear, and demands further investigation.

There is practically no difference in the concentration of K^+
found in Deiters' cells and their glia isolated by Hydén's method
(Hamberger and Röckert, 1964). However incubation of these fractions
at various temperatures after separation led to differential
changes in potassium concentration. In the isolated neurones, the
K^+ content was reduced at all temperatures, while in the glia a
lowering was found only at 4° C. At 20° C the K^+ concentration
remained unchanged, while at 37° C it increased.

An analogous accumulation of K^+ was observed in isolated glial
cells (but not in neurones) by Bradford and Rose (1967), who used
an incubation medium containing K^+ with enriched fractions of
neuronal and glial cells from rat cerebral cortex. If the incubation was carried out under anaerobic conditions (100% N_2) instead
of aerobic (95% O_2 + 5% CO_2) or if tissue glycolysis and respiration
were inhibited by the addition of 10 mM iodoacetic acid and 1 mM
potassium cyanide to the medium, the uptake of K^+ by the slice was
reduced by 75-80%, while that of glia was reduced by 50%. The
value for the neuronal fraction was unchanged.

K$^+$ uptake has also been shown to be higher in glia than in neurones in cerebral cortex of rabbit. Addition of ouabain to the mixture inhibited this uptake in both fractions to roughly the same degree (Haljamäe and Hamberger, 1971).

It is difficult to draw firm conclusions concerning metabolic differences between neurones and glia from these facts, since the rate of diffusion of ions into and out of the cell largely depends on the condition of the cell membrane. The separation of neurones from glia, whether using microdissection, or enriched fractions, inevitably involves some damage to the cells, while the degree of damage to the two cell types is not determined.

Measurements of ionic composition of cultures of normal and tumor astrocytes from hamster carried out by Lees and Shein (1970) using flame photometry show that the mean content of K$^+$ per cell is 140-180 fmol, while that of sodium is 5-10 times lower, being 20-30 fmol. However these authors underline the difficulty of applying these data directly to glial cells in situ.

Liebovitz (1970) has given a theoretical discussion of the biophysical properties of neuronal and glial membranes, including their ionic permeability. The question of the connection between the extracellular space of brain and the glial cells and in particular the problem of ionic composition and transport in neurones and neuroglia has been discussed in detail in the monograph by Franck (1970) which includes an extensive bibliography. This author has also carried out a series of experimental studies on the dynamics of uptake and release of cations (using ^{42}K and ^{24}Na) in cerebral cortex slices as a function of the composition of the medium, the presence of markers of extracellular space and of various inhibitors, the degree of morphological maturity of neuronal and glial cells at various stages of postnatal development, and so on. Franck suggests that there are at least four types of space in brain tissue, differing in a number of properties including cation ratios. One of these spaces corresponds to the extracellular space, a second to the glial cells. The two others represent the cell bodies and processes of the neurones. The two latter spaces are characterized by a much higher content of K$^+$ compared to Na$^+$ (although the exact ratio is not determined) and by a marked sensitivity to electrical stimulation. The glial space, on the other hand, is characterized by a high level of Na$^+$, while K$^+$ content is low. Since the potassium concentration in this space rapidly increased 4-5 times in a K$^+$-rich medium, Franck attributes to it a unique role in regulation of the potassium content of the extracellular space.

Enriched fractions of neuroglia from rabbit cerebral cortex showed a much higher rate of uptake of calcium than neuronal fractions. Passive binding of calcium was also higher in the

enriched glial fraction than in the neuronal or synaptosomal
fractions. High concentrations of external potassium stimulate
calcium uptake into glial cells (as well as into synaptosomes)
but do not affect neuronal uptake (Lazarewics et al., 1974).

Giacobini's work should also be discussed here (Giacobini, 1962,
1964). Using his own modification of the Cartesian diver method,
he determined the carbonic anhydrase activity in Deiters' neurones.
This turned out to be 120 times less than in a clump of glial cells
closely matched for volume; on a per cell basis the carbonic
anhydrase activity was six-fold higher in the glia than in the
neurones. The author associates this property of the neuroglia with
ion transport, suggesting the following sequence of events: the
formation of CO_2 in the neurone; the diffusion of CO_2 from the
neuronal perikaryon to the neuroglia; the carbonic anhydrase
reaction producing H_2CO_3 in the glial cell; the dissociation of
H_2CO_3 in the glia into H^+ and HCO_3^- ions; the exit of HCO_3^- ions
from glia into the blood, and in exchange the entry of Cl^- ions
into the glia from the blood; the entry of Na^+ into the glia,
together with Cl^-, and in exchange the exit of K^+ cations from the
glia into the blood and cerebrospinal fluid.

Amino Acid Metabolism

In vitro experiments showed that uptake of ^3H-2,5-histidine
from the incubation medium occurred only in enriched glial fractions,
but not in neurones (Bradford and Rose, 1967). This observation
was later confirmed by the detailed studies of Hamberger (1971).
The majority of ^{14}C-labelled amino acids (in particular glutamate,
aspartate, threonine, serine, proline and glycine) were actively
taken up against a concentration gradient by enriched glial cell
fractions, while the neurones accumulated amino acids very weakly.
The uptake by both fractions was inhibited by ouabain or 2,4-
dinitrophenol.

The uptake of glutamate by glial cells has attracted considerable attention. Two distinct uptake mechanisms for glutamate have
been detected in nervous tissue incubated in vitro, having high
and low affinities (Henn et al., 1974; Roberts and Keen, 1974;
Hamberger et al., 1975). The Michaelis constants for these two
systems for enriched fractions of cerebral cortex, and for cultured
astrocytes were shown to be in the ranges 12-66 µM and 130-1300 µM
respectively (Henn et al., 1974). Comparable values of K_m (15 and
160 µM) have been found by other authors (Faivre-Bauman et al.,
1974). Autoradiography has shown that the high affinity glutamate
uptake system is localized in the glia (Schon and Kelly, 1974a;
Bowery et al., 1975; Minchin and Beart, 1975). When ^{14}C-acetate
was incubated with spinal ganglia, the label was incorporated
mainly into glial glutamate, in contrast to results with ^3H-glucose

TABLE XI

Amino acid content (in nmoles/mg protein) in neuronal
and glial enriched fractions of rat cerebral cortex
(Rose, 1970)

Amino acid	Neurones	Glia	Original cell suspension
Glutamic acid	14.18 ± 2.08	23.01 ± 2.89	75.41 ± 5.12
Aspartic acid	6.03 ± 1.46	6.33 ± 0.76	25.15 ± 3.09
Alanine	6.44 ± 1.21	2.78 ± 0.13	6.61 ± 0.63
Glutamine	6.94 ± 1.38	4.79 ± 0.60	34.43 ± 3.17
GABA	3.15 ± 1.15	4.10 ± 0.45	9.78 ± 0.63

(Minchin and Beart, 1975). Glutamate uptake into glial cells was inhibited by ouabain, by removal of potassium from the medium, and also by the substitution of choline for sodium (Faivre-Bauman et al., 1974).

When superior cervical sympathetic ganglia were incubated with ^{3}H-amino acids, followed by autoradiography, it was shown that leucine was accumulated mainly by the neurones, while glycine was taken up by the glial cells. However, β-alanine was shown to be taken up mainly by the glia at low concentrations (0.2 μM), while at higher concentrations (1 μM) it was found in the neurones as well (Bowery et al., 1975). However, in other work, spinal ganglia or slices of cerebral cortex were incubated with ^{3}H-β-alanine over a broad range of concentrations. Label was found only in the glial cells. This uptake was strongly inhibited by GABA (Schon and Kelly, 1975).

Rose (1970) carried out a chromatographic analysis of a series of free amino acids (Table XI) showing that in the neurones only the concentration of glutamic acid was lower than in the glial fraction, while the concentration of the four other amino acids in the neuronal fraction was equal to or greater than that in the glia. Table XII shows the more complex picture which was obtained with carbohydrate substrates (Rose, 1968b): after a two hour incubation of enriched fractions with glucose, the content of glutamine and alanine in the neuronal fraction (calculated per mg protein) was 2.3 times higher than in the glial fraction, while for GABA the ratio was 2.0, and for aspartate and glutamine, 1.7. However, after incubation of these fractions with pyruvate only the neuronal

TABLE XII

Free amino acid content (in nmoles/mg protein) and turnover (in counts/min/nmole) in neuronal and glial enriched fractions of rat cerebral cortex after 2 h incubation with uniformly labelled ^{14}C-glutamate or pyruvate (Rose, 1968b)

Conditions of the experiment and type of enriched fraction	Glutamic acid	Aspartic acid	Alanine	Glutamine	GABA
Incubation with glucose					
Amino acid content:					
in neurones	34.3 + 4.4	32.6 + 7.0	38.0 + 4.0	41.6 + 6.3	14.7 + 1.8
in glia	19.5 + 2.0	13.9 + 3.3	18.8 + 3.3	24.2 + 7.7	6.4 + 1.0
Amino acid turnover:					
in neurones	35 + 9	47 + 8	30 + 20	–	17 + 6
in glia	293 + 40	138 + 21	91 + 15	–	97 + 19
Incubation with pyruvate					
Amino acid content:					
in neurones	21.2 + 5.5	35.6 + 5.5	17.1 + 3.8	43.4 + 11.0	2.1 + 0.2
in glia	45.6 + 7.5	36.6 + 4.0	30.0 + 3.3	30.4 + 1.8	2.0 + 0.2
Amino acid turnover:					
in neurones	361 + 88	61 + 18	115 + 17	54 + 6	59 + 31
in glia	633 + 100	250 + 58	109 + 18	233 + 35	212 + 94

TABLE XIII

Compartmentation of glutamate metabolism in
enriched glial fraction of rat cerebral cortex
(Rose, 1970)

Fraction	Ratio of radioactivity in glutamine to that in glutamate after intraperitoneal injection of 10 µCi U-^{14}C-glutamate		
	15 min	30 min	120 min
Neurones	1.01 ± 0.12	0.73 ± 0.21	0.29
Glia	2.13 ± 0.45	1.34 ± 0.22	1.02
Original cell suspension	2.10 ± 0.45	1.64 ± 0.13	1.57

content of glutamine remained higher than in the glial fraction
(1.4 times). The content of aspartate and GABA was practically
identical in the two fractions, while the content of glutamate and
alanine turned out to be twice as high in the glial fraction. At
the same time the incorporation of label from U-^{14}C-glucose or
U-^{14}C-pyruvate into all amino acids of the glial fraction was 2.8
times higher than into amino acids of the neuronal fraction. The
only exception was alanine, whose radioactivity did not differ
between the neuronal and glial fractions after incubation with
pyruvate. Rose justly points out that not all the characteristics
of amino acid metabolism in enriched fractions represent properties
of neurones or glia in vivo. However, he clearly believes that the
lower concentration and higher turnover rate of glial cells is a
real difference between the two cell types.

Nagata et al. (1974), using a slightly different scheme from
that of Rose to prepare enriched fractions, obtained a different
distribution of the same amino acids. They found that the concentration of glutamate in glia was higher than that in neurones, not
lower, although this difference did not reach statistical significance. The concentrations of GABA and of glycine in the glia were

respectively 1.5 and 2.0 times higher than in the neurones.
Interpreting these data, the authors make two points; firstly that
the concentrations of amino acids in the enriched fractions is
3-5 times lower per mg protein than that in a homogenate of cerebral
cortex; and secondly that the glial fraction may be contaminated
with synaptosomes (see Robert et al., 1975) which are particularly
rich in those amino acids which are thought to be neurotransmitters.

The amino acid metabolism of brain differs from other organs
and tissues in being compartmented, that is there are at least two
spatially separated amino acid pools, significantly differing in
metabolic activity. The idea of compartmentation was put forward
by Waelsch (Waelsch, 1959; Waelsch et al., 1964) on the basis of
experiments with glutamate. At early times (from 1 h) after
administration of labelled glutamate to an animal its radioactivity
in brain tissue turned out to be not higher, but lower than that of
glutamine, although glutamate is the direct precursor of glutamine.
This discrepancy was explained by Waelsch as being due to the
existence in brain tissue of two separate pools of glutamate: the
smaller one having the more rapid turnover, while the larger one is
metabolically less active (Berl et al., 1962, 1968; Van denBerg et
al, 1969). In the course of these studies the authors raised the
question as to whether the small, but active pool could be neuronal,
while the larger, less active pool was glial. However this
suggestion was later rejected (Waelsch, personal communication).

This question was later again raised by Rose (1970, 1973).
U-^{14}C-glutamate was injected into the lateral ventricle of the rat;
15-30 min later enriched fractions were prepared from the brain.
The radioactivity of free glutamine extracted from the neuronal
fraction was lower than that of the glutamate, while in the glial
fraction the radioactivity of glutamine was higher than the
glutamate (as was the case in whole brain homogenate). These
results are summarized in Table XIII. The author concludes that
the metabolically active glutamate fraction is localized in the
glial cells (and possibly in synaptosomes). Experiments were also
carried out in octopus. 15-16 min after an intracerebral injection
of L-U-^{14}C-glutamate, the radioactivity of glutamate was markedly
higher than that of glutamine (Rose and Cory, 1970). The authors
suggest that this difference is related to the low numbers of glial
cells in the invertebrate nervous system.

Balazs et al. (1973) have suggested that the small compartment
in nervous tissue is glial, while the large compartment is the
neurones and nerve endings. Garfinkel (1973), on the other hand,
proposed that the large compartment consists of neurones and
dendrites, while the small compartment is made up of the nerve
terminals, and perhaps the oligodendroglia, while the astrocytes
partially resemble the large compartment.

Hamberger, reviewing his own data and other published work (Hamberger et al., 1975) concluded that while the rate of incorporation of amino acids into glial cell protein is lower than for neuronal protein, the glial cells accumulated amino acids from the surrounding medium in quantities exceeding their apparent requirements (Hamberger and Henn, 1973; Hamberger et al., 1975). This could indicate that the glia provide a supply of amino acids to the neurones.

Microchemical determination of glutamate-oxaloacetate aminotransferase activity carried out by Lowry et al. (1956) showed that in the glial capsule of spinal ganglion neurones, this enzyme was six times less active than in the nerve cells of the ganglion, as well as in the motoneurones of spinal cord anterior horn. Comparing enzyme activities in the geniculate body of rat before and after selective neuronal degeneration Utley (1964) concluded that glutamate-oxaloacetate aminotransferase was localized primarily in nerve cells (which agrees well with the above data of Lowry et al.) that glutamine synthetase was present mainly in the glia, and that glutaminase was present to an equal degree in neuronal and glial cells. The latter conclusion is also in agreement with that of Promyslov and Andreeva (1967) who studied the activity of glutaminase in gliomas at various stages of malignancy.

Data from work with enriched fractions suggest that the activity of aspartate α-ketoglutarate aminotransferase is higher in neurones than glia (Rose, 1968b); this is not in agreement with Utley's work on glutamine synthetase using selective degeneration in the geniculate body.

Neurotransmitter Metabolism

A great deal of work has been carried out on GABA in neurones and glia. As shown above, its concentration in the glial fraction was only half that of the neuronal fraction after incubation of the enriched fraction with glucose, but identical in the two fractions after incubation with pyruvate (Rose, 1968b). The activity of glutamate decarboxylase, which breaks down glutamate to GABA, turned out to be only slightly higher in the neuronal fraction than in the glia. If detergent (Triton X-100) was added to the medium, the activities were practically equal in the two fractions. The incorporation of label into GABA during incubation of enriched fractions with U-^{14}C-glutamate on the other hand, is strikingly different: for the neuronal fraction, the specific radioactivity of GABA was less than 11, while for the glia it was 415 ± 72 cpm/nM (Table XIV).

Rose considers the rate of formation and the level of GABA in the glial cell to be no less than in the neurone. This conclusion

TABLE XIV

GABA formation and glutamate decarboxylase activity in neuronal and glial enriched fractions of rat cerebral cortex (Rose, 1968b)

| Fraction | Specific radioactivity of GABA (in counts/min/nmole) after a 2 h incubation with U-^{14}C labelled precursor | | | Activity of glutamate decarboxylase (in µmoles/mg protein/h) | |
	Glucose	Pyruvate	Glutamate	Without Triton X-100 treatment	With Triton X-100 treatment
Neurones	17 ± 6	59 ± 31	11	0.24	0.36
Glia	97 ± 19	212 ± 94	415 ± 72	0.20	0.65
Whole cell suspension	78 ± 10	261 ± 74	482 ± 113	0.16	0.36

is in contradiction to that of Promyslov (1969) who studied glial tumor tissue at various stages of malignancy, and demonstrated only a trace of GABA. He came to the conclusion that this amino acid is not formed in the normal neuroglia. Otsuka et al. (1971) came to the same conclusion from experiments using micromanipulation methods. In large neurones (Betz cells of the cerebral cortex, spinal moto-neurones, vestibular neurones, etc.) the concentration of GABA ranged from 900 to 6,6000 μM/g wet weight of tissue, while in the neuropil surrounding these neurones the value did not exceed 2 μM/g (Table XV).

At the present time, it is difficult to evaluate these different data. The concentration of GABA in enriched glial fractions might easily turn out to be increased by the admixture of synaptic endings, in which this amino acid might be localized in high quantity. In the neuropil however, the general concentration of GABA is naturally lower than its concentration in the glial cell bodies since these cells take up only a part of the volume of nervous tissue, surrounding the neuronal cell bodies. Enough has already been said in the preceding chapter of the difficulties of obtaining data about neuroglia from experiments on glial tumors.

Incubation of spinal ganglia in a medium containing GABA shows that this amino acid is selectively taken up by the glial cells (Schon and Kelly, 1974b). The kinetic properties of this uptake showed that both passive diffusion and active uptake were occurring (Minchin, 1974; Iversen and Kelly, 1975). The release of this GABA was increased by factors causing membrane depolarization (Minchin and Iversen, 1974; Iversen and Kelly, 1975).

Accumulation of GABA by enriched fractions from cerebral cortex has been studied in detail by Hamberger's group (Henn and Hamberger, 1971; Hamberger and Henn, 1973; Sellström and Hamberger, 1975; Hamberger et al., 1975; Sellström et al., 1975). GABA uptake by glial cells was shown to be 2-6 times higher than by neurones, while the synaptosomes were even more active than the glia. Raising the calcium concentration led to an inhibition of GABA uptake in the glial cells, while neuronal uptake was increased. Potassium ions stimulated GABA uptake at low concentrations, but inhibited it at higher levels; this latter effect was greatest in the nerve endings and least in the neuronal fraction. The GABA uptake system in the glia has been shown to have similar properties to that in the nerve endings, having a high affinity for GABA and a requirement for Na^+ ions. Both synaptosomal and glial fractions showed substantial release of pre-loaded GABA in response to high (depolarizing) concentrations of potassium in the medium, but only in synaptosomes was this effect Ca^{++}-dependent. The glial fraction showed a higher GABA aminotransferase activity than either of the other two fractions. However glutamate decarboxylase, the enzyme responsible for GABA formation, showed a high activity in the synaptosomes, while the

TABLE XV

GABA content in various neuronal cell bodies and their
surrounding tissue from cat central nervous system
(Obata et al., 1970; Otsuka et al., 1971)

Type of neurone	GABA content $(10^{-14}$ M) per neurone	Concentration of GABA in Neurone (in mM)	Concentration of GABA (in µmoles/g wet weight) in the tissue surrounding the neurones
Large neurones of the lateral vestibular nucleus	26.7 $\pm$ 5.1	6.2 $\pm$ 0.9	1.8 $\pm$ 0.2
Motoneurones of anterior horn of spinal cord	2.8 $\pm$ 0.7	0.9 $\pm$ 0.2	1.0 $\pm$ 0.1
Cells of cerebellar nuclei	5.0 $\pm$ 1.0	6.0 $\pm$ 1.1	1.8 $\pm$ 0.2
Purkyně cells of cerebellum	3.1 $\pm$ 0.6	6.6 $\pm$ 1.0	1.0 $\pm$ 0.1
Betz cells of cerebral cortex	2.3 $\pm$ 0.3	2.5 $\pm$ 0.3	1.4 $\pm$ 0.1

TABLE XVI

Ratio of the activities of the H and M forms of
lactate dehydrogenase in neurones and neuroglia in various
regions of the central nervous system (Brumberg and Pevzner, 1975b)

Region	Ratio H/M	
	Neurones	Neuroglia
Cerebral cortex (layers II-III of visual cortex)	0.48 $\pm$ 0.026	0.61 $\pm$ 0.048
Cerebellar cortex (Purkyně cell layer)	0.53 $\pm$ 0.042	0.61 $\pm$ 0.049
Spinal cord anterior horn	0.58 $\pm$ 0.038	0.53 $\pm$ 0.032
Spinal ganglia	0.60 $\pm$ 0.044	0.76 $\pm$ 0.048

activity in the glia was low (Sellström and Hamberger, 1975; Sellström et al., 1975; Hamberger et al., 1975).

The existence of a number of histochemical methods for demonstrating cholinesterase activity has resulted in a considerable number of studies in the histochemical literature in which a comparison of cholinesterase activity in neurones and glia is made. The majority of the authors conclude that neuronal cell bodies contain principally acetylcholinesterase, while the glia contain butyrylcholinesterase. (For reviews, see: Giacobini, 1959, 1964; Adams, 1965, 1969; Friede, 1965, 1966, 1967; Silver, 1967; G"l"bov, 1969; Pevzner, 1969b; Valenzuela y Chacon, 1969; Manolov and Davidov, 1976.)

This conclusion has been confirmed by electron microscope histochemistry (Manolov and Davidov, 1976). Data from direct biochemical determination of these enzymes in microdissected cells and in enriched fractions are summarized in Table VI. It is clear from the table that the activity of acetylcholinesterase in various types of neurone is substantially higher than in the corresponding glial cells. However, this difference turns out to be not very striking, being about 1.5-3 fold. Thus Aleksidze et al. (1974) found twice as much acetylcholine in neurones as in glia, using enriched fractions from rabbit cerebral cortex. The activity of butyrylcholinesterase was equal in the two fractions. Further confirmation comes from the work of Pavlin (1974), working with microdissected neurones from putamen and reticular formation. He found that the activity of acetylcholinesterase was 1.5 times higher in neurones than in glia, while the butyrylcholinesterase activity was similar in both cell types. Vernadakis and Gibson (1973) found more acetylcholinesterase in neuronal than in glial enriched fractions from chick brain, while butyrylcholinesterase predominated in the glia. Other work with enriched fractions, however, in this case from rat cerebral cortex, showed no difference in acetylcholinesterase activity between the two fractions (Hemminki et al., 1973).

Choline acetyltransferase has also been studied in enriched fractions. Vernadakis and Gibson (1973) demonstrated the presence of this enzyme in the neuronal fraction from chick brain, but found no measurable activity in the glial fraction. A similar result was found by Arbogast and Arsenis (1974) using 10-40-day old rats. However, Tsukada's group did find choline acetyltransferase activity in glial enriched fractions from rat brain. Between 10 and 40 days after birth the activity increased, but then decreased, until in the adult animal its activity was the same as in the 20-day old animal (Nagata et al., 1976). A similar pattern was also found in the neurones, although the activity was about twice the level found in the glia. By contrast, Hemminki et al. (1973) found practically equal levels of this enzyme in the two enriched fractions from rat cerebral cortex.

While glial cells accumulate GABA more actively than neurones, the uptake of monoamines (dopamine, noradrenaline and serotonin) was practically equal in the two enriched fractions (Henn and Hamberger, 1971; Hamberger and Henn, 1973).

Catechol-O-methyltransferase activity could be demonstrated only in the neuronal fraction from 10-40 day old rats (Arbogast and Arsenis, 1974).

According to Hamberger et al. (1970) monoamine oxidase (MAO) activity is practically the same in neuronal and glial fractions from rabbit cerebral cortex; no significant difference was found when mitochondria separated from these enriched fractions were used. Hemminki et al. (1973) also found similar levels of MAO in the two enriched fractions. However Sinha and Rose (1972a) found that the neuronal fraction from rat cerebral cortex contained almost twice as much MAO activity as the glial fraction. Hazama et al. (1976) arrived at the same conclusion. They prepared enriched fractions from five regions of rat brain; only in the caudate nucleus was the activity equal in the two fractions, while in the other regions (hypothalamus, hippocampus, amygdala (nuclei), and cerebellum) and also in whole brain, the MAO activity was 1.5-3 times higher in neurones than in glia.

DOPA decarboxylase activity was found to be 1.5 times higher in the neuronal fraction of rat brain than in the glia (Hemminki et al., 1973).

The neuromediator (or mediator-like) effects of cyclic nucleotides have attracted a great deal of interest. It has been shown that the activity of cyclic adenosine-2',3'-monophosphate 3'-phosphohydrolase (phosphodiesterase) in glial cell cultures was half that found in a homogenate of rat brain, or in synaptosomes isolated from cerebral cortex. However the affinity of the enzyme for its substrate (the Michaelis constant) was identical in both cases (Zanetta et al., 1972). These data are of course indirect. Direct determination of the activity of this enzyme in enriched fractions of rat brain showed that phosphodiesterase activity was higher in the glial fraction (Desmukh et al., 1974; Nagata et al., 1974, 1976). These results are summarized in Table VI. However, it is possible that this difference between the glial enriched fraction and the neurones might be due to contamination with myelin. In the scheme used by Nagata et al., for example, the myelin fraction is found directly adjacent to the glial fraction (see Fig. 4). Phosphodiesterase is known to be very active in myelin, and is sometimes used as a marker.

Mandel's group have compared tissue cultures from chick embryos, which differ in cell composition. They showed that those cultures which were comparatively rich in neurones also showed high guanyl

cyclase activity. This enzyme catalyzes the formation of cyclic
guanosine-3',5'-monophosphate from GTP. No guanyl cyclase activity
could be detected in cultures which contained only glial cells
(Goridis et al., 1974).

Possible Enzyme Markers and Isoenzyme Ratios

The use of enzymes which are specific for a particular fraction
as markers is a commonplace in the biochemistry of subcellular
fractions. Examples are succinate dehydrogenase for mitochondria,
and phosphodiesterase for myelin. It would be extremely useful to
have analogous enzyme markers specific for neurones and glia. This
would make it possible to deal with the problem of the contamination
of each fraction with material from other structures of the nervous
system, a difficulty which may be considered the Achilles' heel of
the enriched fraction method.

Since Giacobini (1962) demonstrated that the activity of
carbonic anhydrase is 120 times greater in the perineuronal glial
capsule than in the Deiters' cell body (comparing matched volumes
of tissue) this enzyme has been considered a reliable glial marker.
Murai (1973) measured this enzyme in enriched fractions from rabbit
cerebral cortex, and showed that the activity (per mg protein) was
11 times higher in the glial fraction than in the neurones. Similar
work using rat brain gave a figure of 7-8 times (Nagata et al., 1974).

Sinha and Rose (1972b) found that the activity of β-galactosidase was 10 times higher in the neuronal enriched fraction from
rat brain than in the glial fraction. They suggest that this enzyme
may be used as a neuronal marker. However Raghavan et al. (1972)
failed to find a significant difference in the activity of this
hydrolase between neurones and glia.

This discrepancy could be explained by a technical defect, for
example, insufficient purity in the fractions prepared. However,
in our opinion, the existence of a strictly neuronal or glial enzyme
marker is inherently unlikely. The point is that while the metabolically active subcellular fractions are essentially artefacts of
the preparation and represent only fragments of the complete cell,
any complete cell (including neurones and glia) represents a self-
sufficient structural unit. The fact that both neuronal and glial
cells are capable of prolonged functioning in vitro suggests a
considerable degree of functional-metabolic autonomy (see, for
example, Dittman et al., 1973; Kostenko et al., 1974; Svanidze and
Museridze, 1974; Varon, 1975; Veprintsev, 1976). Such properties
as appropriate reactions to stimuli (Svanidze et al., 1973; Svanidze
and Museridze, 1974; Kuz'min et al., 1975) and sensitive regulation
of metabolism of ions and hormones (DeVellis and Inglish, 1973;
DeVellis and Kukes, 1973; Varon and Saier, 1975) are fully preserved

under these conditions. Such functional autonomy indicates a
comparable degree of biochemical autonomy, and it is difficult to
conceive a self-regulating cell, capable of surviving for long
periods, which is completely lacking in a particular metabolic
pathway, or in a particular enzyme activity (this was discussed
above, in the context of oxidative enzymes). It is necessary to
think in terms of an intercellular cooperation of neuronal and glial
cells in the neurone-neuroglia unit (this will be fully discussed
later), and this leads to the idea that there will rather be a
predominance of certain aspects of cell metabolism in neurones or
glia.

   We therefore consider it both more interesting and more relevant
to consider the unequal distribution of isoenzymes between neurones
and glia. So far there is not a great deal of data on this subject.

   Gerhardt (1968) used the topochemical approach to compare the
ratios of activity of the five isoenzymes of lactate dehydrogenase
(LDH) in 23 regions of human brain (post-mortem). The ratios of
neuronal and glial cell numbers in different brain regions were
obtained from published neurohistological data. The authors
concluded that brain regions rich in glial cells contained mainly
the electrophoretically mobile group of LDH isoenzymes (i.e.
predominantly the H-subunit). The electrophoretically less mobile
forms (i.e. predominantly the M-subunit) were found in those brain
regions which contained the most neurones. The H-form of LDH is
considered to catalyze primarily the anaerobic metabolism (see
Brumberg and Pevzner, 1975a). If this is the case, Gerhardt's data
lead to the paradoxical conclusion that aerobic catabolism of carbohydrate is localized in the glia, while anaerobic processes are
concentrated in the neurones.

   Aleksidze and Khaglid (1970), however, came to the opposite
conclusion. They separated the LDH isoenzymes from Deiters' neurones
and their glial capsules by electrophoresis. They determined the activity of the three more rapidly migrating forms, LDH-1, LDH-2, and
LDH-3. The relative activities of these forms in the neurones were
59.1 ± 1.0, 33.4 ± 0.8 and 7.5 ± 1.6 respectively. In the glia,
the ratios were 76.0 ± 1.2, 20.0 ± 1.0 and 3.9 ± 1.8. Thus the
aerobic form of the enzyme seems to predominate in the neurones,
and the anaerobic form in the glia. It is unfortunate that there is
no data on the truly anaerobic forms of the enzyme, LDH-4 and LDH-5.
Another difficulty in interpreting this work is that processes, which
are found mainly in the glial fraction, may differ from cell bodies
of both neurones and glia in the proportion of H- and M-forms. This
is suggested by the histochemical data of Gerebtzoff (1968) and of
Dimova (1971). The work of Korochkin (1972), however, supports the
conclusions of Aleksidze and Khaglid. He separated all five LDH
isoenzymes from microdissected Deiters' cells, hippocampal neurones,
and also from cells called microgliocytes by the author (although

judging by the published photomicrographs, they appear to be oligo-dendroglia). The estimation of the LDH isoenzymes in this work is visual appraisal, without quantitative measurement of the intensity of staining of the bands on the polyacrylamide gel.

Packman et al. (1971) carried out similar work on enriched fractions of rabbit brain. They were able to clearly distinguish only four fractions, as LDH-5 was poorly separated from the origin, and did not appear on the electrophoregram. They found a relatively high level of LDH-1 and LDH-3 compared to LDH-2 and LDH-4 in the neurones, while in the glial fraction the activities of all four forms were equal. They also conclude that neuronal metabolism has a more aerobic character than that of the glia. A problem with this work relates to the tetrameric nature of the LDH molecule, and the ratios of its subunits. It is difficult to see how a high concentration of LDH-1 (H_4) and LDH-3 (H_2M_2) can be combined with a low concentration of LDH-2 (H_3M) and LDH-4 (HM_3). In their paper, the authors present electrophoregrams from brain mitochondria which also show low LDH-2 and LDH-4, which the authors suggest is due to the loss of some of the LDH during separation of the mitochondria. It seems to us that the differences in the isoenzyme spectrum of LDH between neurones and glia could also be due to methodological reasons.

Another study of LDH in enriched fractions, in this case from rat brain, showed no difference in the electrophoretic patterns between the two cell types (Nagata et al., 1974).

We have studied this problem using the cytospectrophotometric method to determine the activities of the H- and M-forms of LDH in cryostat sections of nervous tissue. The method is a modification of that described by Gerebtzoff (1966). The H/M ratios in neurones and neuroglia in various regions of the CNS are shown in Table XVI. On the whole, these data confirm the similarity of the LDH isoenzyme spectrum between neurones and glia. Only in the spinal ganglia did the glial cells differ from the neurones, showing a higher activity of the aerobic form of the enzyme (Brumberg, 1975b).

Packman et al. (1971) also investigated malate dehydrogenase in enriched fractions from rabbit cerebral cortex. They demonstrated the existence of two forms of this enzyme, which were present to an equal degree in the neuronal and glial fractions. However, Venkov and Rusanov (1976) showed that two isoenzymes of malate dehydrogenase were present in mitochondria separated from the neuronal fraction from rat brain, while three forms of the enzyme were present in glial mitochondria. Direct comparison is difficult, because in the former work Hamberger's separation scheme was used, while in the latter case a modification of Sellinger's method was used.

Hirsh (1972) compared the relative activities of two isoenzymes of β-galactosidase in microdissected motoneurones from human spinal cord (post-mortem material). One of the isoenzymes is thermolabile (form A) the other is thermostable (form B). While the ratio A/B was 82:18 in these cells, in the surrounding glia it was found to be 62:38. Interpretation of this work is again made difficult by uncertainty as to the contribution made by processes of both neurones and glia in the glial fraction, as discussed above. So the question of the proportion of isoenzymes in the glial cell bodies themselves is not yet answered.

Protein Metabolism

The dry mass of neurones from various regions of the nervous system is on average almost the same as that of an equivalent volume of the glial capsule (Hydén, 1962, 1963, 1964).

However, Freysz et al. (1967) showed that the total concentration of protein nitrogen in enriched neuronal fractions from rat brain was rather higher than in the glia. Probably this discrepancy is connected with the presence in the glial fraction of various contaminants in particular fragments containing myelin, and other structures rich in lipid which leads to the low content of protein in the glial fraction by comparison with the neurones.

Expressed on a per cell basis, the dry weight of the glial cell was on average 3.5 times and the overall content of protein 2.8 times higher than in a single neurone (Freysz et al., 1968).

A similar comparison was made by Norton and Poduslo (1971), who showed that both dry weight and protein content per astrocyte arc on average three times higher than in a single neuronal cell body (Table XVII). However, the authors themselves do not consider this data conclusive, since they suggest that during the separation proteins may be lost from neurones to a greater degree than from glia. The protein content of a single oligodendroglial cell is several times lower than that per neurone, as can be seen from Table XVII (Poduslo and Norton, 1972).

Satake and Abe (1966) showed that the rate of incorporation of labelled amino acid into protein of enriched neuronal fractions was higher than into protein of total brain homogenate. On the basis of these data it is difficult to conclude with confidence that there is a more rapid biosynthesis of proteins in neurones than in glia, in so far as the concentration of neuroglial cells in the whole cell suspension is lower to an unknown degree than the concentration of neuronal cell bodies in the enriched neuronal fractions. Not less than 1/3 of such a homogenate consists of processes (Kaplan, 1965). In these lipid-rich structures the rate of protein biosynthesis is undoubtedly much lower than in the cellular elements of the brain.

TABLE XVII

Dry weight and content of nucleic acids, protein and lipids in enriched
fractions of neurones, astrocytes and oligodendroglia
(Norton and Poduslo, 1971; Poduslo and Norton, 1972)

Substance	Whole rat brain (concentration in %)	Neurones (rat cerebral cortex)		Astrocytes (rat cerebral cortex)		Oligodendroglia (calf white matter)	
		Concentration (%)	Absolute quantity (pg/cell)	Concentration (%)	Absolute quantity (pg/cell)	Concentration (%)	Absolute quantity (pg/cell)
Dry weight			178		590		≃25
Protein	57.9 – 60.5	55.7	99	51.9	307		≃11
DNA	0.58 – 0.72	4.6	8.18 ± 1.18	1.9	11.2 ± 0.9		5.14 ± 0.75
RNA	1.27 – 1.59	13.6	24.2 ± 2.1	4.9	29.1 ± 4.4		1.95 ± 0.43
Total lipid	27.8 – 33.5	24.1	43.0 ± 3.9	38.9	230 ± 42	29.5	6.66 ± 0.84
Phospholipid			31		163		4.2
Galactolipid			0.9		4.2		0.7
Cholesterol			5.6		32		0.9

Protein content was calculated by the authors as the difference between dry weight and
the sum of lipid and nucleic acid content. The absolute amounts of the various lipid fractions were calculated by us on the basis of the authors' figures of total lipid and the
proportions of the various fractions.

Rose (1968b) incubated enriched fractions from rat cerebral cortex for 2 h with ^{14}C labelled precursors, and then measured the radioactivity of the fractions remaining after three washes in perchloric acid. This residual acid-insoluble fraction is more than 80% protein. The incorporation of U-^{14}C-glucose turned out to be somewhat higher in the glia than in the neurones, while the incorporation of U-^{14}C-pyruvate and U-^{14}C-glutamate into the neuronal fraction was more than twice as high as in the enriched glial fraction.

Analysis of the synthesis of neuronal and glial protein in vivo was first carried out by Blomstrand and Hamberger (1969). In their experiments with rabbits, they injected L-^{3}H-lysine intravenously at a dose of 60 μCi/kg, and after a period varying from 15-360 h prepared enriched fractions by their own method from cerebral cortex and spinal cord. The dynamics of incorporation of label into neuronal and glial proteins were very similar (Fig. 12); but at all times the radioactivity of the protein in the neuronal fractions of brain and spinal cord was approximately twice as high as in the corresponding glial fractions. Again it is hardly possible in this case to conclude that the biosynthesis in the neuronal cell body procedes at a higher rate than in the glial cell. The data of Blomstrand and Hamberger were obtained on a per milligram protein basis in each fraction, and the microphotography presented in their

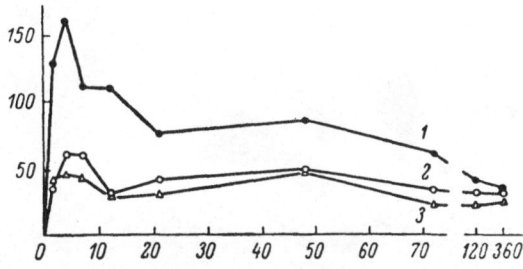

Fig. 12. Radioactivity of proteins of enriched fractions from rabbit cerebral cortex after intravenous injection of ^{3}H-leucine (Blomstrand and Hamberger, 1969). Ordinate-radioactivity of proteins (counts/min/mg protein); Abscissa-time (in hours) after intravenous injection of leucine at a dose of 60 μCi/kg; 1- neurones; 2- glia; 3- myelin.

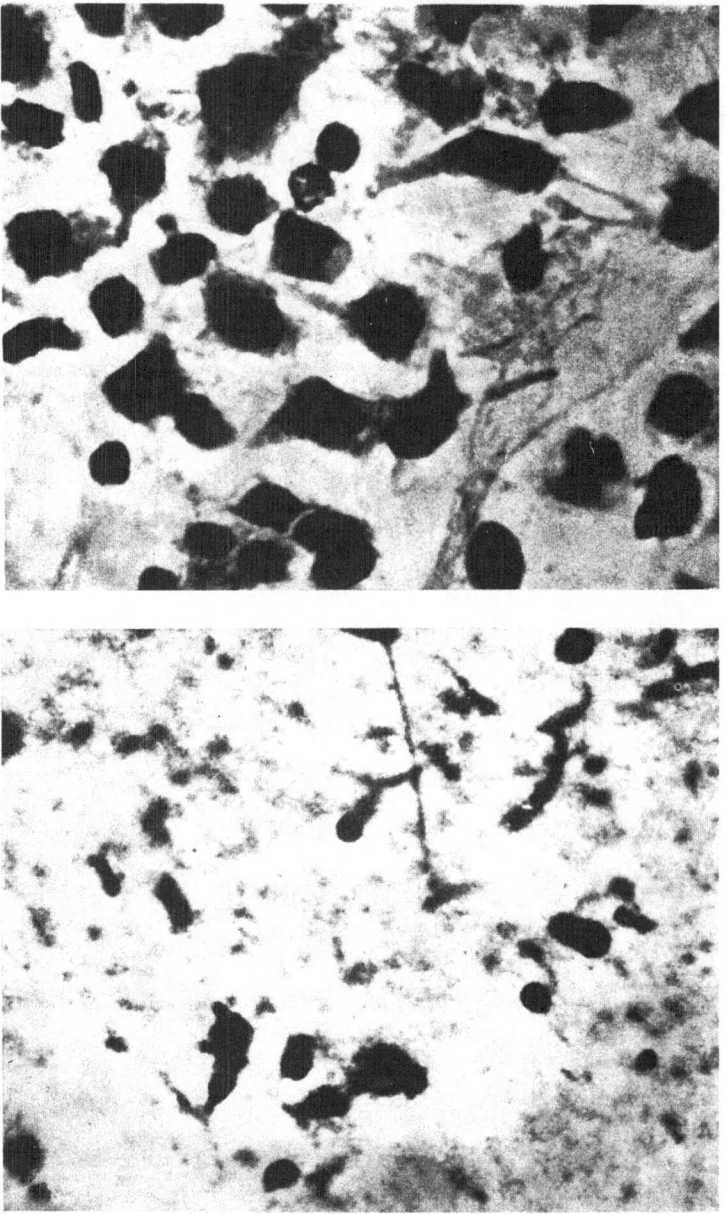

Fig. 13. Microphotography of neuronal (above) and glial (below) enriched fractions from rabbit cerebral cortex. 1000 x magnification (Hamberger et al., 1970). Separated enriched fractions were diluted in 10 volumes 0.32 M sucrose containing 0.1 M NaCl, fixed in 50% ethanol and stained according to Papanicolau.

work (Fig. 13) clearly indicates that in these fractions the density
of the neuronal cell body is considerably higher than that of the
glial cells. These authors later calculated that the DNA/protein
ratio in the enriched neuronal fractions was almost five times
higher than in the glia (Hamberger et al., 1971). The value of this
ratio reported by Norton and Poduslo (1971) was almost threefold,
which is less striking but still sufficiently clear. Thus each
milligram of protein would represent a much higher fraction of the
cell protein (i.e. for one cell body) than in the glia, while for
each milligram of glial protein the portion actually of glial origin
will be much lower in view of the presence of contamination by other
structures besides glial cells (as a rule lipid enriched). The
latter might substantially reduce the overall magnitude of incorporated radioactivity in the whole glial fraction. In fact, as can
be seen from Fig. 10, the incorporation of lysine into glial fraction
protein is much more rapid than incorporation into isolated myelin.
On the other hand when enriched fractions from rabbit are subjected
to subcellular fractionation by differential centrifugation according
to Gray and Whittaker, 2 or 3 h after intravenous injection of
$4,5-^3$H-lysine (100 µCi/kg) the radioactivity of proteins of nuclei
turns out to be identical in neuronal and glial fractions, while in
mitochondria and particularly in microsomes (see Fig. 14) the incorporation of label was higher into neuronal than into glial proteins
(Hamberger et al., 1971).

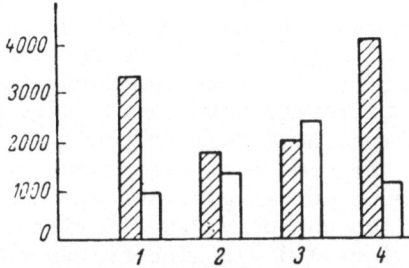

Fig. 14. Radioactivity of proteins in subcellular fractions
of neuronal and glial enriched fractions of cerebral cortex
3 h after intravenous administration of ^{3}H-leucine. (Hamberger
et al., 1971). Ordinate- radioactivity of proteins (counts/
min/mg protein); 1- whole homogenate; 2- nuclei; 3- mitochondria; 4- microsomes; Hatched columns- neurones; Blank
columns- glia.

    In the experiments of Giorgi (1971), 1-^{14}C-glycine was injected
into rat brain. After 2 h, the radioactivity of the proteins
separated from large neuronal and astrocyte fractions was approximately equal and considerably exceeded the values for the other
enriched fractions (small neurones, oligodendroglia, capillary cells).

    Johnson and Sellinger (1971) prepared enriched fractions from
cerebral cortex of rats of different ages at periods of 5-45 min
after injection of U-^{14}C-phenylalanine. At 10 and 18 days of age
the radioactivity of protein in the neuronal fraction was 13-17
times higher than in the glial fraction, and at 43 days of age,
twice as high. It is interesting that the rate of appearance of
label in the glial fraction was higher than in the neuronal fraction
and characteristically biphasic. The authors suggest that this
biphasic character of incorporation of label might possibly reflect
different rates of protein synthesis in oligodendroglia and
astrocytes.

    These differences may be related to the use of intracerebral
injection of label. Experiments using both intraperitoneal
injection (Lisý and Lodin, 1973; Albrecht and Smialek, 1975) and
intraarterial perfusion (Blomstrand et al., 1975) consistently
show a higher incorporation of labelled amino acids into neurones
than into glia. Lisý and Lodin (1973) followed the rate of incorporation during postnatal development of mice. They found that the
specific activity of the neuronal fraction (d.p.m./mg protein) at
11 days of age was 1.85 times higher than the glia; at 20 days,
2.56 times higher; at 31 days, 2.66 times higher, and in the adult
(3 months), 2.32 times higher.

    A similar difference between incorporation into neuronal and
glial fractions has also been shown in vitro for a number of
labelled amino acids (e.g. leucine, glycine, lysine, glutamate, and
phenylalanine). Blomstrand and Hamberger (1970) prepared enriched
fractions from slices of rabbit cortex after a 30-60 min incubation
at 37° in oxygenated Krebs-Ringer medium containing ^{3}H- or ^{14}C-
labelled amino acid. Different amino acids gave markedly different
results: leucine was incorporated into neuronal protein (on a per
mg protein basis) 5-6 times more actively than into glial protein,
while for the other amino acids (glycine, glutamate, phenylalanine)
the rate of incorporation was only 2.5-2.8 times higher. The incorporation of uniformly labelled ^{14}C-glucose was only slightly higher
into neuronal protein than into glia. Thus the results of experiments with glucose and glutamate confirm the data of Rose (1968b)
cited above. Comparison of slices, incubated in 8.2 or 100% O_2
showed that as the partial pressure of oxygen increased, the
incorporation of L-leucine into neuronal protein increased linearly,
while in the glial fraction it remained practically unchanged.
Increase of the ionic concentration of K^+ in the medium from 10 to
100 mM had almost no effect on the incorporation of leucine into

glial protein, while incorporation into the neuronal fraction first
of all increased and then rapidly fell off at a concentration of
K^+ higher than 40 mM. Addition of puromycin to the medium inhibited
the incorporation of leucine into proteins of both fractions to an
equal degree, while cycloheximide and 2,4-dinitrophenol caused a
greater degree of inhibition in the neuronal fraction (Table XVIII).

Takahashi et al. (1970) found a different effect of K^+ on the
synthesis of glial protein. They separated enriched glial fractions
from bovine white matter by a modification of the method of Satake
and Abe (1966) and incubated this with glucose and ^{14}C-leucine.
The radioactivity of the protein after incubation in a medium
containing 105 mM KCl was about 25% lower than in a normal medium.
It is interesting that an even greater inhibition of incorporation
of labelled lysine into protein of the glial fraction was noted when
the medium contained 10 mM glutamate.

TABLE XVIII

Effect of inhibitors on the incorporation of 4,5-^{3}H-leucine
into proteins of neuronal and glial enriched fractions from
rabbit cerebral cortex (Blomstrand and Hamberger, 1970)

Inhibitor	Concentration (in µg/ml)	Relative radioactivity of the trichloroacetic acid insoluble (as % of control)	
		neurones	glia
Puromycin	25	62	57
	50	34	44
	200	31	34
Actinomycin D	0.5	108	100
	50	87	65
Cycloheximide	50	9	20
	150	7	15
2,4-dinitrophenol	1.9	44	56
	9.3	12	18
Chloramphenicol	50	108	103
	200	104	102

Tiplady and Rose (1971) studied the biosynthesis of protein in neuronal and glial enriched fractions incubating in vitro with ^{3}H-4,5-lysine, ^{3}H-tryptophan or ^{14}C-glutamate. In all cases the radioactivity of the neuronal protein was 2-2.6 times higher than in the glia. Addition of a mixture of 19 amino acids inhibited incorporation of label into protein by 40% both in neurones and in glia.

Incubation of enriched fractions of rat cerebral cortex with ^{3}H-leucine also led to a more rapid incorporation of label into both water-soluble neuronal protein (Hemminki and Holmila, 1971) and into insoluble proteins (Hemminki, 1973) compared to the corresponding glial proteins.

In the work of Hamberger et al. (1971), enriched fractions were prepared from slices after a 30 min incubation with ^{3}H-4,5-leucine and subcellular fractions then prepared. It was found that the excess radioactivity in the neuronal fraction was largely located in the microsomal fraction. In the mitochondrial fraction the incorporation in neurones and glia turned out to be equal, while in the nuclear fraction it was 1.5 times higher in the glial fraction (Fig. 15). If the enriched fractions or their subcellular fractions were themselves incubated with label then for whole homogenate and for mitochondria the incorporation was higher in the neuronal

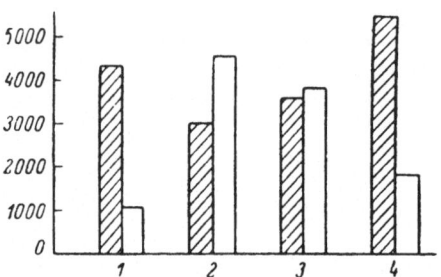

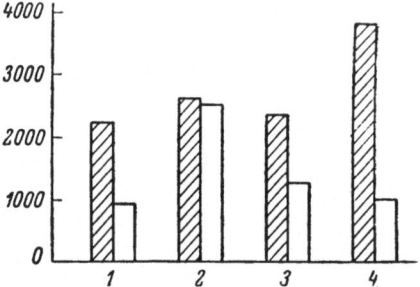

Fig. 15. Radioactivity of proteins in subcellular fractions from neuronal and glial enriched fractions of rabbit cerebral cortex after a 30 min incubation of cortex slices in a medium containing ^{3}H-leucine (Hamberger et al,. 1971). Designations as in Fig. 14.

Fig. 16. Radioactivity of proteins of subcellular fractions from neuronal and glial enriched fractions of rabbit cerebral cortex after a 30 min incubation of the fractions in a medium containing ^{3}H-leucine (Hamberger et al., 1971). Designations as in Fig. 14.

fraction than in the glia, while incorporation into the nuclear protein was the same in the two fractions (Fig. 16). Thus the turnover of all proteins in vitro seems to depend on the degree of separation of the neuronal and glial fractions. The authors themselves emphasize that the incubation conditions do not represent the in situ conditions in nervous tissue, where low molecular weight substances pass from the blood first of all to the neuroglia, and are only then incorporated into macromolecules in the glia. Later, Haglid and Hamberger (1973) compared the kinetics of protein synthesis in enriched fractions of rat brain during postnatal development. At all ages studied, the incorporation of ^{3}H-leucine into the neuronal fraction in vitro was higher than that into the glial fraction.

Løvtrup-Rein (1970) separated cell nuclei by her own method, which separates nuclei from neurones, astrocytes and oligodendrocytes plus microgliocytes. Both in experiments in vitro using a 60 min incubation of the separated fractions with ^{14}C-leucine, and in vivo, 30 min after intracisternal injection of rats with leucine, the incorporation of label (on a per mg protein basis) was highest in the astrocytes, somewhat lower in the neuronal nuclei, and lowest of all in the nuclei of oligodendroglia plus microgliocytes.

Using the same method of preparing nuclei and incubating them for 60 min with ^{3}H-4,5-leucine, Burdman (1970) obtained a somewhat different relationship: he found an incorporation of 2215 c.p.m./ mg protein for oligodendroglia plus microgliocytes, 1463 c.p.m./mg protein for neurones and 1212 c.p.m./mg protein for astrocytes. With intraperitoneal injection and subsequent separation of the nuclear fraction of cortex, the radioactivity of total protein in the neuronal nuclei after 15-240 min was invariably higher than in the glial fractions. The author also separated various types of proteins, by consecutively treating the nuclei with physiological solution, 2 M NaCl, and 0.25 M HCl. Soluble proteins, acid soluble DNP, chromatin acid protein, and a residual protein fraction are thus obtained. Only in the chromatin fraction was the radioactivity incorporated in neuronal and glial fractions equal. In all the other fractions the incorporation of label was 1.5 times higher in neuronal than in glial cells.

There are very few studies of the effects of changes in the functional state of the nervous system on the metabolism of neuronal and neuroglial cells. Blomstrand (1970) subjected rabbits to hypoxic hypoxia, exposing them for 3 or 12 h to an atmosphere containing 8% O_2 and 92% N_2. Cortex slices of these animals were then incubated with ^{3}H-leucine. Incorporation of label into protein was reduced after 3 h and increased after 12 h in the barochamber. Separation of enriched fractions from these slices showed that the effect of 3 h hypoxia was localized predominantly in the glia, but the effect of 12 h was more marked in the neuronal fraction. If

the rabbits were injected intravenously with ^{3}H-leucine, then after
3 h the radioactivity of the neuronal protein under the influence
of 3 h hypoxia was somewhat raised, while that of the glia was
reduced to a much greater degree. However under the influence of
12 h hypoxia as in the experiments in vitro, the incorporation of
label was selectively increased in the neuronal fraction. In other
work Blomstrand (1971) used local X-irradiation of rabbit cerebral
cortex at a dose of 3000 R. A week later the in vitro incorporation
of ^{3}H-leucine into neuronal protein fractions was somewhat elevated,
and in glial fraction, reduced.

When rats were injected intravenously with ^{75}Se-methionine,
and then exposed for 30 min to a hypoxic atmosphere (4% O_2) an
inhibition of incorporation of label into proteins of both enriched
fractions (prepared by Hamberger's method) was noted. A similar
inhibition occurred if the hypoxic hypoxia was combined with cerebral
ischemia (interruption of the carotid artery) or if the rats were
placed in a barochamber with elevated carbon dioxide for 1.5 h.
24 h after the cessation of these conditions, the incorporation of
label into neuronal protein was still somewhat lower than normal,
while that into the glial fraction had returned to normal, or was
even slightly above the normal level (Albrecht and Smialek, 1975).

Jarlstedt and Hamberger (1972) injected rats intraperitoneally
with 40% ethanol, using a dose of 3.2 g/kg. Controls received
saline. After 3 h, cortex slices were prepared, and incubated with
^{3}H-leucine. Enriched fractions were then prepared from the slices.
The incorporation of label into glial protein was reduced in the
ethanol-treated rats, while that into the neuronal protein was
unchanged. The glial localization of the inhibitory effects of
ethanol on protein synthesis has also been shown in experiments in
vitro.

Rats reared in the dark from birth show equal in vivo incorporation of ^{3}H-lysine into proteins of neuronal and glial fractions at
the age of 1.5 months, although in normal animals incorporation under
the same conditions (0.5-2 h after intraperitoneal injection) was
1.5 times higher into neuronal than glial proteins. If such dark-
reared rats are exposed to normal illumination (12 h light - 12 h
dark) the synthesis of neuronal proteins is enhanced, and that of
the glia reduced. If dark-reared rats are subjected to normal
lighting for 4 days, the incorporation of label was found to be
increased only in the glial fraction, while neuronal incorporation
was normal (Rose et al., 1973; Rose and Sinha, 1974).

In recent years, considerable work has been carried out on
individual proteins of the nervous system. Proteins have been
isolated which occur only in the nervous system, or almost exclu-
sively so. One of these proteins (14-3-2) is found mainly in the
neurones, while two others (S-100 and 10B) are localized in the

glial cells. (For reviews, see: Bogoch, 1969; Moore, 1969; Cicero et al., 1970; Cicero, 1973; Calissano, 1973; Palladin et al., 1976.) Both at the neuronal protein 14-3-2 and the glial protein S-100 have been demonstrated in newborn mouse brain, and they increase markedly between 7 and 14 days after birth. 14-3-2 was shown to decline in a number of brain regions of old mice, while S-100 continued to increase (Cicero et al., 1972). A corresponding messenger RNA for each of these proteins has been described in brain polysomes (Zomsely-Neurath et al., 1973).

Various authors have described other proteins in nervous tissue, including the neuronal sialoglycoprotein GP-350 (Van Nieuw Amerongen et al., 1974); glial α_2 glycoprotein (Warecka et al., 1972); the fibrillar astrocyte protein (Bignami et al., 1972; Bignami and Dahl, 1973, 1974; Dahl and Bignami, 1973; Antanadis et al., 1975, 1976; Dahl, 1976); proteins of the glial membranes (Schachner, 1974; Schachner et al., 1975) and membrane antigens D1, D2, and D3 (Jacque et al., 1974).

Protein S-100, in particular, making up around 0.2% of the total brain protein, has acidic properties. It has a rapid turnover and is localized mainly in the glial cell bodies, as well as in the neuronal nucleus. Glial cells retain their ability to synthesize S-100 when they are cultured outside the organism (Benda et al., 1968; Pfeiffer et al., 1970). The rate of synthesis of this protein in astrocytes cultured in vitro is around 0.3 pg/cell/24 h (Lightbody et al., 1970). The thermal denaturation and optical properties of protein S-100 closely resemble those of a polypeptide chain consisting of 25% α-helix and 25% β-helix and 50% random structure (Kessler et al., 1968). The molecular weight of this protein is around 20,000 (Dannies and Levine, 1969; Uyemura et al., 1971). In experiments in vitro, addition of S-100 protein to a full RNA synthesis system on nuclear chromatin or on isolated DNA led to an inhibition of protein synthesis (Bondy and Roberts, 1969).

The report of a high rate of synthesis of S-100 protein (McEwen and Hydén, 1966) contradicts the results of determination of its half-life in rat brain - around 16 days, or approximately the average of the water-soluble proteins of brain tissue (Cicero and Moore, 1970). Determinations of the half-life of S-100 in glial cells in culture also give a figure similar to that for total protein (Labourdette and Marks, 1975). After incubation of mouse brain slices in a medium containing ^{14}C- or ^{3}H-amino acids (Wronski and Von den Decken, 1975) the specific activity of S-100 was only one-tenth of that in other water-soluble proteins, including 14-3-2. However, the possibility that S-100 may be heterogeneous cannot be excluded and different portions may turn over at different rates.

Separation of S-100 protein using high speed centrifugation and electrophoresis on polyacrylamide gels made it possible to obtain two (Gombos et al., 1971; Margolis, 1971), three (Dannies and Levine, 1969) and even five (Uyemura et al., 1971) fractions of this protein, very similar in molecular weight and amino acid composition. Gombos et al. (1971) separated S-100 into two fractions, fast and slow, according to their rates of migration during polyacrylamide gel electrophoresis. A comparison of 21 regions of bovine brain and spinal cord showed as a rule that in the phylogenetically older formations of the central nervous system, the concentration of the rapidly migrating component was higher than in the more recent areas. However, the concentration of the slow-migrating component of S-100 was almost unchanged. The authors suggest that the fast-migrating fraction of S-100 protein is localized predominantly in the oligo-dendroglia, while the slow-migrating component is in the astrocytes.

Immunochemical analysis of enriched fractions from brain, carried out by Tsukada's group (Nagata et al., 1974), supported the sugges-tion of Hydén and McEwen (1966) that S-100 is localized in the glial cells. This is also supported by the immunofluorescent histochemical studies on rat brain by Sviridov et al. (1972). Recently, Hydén's group used a sensitive immunochemical method (peroxidase-antiperoxi-dase) to demonstrate a connection between the S-100 protein and synaptic membrane structures (Hydén, 1974; Haglid et al., 1974; Hydén and Ronnback, 1975; Hansson et al., 1975). The functional implica-tions of such localization for this brain-specific protein will be discussed in the following chapters.

Benda et al. (1970) demonstrated immunologically that protein 10B, described by Bogoch (1969) as being specific to the nervous system, was present only in reactive astrocytes. (They compared various areas of gliosis with regions adjacent to the tumor.) 10B is not found in normal astrocytes, in glial tumors, or in normal neurones.

Interest in these acidic proteins has grown in recent years due to the fact that their metabolism seems to be considerably altered by training, by formation of new behavioral patterns, and so on (Kometiani et al., 1970). Thus Hydén and Lange (1970, 1971), in experiments in which rats were trained to use the non-preferred paw to reach for food, showed that development of a new behavioral pattern was accompanied by the appearance of a distinct peak of the acidic protein S-100 on the electrophoregram of protein separated from the hippocampus (Fig. 17). It is interesting that the overall quantity of protein in the cerebral hemispheres of rat is around 200 mg and from the fraction of S-100 calculated by the authors there should not be more than 0.4 mg of S-100 per brain. Neverthe-less, injection of antiserum against S-100 protein markedly reduced the ability of the animals to learn, without affecting general motor activity.

Electrophoretic and immunochemical analysis, carried out in

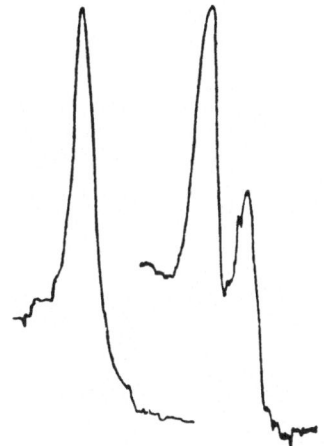

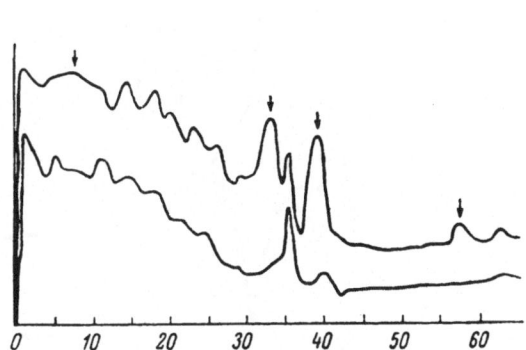

Fig. 17. Electrophoregram of
water-soluble acidic proteins
from rat hippocampus (Hydén and
Lange, 1969). Left- S-100 protein from control rat; Right-
from rat after training to use
non-preferred paw.

Fig. 18. Electrophoregram of
water-soluble proteins from
neuronal and glial enriched
fractions of rabbit cerebral
cortex (Packman et al., 1971).
Ordinate- optical density;
Abcissa- migration distance
from start (in min); Upper
curve- glia; Lower curve-
neurones; Arrows- protein
fractions localized predominantly in the glia.

Hydén's laboratory (Packman et al., 1971) showed that proteins
14-3-2 and S-100 were present in both enriched fractions from brain
(Fig. 18). The authors come to the conclusion that neither of these
proteins is purely neuronal or purely glial; both of them are only
relatively predominant in the respective types of cell. This interpretation agrees with the results of an electrophoretic separation
of soluble proteins (Fig. 19) prepared from isolated Deiters'
neurones and their glial capsule (Aleksidze, 1970). The difference
between neurones and glia were only quantitative. Karlsson et al.
(1973), obtained an increased yield of protein from enriched
fractions by using a detergent such as sodium dodecyl sulphate or
triton X-100. Rather less protein was extracted from glial than
neuronal fractions with triton X-100. When subcellular fractions
were compared, it was found that a protein band, corresponding to
a molecular weight of 37,000 was absent specifically from glial
microsomes. No differences were noted between neurones and glia
in the nuclear or mitochondrial fractions.

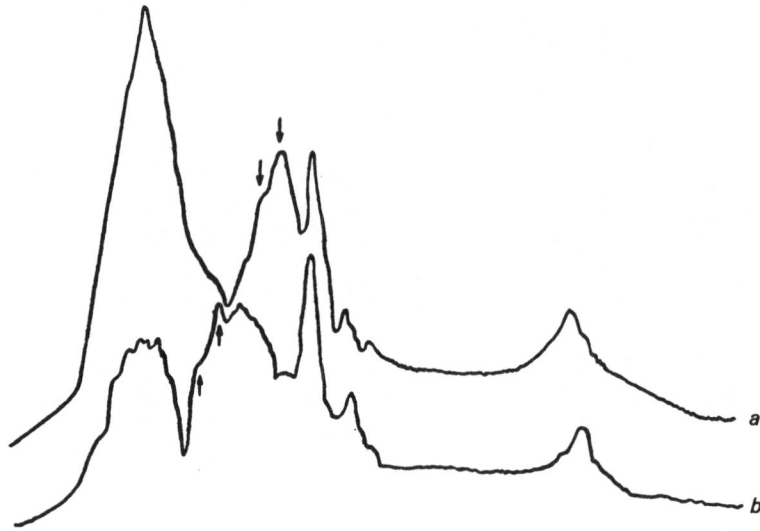

Fig. 19. Electrophoregram of water-soluble proteins from
isolated Deiters' cell bodies and their glial capsule of
lateral vestibular nucleus of rabbit   (Aleksidze, 1970).
a- neurones; b- glia; Arrows- protein fractions localized
predominantly in neurones or glia.

      Some interesting data have been obtained using cell cultures
from nervous tissue. Morphological differentiation of neuroblastoma
cells was activated by humoral factors released into the medium by
glial tumor cells (Monard et al., 1973). These authors suggest that
under normal conditions neuronal differentiation is induced by
factors originating from the glia. Schubert (1973) has confirmed
the ability of cultured glial cells to release protein into the
medium, and shown that this is not passive diffusion, but true
secretion. Varon et al. (1974) have also studied this question,
and offer the tempting suggestion of a connection between this role
of glial cells and the well known nerve growth factor. In fact, the
growth of cell cultures from spinal ganglia of newborn mice is stimulated equally by nerve growth factor and by a homogenate of glial
cells, while antiserum to nerve growth factor inhibits the cultures
and blocks the stimulating effect of glial satellite cells. However,
Monard et al. (1975) showed a number of differences between these
two types of biological factor. In particular there were very marked
differences between their electrophoregrams.

                          Nucleic Acid Metabolism

      DNA content. Relative determinations of DNA were first carried
out by Lapham and co-workers (Lapham, 1962; Lapham and Johnstone,
1963, 1964). In post-mortem human cerebrum and cerebellum,

cytophotometry of Feulgen stained material showed that the over-
whelming majority both of protoplasmic and fibrillary astrocytes
contain the same quantity of nuclear DNA as epithelial cells of the
renal canals. Similar results were obtained for cerebellum of
experimental animals. Thus the bulk of the astrocytes in human brain
are diploid. Only in a small percentage of astrocytes (less than
10%) was a greater amount of DNA found per nucleus, varying from di-
to tetraploid, indicating preparation of the cell for division
(Schwartz et al., 1966). These measurements, however, were carried
out on guinea pigs with experimental ataxia. This effect might
perhaps be produced biochemically by a stimulation of DNA synthesis
in reactive astrocytes.

In the newborn rat, all cells of the cerebellum were diploid,
while at 5-7 days a marked development of polyploidy occurred in
the Purkyně cells. Glial cells of the cerebellum remained diploid
for the whole lifetime (both for animals and for man) (Lapham, 1968;
Lentz and Lapham, 1970). Only the diploid quantity of DNA was
found in glial cells of cat hippocampus (Herman and Lapham, 1968).

Direct determination of the absolute content of DNA using
UV-cytospectrophotometry (Pevzner, 1963b, 1965b) showed that the
mean quantity of DNA per nucleus of a macroglial cell was of the
order of 6-7 pg, i.e. corresponding to the diploid quantity of DNA
in a mammalian cell nucleus. This quantity was found both for
satellite cells of the cervical sympathetic ganglion in cat, and
for macroglia of human cerebrum (autopsy and biopsy material). Only
a very small proportion of human astrocytes were found to contain
tetraploid amounts of DNA (Pevzner, Tomina and Chaika, 1964).

Massive polyploidy was observed cytophotometrically in Feulgen
stained neuroglia in cultures of glial cells from visual cortex
of cerebral cortex from newborn rat (Svanidze, 1965). However, this
polyploidy was noted only in the first few days of culture. The
degree of polyploidy subsequently became progressively less.

On the basis of determination of the relative content of DNA in
Feulgen stained preparations, a considerable number of polyploid
cells were found in neurones of the visual cortex of adult rats:
50-60% of the total number of neurones in layer II-III, 50% in layer
IV and 60% in the deeper layers of the cortex. Polyploidy was shown
also in the glial cell satellites surrounding these neurones
(Svanidze, 1967; Svanidze and Berishvili, 1970). A very important
finding in our view is that the polyploid neurones were associated
with oligodendroglia which also continued a polyploid content of
DNA. Thus the polyploidy of these neurones complements that of their
satellites. If we take the point of view of Brodskii (1964, 1966)
on polyploidy as a form of physiological regeneration of the organ,
then we can draw the conclusion that such a functional regeneration
in nerve tissue involves not only the neurone, but the whole

neurone-neuroglia unit. It is interesting that in newborn rats this
correlation of polyploidy between neurones and their satellites is
less well-defined, appearing to be a property only of the highly
differentiated nervous tissue and expressing itself only during
postnatal development. Comparison of cerebral cortex cells from
animals at different levels of the phylogenetic scale showed that
the complexity of the morphological structure of the cortex
correlates with the existence of a high number of polyploid neurones,
while in the astrocytes of layer I and in cell satellites surrounding
cortical neurones an increasing number of cells are diploid
(Svanidze and Berishvili, 1970; Svanidze and Museridze, 1974).
However, the correlation is not perfect - in the adult cat, for
example, all the neuronal and glial cells of the visual cortex are
diploid (Museridze and Svanidze, 1975), while in human cerebellum
(post-mortem material) Mann and Yates (1973) showed a substantial
number of polyploid glial cells.

It is interesting that even in the gigantic neurones of
gastropod molluscs, which show a degree of polyploidy of the order
of several thousands, the perineuronal neuroglial cells contain no
more than diploid or tetraploid quantities of DNA (Kuhlmann, 1969).

Rose (1965, 1967), found that in the enriched neuronal fraction
the content of DNA per unit of dry weight was higher than in the
glial fraction. A similar finding was reported by Freysz et al.
(1967) with enriched fractions obtained by a slightly modified
method. According to their data, in rat cerebral cortex the content
of DNA per gram dry weight for the neuronal and glial fraction was
respectively 2.46 and 0.73 mg. However, calculations of this sort,
it goes without saying, are not very significant in that it is
difficult to calculate the proportion of cell bodies in the whole
mass of each fraction. Freysz used a Thoma chamber to count the
number of cell nuclei per unit volume of the fraction being studied,
and calculated that a single nerve cell body of cerebral hemisphere
of adult rat contained on average 0.625 ± 0.056 pg DNA phosphate
while the glial cells contained 0.636 ± 0.029. If it is assumed
that phosphorus makes up 1/10 of the DNA molecule, then these data
indicate that both in neurones and glia of cerebral cortex of rat
brain the DNA content is 6-7 pg DNA, i.e. the diploid quantity.

Norton and Poduslo (1971) obtained values for DNA of 8.18 ± 1.18 pg per neurone, and 11.2 ± 0.9 pg per astrocyte (Table XVII).
It seems to us that this result probably does not differ signifi-
cantly from that of Freysz and his co-workers. If we take the
criterion of Student's

$$t = \frac{M_1 - M_2}{\sqrt{m_1{}^2 + m_2{}^2}}$$

then from the data of Norton and Poduslo,

$$t = \frac{11.2 - 8.18}{\sqrt{0.9^2 + 1.18^2}} = \frac{3.02}{\sqrt{2.2}} = 2.02$$

Given the number of determinations, 6, this corresponds to $0.1 > p > 0.05$. Although these authors work on the assumption of the significance of this difference, they do not consider that astrocytes really contain more DNA than neurones. In their opinion, in the separation procedure there is breaking off of processes from the neuronal cell body which causes it to lose part of its DNA and protein, while in the astrocytes this loss, if it exists, occurs to a lesser degree.

Poduslo and Norton (1972) have also developed a method for preparation of an enriched fraction containing mainly oligodendroglia. In these cells the DNA content was on average 6.2 pg/cell. Fewster et al. (1973) used a slightly different method for preparation of oligodendroglia from white matter and also showed a DNA content of 6-8 pg/cell. These results are in good agreement with our cytospectrophotometric determinations (Pevzner, 1963b, 1965b; Pevzner, Tomina and Chaika, 1964). It is gratifying that data obtained by different authors using methods differing in principle should allow one to arrive at the same conclusion.

DNA turnover. While neuroglial cells have the same origin as neurones, i.e. from the neural tube, they seem to be less differentiated than nerve cells, and are more cambial. They retain the capacity for cell division into postnatal development (for literature, see: Privat, 1975). It is natural, therefore, that glial cells, unlike neurones, should show turnover of nuclear DNA. This was first demonstrated by Koenig (1958) and by Leblond and his co-workers (Walker and Leblond, 1958; Messier and Leblond, 1960) using autoradiography after labelling with tritiated thymidine. Incorporation of ^{3}H-thymidine into DNA has since been demonstrated by other authors (for review, see: Gracheva, 1968).

The autoradiographic studies of Gracheva (1968) showed that in cerebral glial cells in newborn rat the normal duration of the mitotic cycle was 26 h - phase G_2 lasted 2 h, phase G_1 (plus mitosis) lasted 12.3 h, and phase S lasted 11.7 h.

Much more intensive DNA synthesis was found in the glial cells of the subependymal layer (Smart, 1961; Smart and Leblond, 1971; Gracheva, 1969; Altman, 1972) which form the basis of the new population of neuroglia in the adult animal.

DNA polymerase activity is somewhat lower in enriched fractions of glial nuclei than in neuronal nuclei (Suzuki and Kato, 1973; Szijan and Burdman, 1974). The kinetic parameters of the enzyme,

such as the optimum pH, the substrate specificity and the dependence
on divalent cations are the same for neuronal and glial nuclei
(Szijan and Burdman, 1974). In this work, the glia were studied as
a whole, and not subdivided into their different types. This subdivision was carried out by Stamblova et al. (1973), who used zonal
ultracentrifugation to separate nuclei from rat brain into five
fractions. These were labelled 1-5 as follows: fraction 1 -
59% neurones, 41% astrocytes; fraction 2 - 19% neurones, 81% astrocytes; fraction 3 - 82% astrocytes, 18% oligodendroglia; fraction 4 -
8% astrocytes, 92% oligodendroglia; fraction 5 - 100% oligodendroglia. DNA polymerase activity was highest in fraction 1 and
progressively decreased towards fraction 5. This probably indicates
that astrocytes have a greater activity of this enzyme than oligodendroglia. DNase activity is also distributed in a similar way,
with a declining activity from fraction 1 to fraction 5 (Stambolova
et al., 1973).

Molecular hybridization has made it possible to detect much
finer qualitative differences between neuronal and glial DNA. Soga
and Takahashi (1976) found differences between enriched fractions
of oligodendroglial nuclei and fractions containing a mixture of
neuronal and astrocyte nuclei in transcription of a unique sequence
of DNA, but no difference in transcription of a repeated sequence.
Differences were found in transcription of a repeated sequence of
DNA between unseparated nuclei from whole brain and nuclei obtained
from a variety of other organs (including liver, kidney and spleen).
This presumably reflects the existence of several types of cell
with different genomes in the nervous system. The data of Soga and
Takahashi are very interesting, although at present it is impossible
to draw conclusions concerning the functional significance of the
differences they have shown between neurones and glia, or between
astrocytes and oligodendroglia. Clearly, further work is required
in this direction.

Changes in DNA metabolism correlated with changes in the
functional state of the nervous system. Under normal conditions,
in particular when stimulation of the neurones is adequate, the
quantity of DNA in the neuronal nucleus is rather constant (for
reviews, see: Pevzner, 1963a, 1966a; Brodskii, 1966). In all
probability, constancy of DNA content is also a characteristic of
the glial cells. Watson (1965a) carried out a comprehensive series
of experiments using autoradiography with tritiated thymidine. He
showed no changes in incorporation into glial DNA in hypothalamic
nuclei of rats chronically given salt water instead of ordinary
water, in the vestibular nucleus after rotation of the animal, or in
anterior horn of spinal cord or spinal ganglia after muscular
excercise. However, the proportion of glial cells showing incorporation of ^{3}H-thymidine may increase in some cases. For example,
Altman and Das (1964) placed 23-26 day old rats in a large cage with
a range of different "playthings" (ladders, wheels, shelves and so

on). In about 4.5 months the increase in brain weight of these
animals was 9.7% higher than in rats kept for the same period in
individual cages without such stimuli, while in the neocortex,
but not in other brain regions, the number of labelled cells was
increased by 59%.

Autoradiographic data shows that DNA synthesis is markedly
increased during early postnatal development, when intense myelination is occurring in the brain (Wender et al., 1974). ^{3}H-thymidine
incorporation into oligodendroglia is much increased at this stage
of development, while incorporation into astrocytes does not
increase (Burdman, 1972; Stambolova et al., 1973). In the adult
animal, the normal rate of DNA synthesis is very low, but it can be
markedly increased by the pathological processes accompanying
reactive changes (particularly proliferation) in the neuroglia. An
increase in incorporation of labelled thymidine into glial cell
nuclei was shown to occur as a result of mechanical stimulation of
brain tissue (Altman, 1962, 1963; Koenig and Barron, 1962; Reznikov,
1974; Reinis, 1975), following general X-ray irradiation of the
animal (Gracheva, 1963, 1968) and during the development of reparative processes after section of a peripheral nerve (Sjöstrand, 1965;
Watson, 1965b).

An increase of glial DNA synthesis after injury to the nerve
occurs in Schwann cells. This was shown by Neimierko and
Oderfeld-Nowak (1968; Oderfeld-Nowak and Niemierko, 1969) using
cytophotometry of a preparation of sciatic nerve of rat, made as
early as 2 h after cutting the nerve.

Tumor growth of glia is also characterised by increased DNA
synthesis. In the cells of a glial tumor of the brain, an increase
in DNA content (Pevzner, Tomina and Chaika, 1964) as well as in
intensity of incorporation of labelled adenine was shown (Promyslov
et al., 1966).

RNA content. Korey (1957, 1958) showed that the concentration
of RNA per mg total nitrogen in an enriched glial fraction from
white matter was 1.5 times higher than in whole brain homogenate.
The DNA/RNA ratio was 22.8 for neuronal fraction while for homogenate of gray and white matter of brain it was 1.3 and 2.2 respectively (Korey et al., 1958; Korey and Orchen, 1959). Even if we
assume that all cells of the neuroglia of cortex are tetraploid
(10-12 pg DNA/cell) then a DNA/RNA ratio of 22.8 would indicate an
RNA content in the neuroglial cell of around 0.5 pg RNA. However,
these studies of Korey were the very first experiments on the
separation of enriched glial fractions, and the degree of purity
of the glial cells was not very high. In particular the glial
fraction was contaminated by numerous fragments of the myelin sheath
(that is, lipid-rich structures) and also with leucocytes and cells
of the connective tissue sheath (i.e. structures with a rather high

content of DNA by comparison with RNA).

Rose (1965, 1967) used a much more highly developed separation scheme and found that the DNA/RNA ratios in glial cells and in the neuronal cell bodies of cerebral cortex of rat were almost identical (about 1.0 and 1.3 respectively). This difference from the data of Korey may be partly explained by differences in the species and age of the tissue under study (Korey used lamb white matter, Rose used cerebral cortex of adult rat) and by morphology (in white matter Schwann cells and oligendroglia predominate, while in the cortex there are more protoplasmic astrocytes, which contain more RNA than the other types of glia). However the most important factor is undoubtedly the different degree of purity of the fractions.

Hydén and his co-workers found that microdissected Deiters' neurones of lateral vestibular nucleus of rabbit contained on average 1545 pg of RNA, while an approximately equal volume of glial cells contained only 123 pg (Hydén and Pigon, 1960; Egyhazi and Hydén, 1961; Hydén and Lange, 1961). In neurones of human globus pallidus (post-mortem material) a mean of 116 pg RNA was found, while in the glial capsule the figure was 17 pg RNA (Gomirato and Hydén, 1963; Hydén, 1964). Thus the concentration of RNA per unit volume seems to be 7-12 times higher in neurones than in glial cells.

Such a contrast with the cytophotometric data and with experiments using visual estimation of neurones and glial cells in histological slices stained with dyes specific for nucleic acids (see, for instance: Pevzner, 1964, 1966, 1969b; Brumberg, 1968a; Geinisman et al., 1970b; Pevzner and Saudargene, 1971; Rubinskaya, 1971) can hardly be explained by the inevitable losses of RNA in fixation and histological preparation of nerve tissue: these losses are not large enough (for reviews, see: Brodskii, 1960, 1966; Pevzner, 1963a) and in any case could hardly differ to such a degree between neurones and glia. The main reason is undoubtedly the presence in the sample of glial capsule of other structures beside glial cells, mainly processes containing little RNA, which thus markedly reduce the overall amount of RNA found.

Deiters' neurones isolated from rat brain by Hydén's method contain 648 pg RNA, while a similar volume of glial capsule contains 330 pg (Grennel et al., 1968). Thus in this case the concentration of RNA per neurone turned out to be only twice as high as in a sample of perineuronal glia.

Norton and Poduslo (1971) found that the RNA content of enriched neuronal fractions was 3 times higher than in glia. However the concentration of neuronal cell bodies is much higher than that of glia in the corresponding fraction (see Chapter 2). Therefore, it is quite natural that calculations of the amount of RNA on a per cell basis give different results - a neuronal cell body contains on average 24.2 pg, while astrocytes contain 29.1 pg (Table XVII).

We have made UV-cytospectrophotometric measurements on the glial cell satellites surrounding neurones of the cervical sympathetic ganglion, anterior horn of spinal cord, and the spinal ganglia (Pevzner, 1965b; Brumberg, 1968a, 1968b; Brumberg and Pevzner, 1968, 1971). The quantity of RNA in these glial cells is around 3-4 pg, which is several times lower than the figure obtained by Norton and Poduslo (1971). However in the work of these authors, the glial enriched fraction contained only astrocytes of cerebral cortex (and clearly, therefore, predominantly protoplasmic astrocytes, in contrast to the fibrillary astrocytes of white matter). By comparison with these large cells, the cell-satellites studied by us, principally oligodendroglia, were of a considerably smaller size. It is logical, therefore, that their quantity of RNA is much less than in protoplasmic astrocytes. The RNA content of oligodendroglia prepared according to Norton and Poduslo (1972) or Fewster et al. (1972) was shown to be 1.5-2 pg/cell.

RNA composition. Fractionation of RNA was first carried out by Volpe and Guiditta (1967). Sedimentation analysis of RNA obtained from enriched fractions of rabbit cortex showed on the whole a similar ratio of the various RNA fractions (4S, 18S and 28S) in neuronal and neuroglial cells.

Løvtrup-Rein (1970b) separated cell nuclei from brain tissue with her own method, obtaining three fractions: astrocytes, neurones, and oligodendroglia plus microglia. In all three fractions 38S and 45S RNA were present, while in the neuronal fraction 35S RNA was also detected. RNA composition was most similar in neurones and astrocytes (Løvtrup-Rein and Grahn, 1970). The polysome composition (i.e. proportion of tri- and polysomes), RNA:protein ratios, etc. are also similar in neuronal and glial enriched fractions from rat brain (Løvtrup-Rein and Grahn, 1974). It seems that while neuronal and glial RNA may show characteristic differences in their metabolism, their physico-chemical properties are closely similar (Giuditta et al., 1972; Giuditta, 1974).

The data on the base ratios in RNA in the neurone-neuroglia unit are highly contradictory. Much attention has been attracted by the report of Hydén and Egyhazi (1962, 1963) that in the lateral vestibular nucleus of medulla oblongata of rat there is an adenine-rich glial RNA with a nucleotide composition approaching that of DNA-like nuclear RNA. However, when various neurone-neuroglia units are studied this conclusion is not borne out (Table XIX). In the same nucleus in rabbit, for example, the RNA of Deiters' cells shows no difference from glial RNA in adenine content, but has a higher cytosine level. In spinal anterior horn motoneurones of rat, the RNA composition shows a predominance of guanine and uracil (Slagel

TABLE XIX

Nucleotide composition of RNA of isolated neurones and their
glial capsule from various regions of the central nervous
system (after Brumberg and Pevzner, 1971)

Region	Base	Neuronal RNA	Glial RNA	Difference between neurones and glia (%)
Lateral	Adenine	19.7 ± 0.37	20.8 ± 0.28	
vestibular	Guanine	33.5 ± 0.39	28.8 ± 0.64	-14
nucleus of	Cytosine	28.8 ± 0.36	31.8 ± 0.27	+10
rabbit	Uracil	18.0 ± 0.18	18.6 ± 0.55	
Lateral	Adenine	20.5 ± 0.54	25.3 ± 0.16	+23
vestibular	Guanine	33.7 ± 0.33	29.0 ± 0.24	-14
nucleus of	Cytosine	27.4 ± 0.34	26.5 ± 0.43	
rat	Uracil	18.4 ± 0.26	19.2 ± 0.27	
Nucleus of the	Adenine	21.1 ± 0.63	28.1 ± 1.30	+34
hypoglossal	Guanine	24.8 ± 0.60	23.5 ± 1.47	
nerve of	Cytosine	31.9 ± 0.53	21.8 ± 1.15	-32
rabbit	Uracil	22.2 ± 0.53	26.6 ± 1.83	+20
Anterior horn	Adenine	21.6 ± 0.46	25.8 ± 0.82	+19
of the spinal	Guanine	30.4 ± 0.47	27.3 ± 1.83	
cord of rat	Cytosine	24.9 ± 1.22	23.4 ± 1.59	
	Uracil	23.3 ± 0.50	21.6 ± 1.52	
Globus	Adenine	18.3 ± 0.42	19.0 ± 0.78	
pallidus of	Guanine	30.5 ± 0.44	29.1 ± 0.15	
human	Cytosine	35.3 ± 0.60	33.7 ± 0.72	
	Uracil	15.9 ± 0.36	18.2 ± 0.36	+14

NOTE. The base composition of RNA is expressed as percentage of
the total in terms of molecular weight. Differences between
neurones and glia are only recorded where they reach statistical
significance (p < 0.05).

et al., 1966). RNA of neurones of the hypoglossal nerve nucleus
contains substantially less adenine and uracil than glial cells,
while considerably exceeding them in cytosine (Daneholt and Brattgård, 1966). Finally, in human globus pallidus (post-mortem material)
the neuronal RNA differed very little from the glial capsule
although containing slightly less uracil (Gomirato and Hydén, 1963).

In the nerve ganglia of the gastropod mollusc <u>Limnaea</u> <u>stagnalis</u>, the RNA composition in the neuroplasm of a giant axon was practically identical to that of the glial cell body (Uzorin, 1972).

In this connection it is necessary to mention axonal RNA. Its base composition has not been studied in detail. According to one report it differs from the nucleotide composition of neuronal RNA (Edstrom, 1964b), while according to another it is close to the ribosomal RNA of the neurone (Koenig, 1965). It is perhaps possible that the differences described above in nucleotide composition for different neurone-neuroglia units could be explained not by real topochemical differences in these systems, as much as by different proportions of glial and axonal RNA (and their nitrogenous bases) in the sample of glial capsule.

<u>RNA turnover</u>. Incorporation of labelled precursor into neuroglial RNA has been clearly shown by autoradiography (for review, see: Gracheva, 1968). More recently, interesting autoradiographic studies on gigantic neurones and their glia in invertebrate ganglia have appeared (Bezruchko et al., 1970; Kuz'min et al., 1975). These studies indicated a rapid turnover of RNA in glia and a marked increase in synthesis both of neuronal and glial RNA as a result of electrostimulation. An important regularity was observed by Veprintsev's group (D'yakonova, 1970; Bocharova et al., 1972) who showed autoradiographically that during electrostimulation of the gigantic neurones of snail the synthesis of glial RNA continued while the synthesis of RNA in the neurones themselves was inhibited. However, data of this sort are difficult to appraise quantitatively.

Quantitative determination of RNA turnover in the neurone-neuroglia unit was first carried out by Daneholt and Brattgård (1966). They separated neurones and their surrounding glial capsule from the hypoglossal nerve nucleus using Hydén's method, and found that incorporation of tritium-labelled adenine and cytosine into adenine, guanine and uracil of glial RNA was approximately 2.5 times higher than in the same bases of the neuronal RNA, while the rate of turnover of cytosine in neurones and glia was practically identical. Calculation showed that during a 4 h experimental period around 30% of glial RNA had turned over, compared to only 15% of total neuronal RNA.

Volpe and Giuditta (1967) injected rabbits in the sub-arachnoid space with $6\text{-}^{14}C$-orotic acid and separated radioactive 28S RNA from enriched fractions prepared by the Rose method at various times. In the first hour, the highest specific activity was in the glial fraction, while at 3-6 h the neuronal activity was higher. After 14 h a predominance of turnover was again found in the enriched glial fraction.

However, different results were obtained by Flangas and Bowman

(1970) in experiments using intracisternal injection of ^{3}H-cytidine
or orotic acid into rats. In enriched fractions, again prepared by
the Rose method, the radioactivity of total RNA in neurones in the
course of several experiments (from 0.5 to 16 h) was consistently
higher than in the glial fraction. The cause of this divergence
from the results of Volpe and Giuditta is not clear. It may have
something to do with the use of different animals, the exact
fractions of RNA used, or the isotope in question.

Four hours after intraperitoneal injection of tritium-labelled
cytidine or uridine, the radioactivity of RNA in nuclei of the
neuronal fraction and of the astrocyte fraction was 4 times that of
the cell nuclei of the oligodendroglia plus microglia fraction
(Løvtrup-Rein, 1970b). However, no significant differences were
observed between the nuclei of neurones and astrocytes. Austoker
et al. (1973) showed that 30 min after intracisternal injection of
^{3}H-uridine the radioactivity in RNA was higher in astrocyte nuclei
than in nuclei of neurones or of oligodendroglia.

While the question of the relative rates of RNA synthesis in
neurones and glia in vivo remains open, studies in vitro show more
agreement. Jarlstedt and Hamberger (1972) and Yanagihara (1974b)
have compared enriched fractions of neuronal perikarya and glial
cells, while Kato and Kurakawa (1970), Thomson (1973) and
Banks-Schlegel and Johnson (1975) used isolated neuronal and glial
nuclei. A higher rate of incorporation of label (orotic acid,
uridine, cytosine, UTP or GTP) into neuronal RNA than into glial
RNA was found. The rate of RNA synthesis differs between the
various types of glial cells. Austoker et al. (1973) incubated
fractions containing different proportions of neurones, astrocytes,
and oligodendroglia with ^{3}H-uridine, and showed a higher incorporation into astrocytes than into the other two fractions. These
results are in agreement with the in vivo data.

RNA Metabolism as a Function of Changes in
the Functional State of the Nervous System

As has already been emphasized in the introduction, credit for
the definition of the problems of functional biochemistry of the
nervous system undoubtedly rests with Holger Hydén (Goteborg,
Sweden). Since 1958, he has published many original pioneering
works, which drew the attention of many neurochemists to the study
of the neuroglia. Our own study of the quantitative cytochemistry
of the neurone-neuroglia system was begun in this way (see, for
example: Pevzner, 1963b, 1963c, 1964, 1965e). A great deal of very
interesting work by other authors has since been published. The
later works (mainly carried out in the early seventies) will be
discussed in subsequent chapters together with our own experimental
data, and will lead to an overall appraisal of the functional-

biochemical properties of the neurone-neuroglia unit. In the
present chapter, we will deal with those studies of glial RNA
metabolism in various functional states which have been carried out
in Hydén's laboratory.

These studies may be divided into three groups. In the first
group, acute influences on the experimental animal are used. For
example, a number of nervous system stimulants have been shown to
produce an increase in RNA content of Deiters' neurones of the
lateral vestibular nucleus within one hour after injection, while
the RNA content of the surrounding glial cells was decreased (See
Fig. 20). Among the pharmacological agents producing this effect
were tranylcypromine, imipramine, and 2-amino-1,1,3-tricyanopropene.
The influence of the latter (malononitrile dimer) was studied in
more detail. It was shown that in the rabbit administration of this
compound caused a 26% increase in the quantity of RNA in the Deiters'
cell body, while in a sample of glial cells there was a 45% reduction. The nucleotide composition of the RNA was also determined.
In the neurones, the relative content of cytosine was markedly

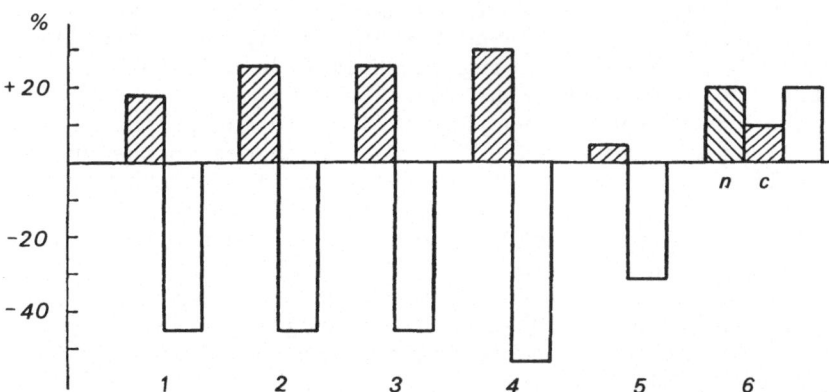

Fig. 20. Changes in RNA content of isolated Deiters' cell
bodies and their glial capsule from lateral vestibular nucleus
after different kinds of stimulation   (Hydén, 1964; Hydén
and Lange, 1966). Ordinate- percentage RNA changes by
comparison with controls; Hatched columns- neurones; Plain
columns- glia; 1- changes in RNA content per cell 1 h after
an injection of Imipramine, prep. G-31406 (4 mg/kg); 2-
1 h after an injection of Imipramine, prep. G21169; 3- 1 h
after injection of tricyanoaminopropene to rabbits (20 mg/
kg); 4- 1 h after injection of phenylcyclopropylamine (0.3
mg/kg); 5- 7 days after daily rotation of the rabbits for
25 min/day; 6- 3-5 days after training of rats to balance up
a narrow wire; n- nucleus; c- cytoplasm.

lowered with a corresponding increase in guanine. However in the neurones, by contrast, the content of guanine was sharply lower, while cytosine was increased; the concentration of adenine and uracil in neuronal and glial RNA was practically unchanged (Egyhazi and Hydén, 1961; Hydén, 1964; Hydén and Lange, 1966).

The second group concerns experiments with repeated long-term influences, which the authors designate as passive stimulation. The favorite stimulation of this kind seems to be rotation of rabbits, fixed in a cage (25 min daily for 7 days). As a result of this rotation, changes in the Deiters' cells of the vestibular nucleus occurred which were of the same type as those occurring with malononitrile dimer, although quantitatively less marked: a 4.5% increase in the content of neuronal RNA, and a 31% reduction of the RNA content of the glial sample. However the nucleotide composition of both neuronal and glial RNA remained unchanged (Hydén, 1960, 1962; Hydén and Pigon, 1960).

The third group comprises the series of experiments in which animals were repeatedly subjected to learning situations. Thus rats were trained to get food which was on a small raised platform, which could only be reached along a thin steel wire, 90 cm in length, stretched at an angle of $45°$. Training was for 45 min per day for 4 days. At the start of training the animal typically succeeded in climbing up the wire and bringing the food down only 3-5 times, while after 4 days 20 such trips could be made. At the end of the training the RNA content in the Deiters' cells and in their glial cell-satellites was increased (Fig. 20). A change in the nucleotide compositions of these RNA's was also noted. At the beginning of training, when there was a daily increase in the number of successful attempts, increase in the nucleotide composition of RNA of these cells became closer to that of DNA (Table XX). At the end of training, however, when the performance of the animals had basically stabilized, the DNA-like RNA was again replaced by RNA with a base composition characteristic of ribosomal RNA (Hydén and Egyhazi, 1962, 1963).

Another line of Hydén's work is devoted to the analysis of cells of globus pallidus (biopsy material) from persons with Parkinson's disease (Gomirato and Hydén, 1963; Hydén, 1964). A striking change was noted in RNA of glial cells in Parkinsonism. The suggestion is made that the pathogenesis of this disorder is linked with the activation of undesirable genes causing principally a biochemical defect in the neuroglia.

One of Hydén's theoretical constructs is the suggestion of the role of RNA in the mechanism of memory (Hydén, 1959b, 1960, 1962). This was followed by the hypothesis that neuroglial RNA is the short term substrate of memory, while neuronal RNA is the long term substrate (Hydén and Egyhazi, 1963; Hydén, 1964, 1967a). We do not

TABLE XX

Nucleotide composition of RNA of isolated Deiters' cell nuclei and their glial capsule from lateral vestibular nucleus of the rat after training
(Hydén and Egyhazi, 1962, 1963)

Base	RNA of neuronal nuclei			RNA of glial capsule			DNA of whole brain homogenate of rat
	Control	Training	Δ RNA	Control	Training	Δ RNA	
Adenine	21.4 ± 0.44	24.1 ± 0.39	38.1 ± 3.25	25.3 ± 0.16	28.3 ± 0.45	43.0 ± 2.83	28.6
Guanine	26.2 ± 0.45	26.7 ± 0.87	28.8 ± 5.75	29.0 ± 0.24	28.8 ± 0.31	27.0 ± 0.24	21.4
Cytosine	31.9 ± 0.77	31.0 ± 0.97	26.7 ± 6.95	26.5 ± 0.43	24.3 ± 0.36	14.0 ± 3.12	20.4
Uracil	20.5 ± 1.01	18.2 ± 1.11	6.4 ± 8.08	19.2 ± 0.27	18.6 ± 0.21	16.0 ± 1.88	28.4 (Thymine)

The bases in RNA and DNA are expressed as molar ratios (in %). RNA is the additional RNA appearing in the neurones and glia as a result of training. The nucleotide composition is calculated from the difference between the proportions of the bases in RNA before and after training.

consider it possible to go into these ideas in detail, in so far as
their speculative character is acknowledged by the authors themselves. The problem of the biochemistry of memory at the present
time is still obscure and contradictory. Even the definition of
memory is formulated in different ways by different authors. Numerous attempts have been made to connect RNA metabolism and memory
formation, but such work cannot be conclusive until the processes
of memory and information storage can be separated from the general,
or nonspecific, functioning of the nerve cell. A detailed discussion
of this problem, and full bibliographies of the participation of
cellular metabolic processes in memory mechanisms, may be found in
the monographs of Brodskii (1966), Meerson (1967), Beritashvili
(1968), Ungar (1970), Adam (1971), Gaito (1969, 1972), and Ashmarin
(1975). It may be added here that the experimental work of Hydén
involved very varied forms of stimulation to the nervous system.
The hungry animals clambered to the food platform, greedily seized
the piece of food, clambered down the wire, hastily ate the food,
and once more "walked the tight rope" up to the platform. Such a
difficult, complex performance undoubtedly involves a great number
of neural circuits in the cortex, as well as in the non-specific
regions of the nervous system. To dissect out a function specifically connected with the mechanism of memory is hardly possible.
This question is critically examined in the review by Squire and
Barondes (1972).

    We have also recently carried out some studies on metabolism of
macromolecules in neurones and glia at various stages of classical
conditioning and consolidation of long-term memory (Pevzner et al.,
1973; Klement'ev et al., 1975). Not only does this material not
allow us to draw any firm conclusions about the participation of
glial cells in the biochemical mechanisms of memory, but it leads
us to conclude that the biochemical basis of memory is a problem at
a much higher order than neuronal macromolecule synthesis. That is
to say that the organization of memory processes is a more complex
hierarchical system than the neurone-neuroglia unit. At present,
it seems that there is an abundance of differing hypotheses, but a
scarcity of well-documented facts. The problem is very ill-defined
and speculative, and requires much more systematic work. Therefore,
while acknowledging the importance of the unique methodological
side of Hydén's studies, and recognizing his unquestionable services
to the solution of the problems of the metabolic unity of the
neurone and neuroglia, we consider the theories constructed by him
on the role of the neuroglia in the mechanism of memory to be on
much weaker ground, and considerably less well-argued.

                        Problems of Our Own Studies

    If one tries to sum up the basic biochemical and histochemical
data concerning neuroglia expounded in the present chapter, then it

is necessary to emphasize that the biochemical apparatus of the glial cell is in general at least as complex as that of the neurone. The activity of the major enzymes, the overall rate of oxidative processes, the turnover of macromolecules in neuroglia - all are of the same order as in nerve cells. It is true that histochemical studies have shown that many of the oxidative enzymes (in particular the Krebs cycle dehydrogenases) are less active in glial cells than in neurones. However, among the histochemists who have studied this question, not all are in agreement, while the biochemical data using micro-methods clearly indicate the presence of the basic oxidative enzymes in the glia.

A wide range of enzymes have been determined in neurones and glia. The results are summarized in Table VI, and it can be seen that they are often inconsistent. A given enzyme is often found to be more active in the neurones than in the glia by one author, while another author finds the opposite result. Similar discrepancies are also found in comparisons of carbohydrate, lipid and mineral metabolism between the two cell types. Since the early sixties, we have been interested in dealing with similar problems concerning neuroglial RNA metabolism.

Nucleic acids have been relatively well studied, both because of their particular metabolic importance in the cell, and because of their useful optical properties. However, this work has several limitations. While other biochemical components have been studied by several authors using different methods (which has produced a great diversity of accumulated material, but allows a more objective appraisal of the reliability of the data obtained), the nucleic acids in glial cells from 1958 to the present time have been studied principally by Hydén's group, using a single methodological approach. Only in the last few years have some studies appeared, concerned with the analysis of nucleic acids in enriched fractions of neuronal and glial cells. The latter work however, has not yet dealt with the problems of functionally conditioned changes in nucleic acid metabolism in the neuroglia. Thus all the data in the literature on the functional biochemistry of the neuroglia have been obtained with methods which involve the mechanical separation of the nerve cell body from the surrounding glia (and also from the axon and dendrites of the neurones).

As was shown in Chapter 2, in situ quantitative histochemistry may be used to estimate the change in total content of intracellular components, such as nucleic acids, or proteins while preserving the morphological relationship between the neurones and their glial cell-satellites.

Another important consideration is that Hydén and his colleagues have nearly always studied the Deiters' cells of the vestibular nucleus of medulla oblongata of rabbit and their glia. To draw

general conclusions concerning the properties of the neurone-
neuroglia unit it is necessary to investigate various types of
nerve cells with different functions.

Finally, in the works of Hydén, as a rule, dynamic analysis of
the process under study is lacking. In most cases a comparison is
made only between controls and a single gourp of experimental
animals at the end of the stimulation. Such a comparison may yield
significant results for a particular question, but interesting
relative changes are likely to remain unnoticed which could be
demonstrated by analysis of intermediate stages in the process.

These are the considerations that have led to the selection of
the particular theme and methodological approach of our study. We
have considered it important to concentrate our attention on the
class of molecule most studied in the field of functional neurochemistry, nucleic acid, and in particular, RNA.

We know that in the cell, nucleic acids are present mainly in
the form of nucleoprotein complexes. The state of these components
in the neurones is considered by Shabadash (1958, 1963, 1966) to be
a very good indicator of the state of functional activity of the
nerve cell. However at the present time reliable methods for the
quantitative cytochemical estimation of cell nucleoprotein do not
exist; the study of these complexes in the single cell with differential analysis of polynucleotide and polypeptide components is a
problem for the functional biochemistry of the future. Meanwhile
a large literature on nucleic acids has accumulated. The usefulness
of quantitative cytochemistry of nucleic acids in individual cells,
including neurones, depends on the fact that the RNA (mainly ribosomal RNA) of the nerve cell is well preserved during fixation and
subsequent histological preparation of nerve tissue, and at the
same time shows changes when the functional condition of the
nervous system changes. The biological importance of the nucleic
acids is indicated by their role in such cardinal processes of cell
function as storage and expression of hereditary information, biosynthesis of cell protein, development of the functions of specialized cells, and so on. Therefore, as it is impossible to study a
great number of intracellular components, we decided to work with
the quantitative estimation of nucleic acids (principally RNA) and
with the determination of total cell protein in the neurone-
neuroglia unit.

Ideally, one would like to study the influence of a great number
of different factors on neurones and glia of very different types.
However this would require an excessively long time because of its
complexity. So it is necessary to compromise. We chose to analyze
the influence of a series of different factors on the metabolism of
a limited number of types of neurone-neuroglia unit, but under
conditions such that a comparison of two or three units would

include neurones differing substantially in their functional
properties. We used motoneurones of spinal cord anterior horn,
neurones of the sensory spinal ganglia, Purkyně cells of the cerebellum, supraoptic nucleus of hypothalamus, red nucleus of midbrain,
sympathetic neurones of lateral horn of the spinal cord, and
neurones of the superior cervical sympathetic ganglia, in each case
together with the corresponding glial cell-satellite.

Thus, changes in nucleic acid (and also of protein) metabolism
in the neurones and their surrounding glial cell-satellites could
be compared under different functional states, in particular during
spreading excitation of a degree not involving fatigue; during
exhaustion of the animal; during acute stimulation, proceeding to
exhaustion of the nervous system; under conditions hampering the
normal metabolic processes (e.g. hypoxia, or administration of
antimetabolites); and with changes in hormonal balance. It is
important to keep track of the reparative stages of nucleic acid
metabolism at various times after the cessation of the influence.
This could help the analysis of synthetic processes in nucleic acid
metabolism in the neurone-neuroglia unit.

Thus the basic problem of the present study is to follow the
dynamics of the changes of nucleic acid (and in some cases, protein)
content in the neurone-neuroglia unit of several experimental
animals (cat, rat, mouse, ground squirrel) during different
functional states of the nervous system. Analysis of the dynamics
thus permits comparison of the properties of nerve and glial cells
with the aim of understanding the role of the neuroglia in neuronal
function.

CHAPTER 4

CONDITIONS FOR

CYTOSPECTROPHOTOMETRIC DETERMINATION

During the 40 years since the publication of the work of
Caspersson (1936) on the principles of cytospectrophotometry, the
method has been repeatedly subjected to analysis and modification,
both by Caspersson himself, and by a series of investigators in
different countries. Books by Mellors (1955), Brodskii (1966), and
Wied (1966) review in detail a number of methodological problems
including its original variants and optical schemes for cytophoto-
meters. Therefore in the present chapter, we may limit ourselves
to describing only very basic specific questions.

General Principles

It has already been said in Chapter 2 that the cytospectrophotometric method is based on the use of the Bouguer-Lambert-Beer
law, which can be expressed as follows:

$$C = \frac{E}{\chi \ell}$$

where C = concentration of optically absorbing substance in the
layer,

ℓ = specific extinction coefficient (absorbency index),

E = optical density (extinction) i.e. the logarithm of
the ratio of the intensity of the incident light to
the intensity of the light emerging from the
absorbing layer.

Photometric determination of concentration is possible if it is
shown, first that the Bouguer-Lambert law is valid, i.e. that there
is a linear relationship between the optical density and thickness

of the absorbing layer; secondly that Beer's law applies, i.e. that
there is a linear relationship between the optical density and the
concentration of the absorbing substance.

The existence of such a dependence for solutions of nucleic
acids has been shown over a wide range of concentrations. This
dependence is valid in particular under conditions of monochromatic
light, and homogeneous distribution of the absorbing material. The
first condition depends only on the apparatus and with cytophometry
is easily realized. The second condition is practically impossible
for photometry of cells, since the nucleic acid (particularly in
neurones) is always distributed in the form of more or less large
accumulations or granules. Errors in the determination of extinction because of the heterogeneity of the absorbing substance are
inevitable. The question is the magnitude of this error. A mathematical analysis of this problem may be found in the works of
Brodskii and Peizulaev (1955), Brodskii (1956, 1965, 1966) and
Agroskin et al. (1960).

At the present time there are two generally accepted variants
of cytospectrophotometry allowing considerable reduction in the
distribution error. The first variant is to scan the irregularly
shaped absorbing cell with several lines, followed by the integra-
tion of the lines and calculation of a mean absorption for the whole
cell (Caspersson, 1955; Brodskii, 1956). This method is in essence
a development of a very early variant of cytophotometry, spot
photometry (Caspersson, 1936, 1940,1950): in this case, a series
of spots were used to measure photometrically a small area of the
cell, in which the distribution of the substance could be assumed
to be homogeneous.

Another method, differing in principle has also arisen for
overcoming distributional errors (Ornstein, 1952; Patau, 1952;
Mendelsohn, 1958; Agroskin et al., 1960). Photometric measurements
are made of a large part of the cell, and in a series of cases, the
whole cell, thus obtaining a magnitude of absorption including the
error. Then the same part of the cell was measured photometrically
at a different wavelength. Knowing the difference of the value of
absorption of the substance at the two wavelengths with homogeneous
distribution, and comparing their actual difference by two-wavelength
photometry of the cell, the influence of the non-homogeneous distribution of the substance in the cell can be calculated and the real
extinction determined. There exist special tables (Mendelsohn,1958)
and graphs (Agroskin et al., 1960; Sherudilo, 1968) considerably
facilitating this calculation.

There was shown to be a good correspondence between the results
of scanning and of two-wavelength cytospectrophotometry (Mendelsohn
and Richards, 1958; Agroskin, 1960; Tomina, 1970). In the present
studies of nucleic acids, two-wavelength UV-cytophotometry was used.

As was shown in Chapter 2, a cytophotometric determination using a perfect method of eliminating possible errors has nevertheless a basic source of error estimated at $\pm$ 5-10%. However, while the individual measurements (in particular with experiments for estimation of absolute quantity of a substance) are highly variable, the mean magnitude can be obtained with sufficient accuracy by making a large number of determinations.

Object of Study

Experiments were carried out on adult animals, mainly on rats (body weight 180-200 g) and in some cases on cats (2.5-3 kg), mice (25-30 g), and ground squirrels (300-400 g). Rats, mice and ground squirrels were killed by decapitation without anesthesia. The brain was quickly removed, the appropriate parts dissected out and the samples immediately plunged into cold fixation medium. This step took a few minutes. Here it may perhaps be feared that the content of the chemical components being studied might be changed as a consequence of either the shock of killing by decapitation, or by anoxia in the dissected brain tissue.

However, nucleic acids and proteins are macromolecules, forming part of the basic cellular structures (endoplasmic reticulum, cytosol, various cell membranes, etc.). It is difficult to imagine that such molecules would suffer a sufficiently rapid quantitative change, as can happen for example with such labile substances as ammonia, ATP, and creatine phosphate. Therefore shock is hardly likely, in a period not longer than a few seconds, to produce a significant change in the quantity of nucleic acids.

A more serious danger is that of post-mortem anoxic changes. Analyzing the effect of acute hypoxia (both hypoxic and circulatory) we earlier showed a marked lowering of RNA levels in cat cortical neurones (Vladimirov et al., 1961; Pevzner, 1962b; Baranov and Pevzner, 1963a). However this change was observed 1 h after the start of hypoxia. The sutdies of Kreps and Chenykaeva (1955) showed convincingly that even severe hypoxia lasting 5 min had practically no influence on cytoplasmic RNA content in the neurones although considerably less severe hypoxia for 2 h resulted in marked changes.

An analysis of post-mortem changes led Jarlstedt (1962) to the conclusion that there is a high degree of stability in the level of RNA in Purkyně cells of cerebellum, while Moore et al. (1968) came to the same conclusion with respect to the content of brain protein.

In the cat, dissection of the required part of the nervous system was carried out under urethane anesthesia administered intravenously in a dose of 1.2 g/kg. Anesthesia can change the content

of nucleic acid in the neurones of brain (Danilova, 1958) and spinal
cord (Brodskii, 1957). However the dose and the duration of the
anesthesia in our experiments were identical in control and experimental animals.

Fixation and Histological Treatment

Samples of brain tissue were fixed for 1-2 h at 0-4°C in
Brodskii's fixative (10% formalin - 96% ethanol - glacial acetic
acid in the proportion 3:1:0:3 by bolume) or in some cases in
Carnoy's fixative (ethanol - chloroform - acetic acid in the proportion 6:3:1 by volume). It is generally accepted that these
fixatives are the best for quantitative cytochemical studies of
nucleic acid and protein. This was concluded from a series of
studies (for reviews, see: Brodskii, 1960, 1966; Pevzner, 1963a)
which showed that many widely used cytological fixatives were
unsuitable and which recommended the mixtures indicated above for
quantitative analysis of nucleic acids and proteins in cells of
histological preparations.

The fixed pieces of nervous tissue were dehydrated in a series
of ethanol solutions of increasing concentration, transferred into
absolute ethanol, and then through a series of intermediate mixtures
with chloroform and embedded in paraffin wax of melting temperature
56°C.

Sections were prepared on a rotating microtome (Reichert,
Austria). This microtome was considered one of the best from the
point of view of the reproducibility of section thickness. The
thickness of sections was set at 5 or 7 µm depending on the cell
type.

Sections were cleared of paraffin in xylene and brought through
a series of ethanol solutions of decreasing concentration to water.
In the case of UV-cytospectrophotometry, the sections were then put
into aqueous glycerol and photometry carried out in this solution.
For visual cytophotometry, sections were stained with the relevant
dye, brought through a series of alcohol solutions of increasing
concentrations, and embedded in Canada balsam.

Ultraviolet Cytospectrophotometry

Unstained sections were examined photometrically on quartz
slides using a single-beam three-wavelength UV-cytospectrophotometer
constructed by L.S. Agroskin. The optical scheme of this apparatus
is shown in Fig. 21.

The light beam passes first through the monochromator, then

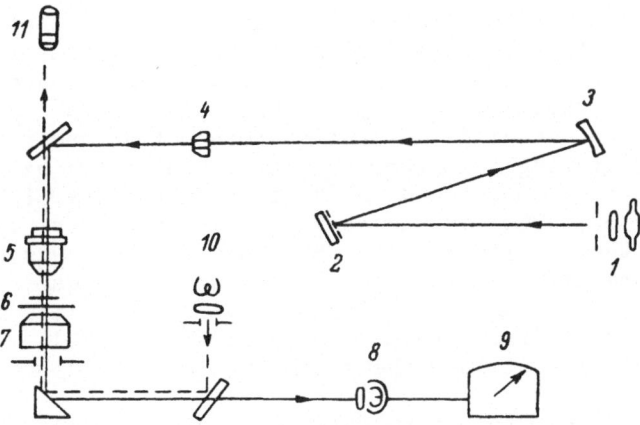

Fig. 21.   Optical arrangement of single-beam two-wavelength aperture UV-cytospectrophotometer built by Argoskin   (Pevzner, 1966c). 1- light source (lamp SVD-120A); 2- monochromator (diffraction grating); 3- focussing mirror; 4- variable aperture; 5- objective; 6- object; 7- condensor; 8- photomultiplier; 9- pen recorder; 10-)I-19 illuminator; 11- eyepiece. Solid line- path of UV-light (for photometry); Dotted line- path of visible light (for visual observation).

through a variable aperature of the appropriate dimension   We selected a size that would project a probe into the microscope field of 6-6.5 µm diameter with a 58 x 0.80 water immersion objective. This allowed measurements both of substantial regions of the neuronal cytoplasm, and of whole glial cell bodies.

The measurement consists of comparing the intensity of the light beam after passing through a part of the preparation free of tissue (i.e. the blank reading) and after passing through the required cell. In both cases one takes the measurement with a galvanometer at two wavelengths: at 265 and 280 nm. These two wavelengths are selected because the absorption by a solution of nucleic acid at 265 is approximately twice that at 280. This ratio of 2:1 is mathematically an optimal value for calculating the correction of distributional error (Mendelsohn, 1958; Agroskin et al., 1960).

The construction of Agroskin's cytophotometer also allows measurements to be made at a third wavelength, 313 nm, which is used in the Russian instruments for estimating non-specific light loss (Brodskii, 1956; Sharobayko, 1958; Pevzner, 1959; Agroskin et al., 1960). However, UV light is absorbed by protein in the fixed preparation as well as by nucleic acid. For the calculation of the

former, all cells were subjected twice to photometry: before and
after acid extraction of nucleic acids (for conditions of extraction,
see Chapter 5). From the difference, a value for the absorption was
obtained, which corresponded only to the nucleic acid. At first we
calculated the non-specific absorption both before and after
extraction, estimating this from the magnitude of the extinction at
313 nm. However we soon found out that in photometry of both
neurones (Pevzner, 1960) and neuroglial cells (Pevzner, Tomina, and
Chaika, 1964) the magnitude of the non-specific loss was practically
equal before and after extraction of the nucleic acids. This fact
(and its detailed discussion) is presented in Brodskii's monograph
(1966), and also confirmed by Geinisman et al. (1970b). Therefore
we subsequently discontinued the determinations at 313 nm.

Two-wavelength cytospectrophotometry actually measures not the
absorption, but the transmission (the ratio of the intensity of
the light passing through the cell to that of the light passing
through the blank). At each of the two wavelengths a quotient was
calculated by division of the transmission of the cell before
extraction by that of the cell after extraction; a mean extinction
at 280 nm by nucleic acid was then found for each cell using the
tables of Mendelsohn (1958). Knowing the thickness of the section
and the specific index of absorbancy for nucleic acids (10,000 at
280nm) we calculated the concentration of nucleic acid per cell in
g/cm^3 (which is numerically equal to $pg/\mu m^3$) from Lambert-Beer's law.

Visible Cytophotometry

Two-wavelength cytospectrophotometry was also used in the
visible region for measuring nucleic acids in sections stained with
gallacyanin chrome alum according to Einarson (1935, 1951), protein
amino- and imino-groups (that is, essentially total protein)
stained with Procion Brilliant Blue RS according to Ivanov (1961),
and protein SH-groups stained with dihydroxydinaphthyldisulphide
according to Barrnett and Seligman (1954).

Absorption curves were constructed for each of these stains in
cells of nervous tissue, allowing two wavelengths to be selected for
each. The first was the wavelength at maximum absorption, the
second at half-maximal absorption (Fig. 22). In particular the pair
of wavelengths used for cytophotometry of nucleic acids was 550 and
465 nm, for total protein, 527 and 588.5 nm, and for protein
SH-groups, 460 and 515 nm.

All the measurements were carried out on a cytophotometer
constructed by L.S. Agroskin, built according to the same scheme as
described in Fig. 21, but with glass rather rather than quartz
optics. Details of the construction of the apparatus are given by
Grinevičius et al. (1966).

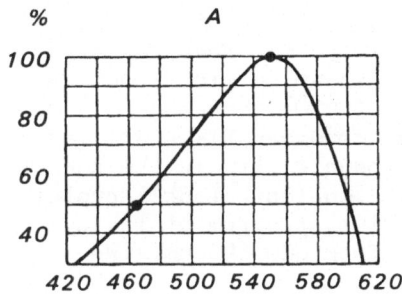

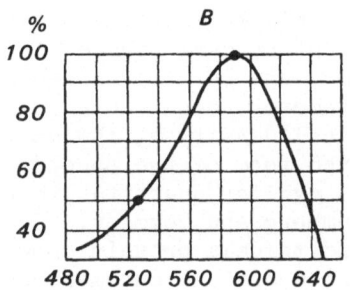

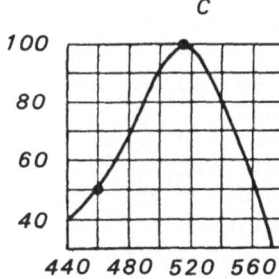

Fig. 22. Absorption spectra of cells in slices stained with chrome-gallocyanin (A), procion brilliant blue RS (B) and dihydroxydianaphtyldisulphide (C). (Pevzner and Saudargene, 1971). Ordinate- optical density as a percentage of maximal absorption; Abscissa- wavelength (in nm); Closed circles- wavelengths selected for two-wave length cytospectrophotometry.

In several series of experiments, both total and basic proteins were determined, using amino black 10B for staining. In this case, total protein was determined by staining sections for 20 min in a 0.03% solution of amido black in Michaelis' buffer at pH 5.3 (Geyer, 1960). Basic proteins were stained according to Alfert and Geschwind (1953) as modified by Gershtein and Vavilov (1969). In this method nucleic acids were first extracted by incubating the sections in 1% trichloroacetic acid for 8 min at 60°C, and then treating with cold TCA and distilled water as described by Ehrenpreis (1965) followed by staining for 10 min at room temperature in a 0.125% solution of amido black 10B at pH 8.2 (Gershtein and Vavilov, 1969). In some cases the basic proteins were also stained, using fast green FCF at pH 8.2 using the method of Alfert and Geschwind (1953) as modified by Bloch and Goodman (1953) and Ehrenpreis (1965). These sections were assayed by two-wavelength photometry at 595 and 650 nm.

Sections stained with amido black were subjected to cytophotometry in the MUF-5 double-beam recording microspectrophotometer. The scheme of this instrument has been described by Bakharev et al. (1964). The absorption spectrum of the cells stained with

amido black was first determined. Detailed analysis of the
absorption curve showed that both in the MUF-5 microspectrophotometer, and in a low-noise high sensitivity shearing microspectrophotometer constructed by Agroskin et al. (1970) there was a high
optical density at the absorption maximum, considerable stability
in the form of the spectral curve and reproducibility in the use of
this dye for quantitative estimation of the content of cell protein
(Brumberg and Pevzner, 1972). The absorption spectrum of the dye
bound to cell protein showed a broad peak from 600-640 nm. We
carried out all our measurements at 620 nm.

The preference for visible cytophotometry over UV-cytospectrophotometry is due to the absence of an unavoidable extraction step,
to the clear visual pictures of the cells being studied and to the
broader range of chemical components that can be studied cytophotometrically. However, with this type of photometry of cells, as a
rule, the specific absorbency index is not known, and so the calculation of the absolute amount of the substance being measured is
not possible. However for comparison of control and experimental
groups of animals, it is sufficient to know the extinction, since
both the specific index of absorption of the dye and the thickness
of the slices can be considered considered constant with errors not
exceeding the general errors of the cytospectrophotometric method.
The linear character of the dependence between concentration of
substance and extinction of the stained cells allows the estimation
of changes in the concentration.

Calculation of Cell Volume

If the volume of the cells compared is virtually unchanged,
measurement of extinction or concentration is sufficient for a
quantitative appraisal of biochemical changes. However, in many
experiments, especially with acute, rapidly acting influences on
the organism, changes in such factors as ion flow, the state of
hydration of proteins, the permeability of the cell membrane, etc.,
can lead to marked differences in cell volume. Such changes have
been repeatedly described in works on the functional morphology of t
the nervous system (for literature, see: Geinisman, 1966;
Khaidarliu, 1967a; Khesin, 1967). These shifts can occur, either
as a result of real changes in the volume of the living cell, or of
changes in the composition and form of the cell which affect the
cells unequally during the histochemical treatment. Changes in the
volume of a cell with unchanged quantity of a particular substance
in it lead, naturally to a reciprocal shift in the concentration
of the substance within the cell. In this case measurement only of
the concentration results in a false view of the effect of the
particular factor under study on the content of the measured
substance.

The difficulty in making measurements of the volume of a neurone is due not only to the irregular form of the cell body, but to the necessity of reproducing the three dimensional configruation on the basis of their surface projections. Evidently, planimetry of the cell, that is measurement of its surface area in the medium, is insufficient because the concentration of substance is in inverse proportion to the volume, and not the area. It is not difficult, for example, to imagine the situation where the cell does not undergo biochemical changes, but only a twofold lowering in the volume of the cell occurs. The concentration of substance being studied in this case will evidently be increased twofold. However the linear dimensions of the cell will be reduced by $2^{\frac{1}{3}}$, and thus the cross-sectional area of the cell will be reduced by $2^{\frac{2}{3}}$, i.e. about 1.6 times. Therefore if we accept initial magnitude of the volume, area, and concentration as 1.0, then the influence of the factor being studied will make the volume of the cell equal to 0.5, the area 0.625, and the concentration 2.0. Multiplying the concentration not by the volume, but by the area, gives 1.25, that is, the conclusion would be drawn that the given factor caused an increase in the quantity of substance in the cell of 25%.

Therefore, in all determination of the volume of the cell, we use the linear dimensions to make the calculation taking the cell body as a regular geometrical shape. From the great number of suggested shapes (for reviews, see: Pevzner, 1963a; Khesin, 1967), we, as a rule, used the formula of the ellipsoid of rotation. In a series of cases, as special measurements show (Pevzner, 1965b; Geinisman et al., 1969) nerve cells in fact correspond well to the rotational ellipsoid, although in some cases (e.g. spinal cord motoneurones) this approximation is not adequate, and it is necessary to use either the dual-cone (Khaidarliu, 1967a) or the three-axis ellipsoid (Geinisman et al., 1969).

We measured the linear dimensions of the cells either directly, using a microscope fitted with the ocular micrometer MOV-1-15[x], or by first throwing an image of the cell onto a screen using projection microscope PMR-1 or drawing apparatus RA-4. In some experiments (carried out in collaboration with L.L. Litinskaya and Yu.R. Khrust) measurements were made with a rapid-scanning integrating microplanimeter (Ivanitskii et al., 1967; Khrust et al., 1971; Litinskaya et al., 1976).

For calculation of the volume of cytoplasm of neurones the volumes of the whole cell and of the nucleus were calculated, and the difference taken. These calculations were carried out on the digital electronic computers Ural-2, Dnepr-21, and Minsk-22.

CHAPTER 5

INVESTIGATION OF SEVERAL

METHODOLOGICAL QUESTIONS

Thickness of the Histological Sections

The optical density of an absorbing substance is in direct proportion to the thickness of the absorbing layer. With normal spectrophotometry of solutions, the thickness of this layer is determined by the dimensions of the cuvette, and therefore has practically no effect on the accuracy of the measurements. With cytophotometry, however, the thickness of this layer is the thickness of a histological section. Therefore it may turn out to be one of the principal sources of error in photometry of cells.

The variability of the thickness of paraffin sections depends on a number of factors: the quality of the microtome knife, the properties of the paraffin wax, the habits of the individual workers, and lastly, the mechanical conditions of the microtome. The latter by its nature cannot guarantee absolute stability of the thickness of serial sections.

A knowledge of the thickness of the sections is necessary in those cases where the concentration is to be calculated from optical density (extinction) (E) according to the formula

$$C = \frac{E}{\chi \ell}$$

Direct measurement of the thickness of the section is a very difficult problem. A great number of different proposals have been made for its solution (Hallén, 1955, 1962; Brattgård, 1956; Merrian, 1957; Hale, 1958; Zaks, 1958; Sharobayko, 1958; Menzel, 1959; Lipp and Gubisch, 1961; Treff, 1963; Lodin, 1964; Chaubal et al., 1967;

Kuznetsova and Brodskii, 1968). Indeed, the very diversity of methods is an indication of the absence of a really satisfactory method.

Even if we do not have a method for monitoring the thickness of each section, we considered it necessary to obtain an indication of the variability of the thickness and of the relationship of the actual to the nominal thickness of the sections. We measured the thickness of sections on the Linnik MIS-11 dual microscope (Zaks, 1958; Agroskin et al., 1960). The aperture of this microscope was specially adapted by N.V. Korolev so that its projection in the microscope field was a narrow horizontal slit. In this microscope the reflection of the aperture from the microscope slide is observed. If part of the projected light passes through the paraffin section, then a break in the projection will be seen in the microscope field (Fig. 23). The thicker the section, the bigger the gap between the reflection of the diaphragm from the microscope slide, and from the upper surface of the paraffin section. This gap may be measured with the screw ocular micrometer MOV-1-15[x] fitted onto the micro- scope, and knowing the dimensions of the optical system it is possible to work out the thickness of the section.

We measured a sample of 100 serial sections. There was no selection under the microscope, that is, no attempt was made to discard sections which sharply differed in thickness.

As can be seen from Fig. 24 the distribution of the section thickness has a characteristic unimodal form, in which the mean thickness turned out to be not 5 μm (as was indicated on the micro- tome scale) but 1.5 times higher. This is in agreement with the data of Merriam (1957) and Hallén (1955, 1962) showing that the actual thickness of sections exceeds the nominal thickness, the degree of this excess being greater the thinner the section. So we

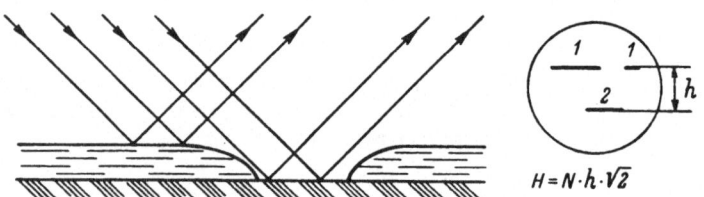

$$H = N \cdot h \cdot \sqrt{2}$$

Fig. 23. Determination of thickness of paraffin-wax sections using the dual microscope MIS-11. Left- scheme of the light path with reflections from the slide and from the surface of the section. The incident angle is 45°. Right- diagram of the microscope eyepiece, and the principle of the calculation; 1- light reflected from the upper surface of the section; 2- light reflected from the slide; H- thickness of the sec- tion; h- displacement between the reflections; N- magnifica- tion of the microscope.

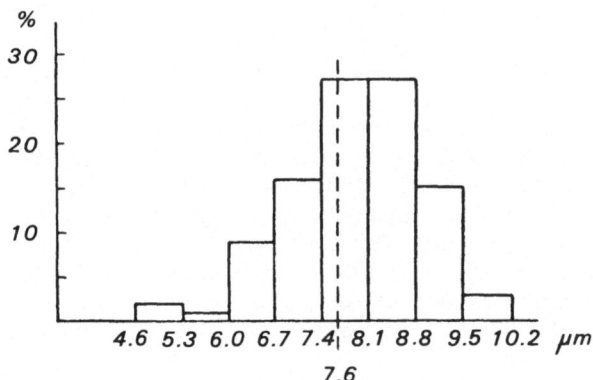

Fig. 24. Distribution of thickness of paraffin-wax sections (Pevzner, 1965). Ordinate- proportion of sections (in %); Abscissa- thickness of sections (in μm); Vertical dashed line- mean thickness of the sections.

calculated that the actual thickness would be 7.5 or 10 μm when the scale of the microtome indicated 5 or 7 μm.

The histogram in Fig. 24 indicates that with a mean thickness of 7.5 μm, the great majority of sections (almost 80%) are grouped in the zone of thickness between 6.7 and 8.8 μm, i.e. their deviation from the mean does not exceed 15%. A similar magnitude of thickness variation of paraffin sections was found by Sharobaiko (1958), who estimated the thickness by the value of the light dispersal (measured by optical density at 313 nm). In our group of sections (around 100) the standard error of the mean (arithmetic) did not exceed 2.5%.

It should not be forgotten that during the normal work, as described above, we regularly examined the sections under the microscope. Usually after some experience it is easy to distinguish sections that are very thick or very thin, and exclude them from the work.

Therefore, in contradiction to the pessimistic statements that even with a perfect microtome quantitative cytochemical determination in sections is impossible because of the great inequality of their thickness, we suggest that with good regulation of the operation of the microtome, and with visual selection of the sections it is possible to make photometric measurements whose variability will not exceed the limit of accuracy of the cytophotometric method (i.e. 5-10%).

Selective Extraction of RNA from Slices of Nerve Tissue

A whole series of histochemical stains including gallocyanin,
toluidine blue and azure show up both RNA and DNA. Both types of
nucleic acid have similar UV absorption properties. Besides this
the UV absorption in fixed cells overlaps that of proteins containing aromatic amino acids. All this indicates the need for
special techniques for the separate determination of RNA and DNA.
It is particularly important for functional-biochemical studies to
be able to selectively reveal RNA, the metabolism of which, unlike
that of DNA, may undergo marked changes in nerve cells as a result
of variations in the functional state of the nervous system.

One of the most reliable and widely used methods for determination of RNA in histological sections is selective extraction. This
approach is often used for biochemical analysis of RNA in tissue
homogenates (Schmidt and Thannhauser, 1945; and see reviews:
Georgiev and Mant'eva, 1962; Munro and Fleck, 1966). However, these
schemes of separation, while fully satisfactory for homogenates are
far from being so for treatment of fixed histological sections.

It seems that the optimal method for selective extractions of
RNA (or DNA) from sections is treatment with specific enzymes - RNase
or DNase. However, experience shows that enzymic extraction may be
incomplete (Brodskii, 1958; Koenig, 1958; Deitch, 1960; Jobst and
Sandritter, 1965; Vejlsted and Pakkenberg, 1972). Evidently the
fixation and subsequent histological treatment can considerably
change the state of the nucleoprotein complex in the cell, in such
a way that the conditions of enzymic hydrolytic reaction in the
section are not at all the same as those in the homogenate.

It is also impossible to ignore the fact that pure preparations
of nucleases are still scarce and expensive. Therefore during the
years of the development of cytochemistry, numerous different
empirically determined methods of non-enzymic extraction of RNA and
DNA have been suggested (for reviews, see: Gurr, 1959; Pearse,
1960). Many of these have not stood the test of time, but some are
still currently used.

The most popular methods involve acid extraction with trichloroacetic (Pollister and Ris, 1947; Pollister et al., 1951; Brodskii,
1955, 1956), hydrochloric (Gutiérres, 1960; Andreeva et al., 1963;
Menzies, 1963) and in particular, perchloric acid (Ogur and Rosen,
1949, 1950; Sulkin and Kuntz, 1950; Koenig and Stahlecker, 1951,
1952; DiStefano, 1952; Brodskii and Suetina, 1960; Brodskii, 1966).

While trichloroacetic acid removes both types of nucleic acid,
perchloric acid turns out to be very suitable for differential
extraction of RNA and DNA. Unfortunately, the selection of optimal
conditions for removal of RNA in the presence of DNA has usually

been carried out only by qualitative, visual methods. A gratifying
exception is the work of Brodskii and Suetina (1960) in which
quantitative estimation with UV spectrophotometry was used to study
the removal of the acid soluble fractions, RNA and DNA from smears
of bone marrow by treatment with perchloric acid at different
temperatures. These authors confirmed that the cell proteins are
conserved during the extraction (Brodskii and Suetina, 1960:
Brodskii, 1966). This is not of crucial importance for visible
cytophotometry of nucleic acids, but of great importance for UV-
cytophotometry, in which removal of part of the protein can lead to
considerable errors in estimation.

Naturally, for each type of tissue, the conditions for selective
extraction of RNA are likely to be different. For example, Koenig
and Stahlecker (1952) showed that under identical conditions of
fixation and subsequent histological treatments, 10% perchloric acid
at 20ºC removed RNA from nervous tissue sections in 12 h, and from
liver sections in 3.5 h.

Therefore we have tried to sort out the conditions for selective extraction of RNA with full conservation of the initial DNA
using quantitative estimation of the completeness of the extraction
by cytophotometry, and to confirm that the content of cell protein
under these conditions remained unchanged.

We had already established (Brumberg and Pevzner, 1966) that
the extraction scheme suggested by Koenig and Stahlecker (treatment
of sections with 10% PCA for 15 min at 37ºC) was unsuitable for our
purpose, as a substantial removal of DNA occurred while the extraction of RNA was still incomplete.

Ogur and Rosen's (1949, 1950) extraction method turned out to
be much more promising since these authors recommend treatment at
a much lower temperature (23-26ºC) which allows the extraction time
to be extended to several hours. To test this scheme of extraction
of RNA we used visible cytophotometry, estimating RNA content in
the cytoplasm of neurones of sympathetic ganglia of cat in sections
stained with gallocyanin chrome alum according to Einarson (1935,
1951; Pearse, 1960) and the content of DNA in nuclei of these
neurones in sections stained with basic fuchsin-sulfurous acid
according to Feulgen and Rossenbeck (1924; Pearse, 1960). The
corresponding absorption at 460 and 580 nm for RNA, and at 490 and
530 nm for DNA was measured with the two-wavelength cytospectrophotometer constructed by Agroskin (1964).

The data obtained (Table XXII) show that Ogur and Rosen's
extraction scheme does not ensure the complete removal of RNA. Even
after 6 h of extraction at 23-26ºC in sympathetic neurones, around
one quarter of the cytoplasmic RNA remained. Another problem is the
increase in optical density of DNA which occurs after staining by

the Feulgen method. This is possibly connected with the shrinkage
of the nucleus, and possible, with a change in the properties of the
DNA molecule, in particular with the degree of hydrolysis caused by
the hot hydrochloric acid used in the Feulgen reaction in sections
already treated with PCA.

Ogur and Rosen also suggested a scheme of RNA extraction using
10% PCA at even lower temperature (0-4°C) with treatment lengthened
to two or three days.

A test of this method of extraction showed that in sections
treated for 72 h, the RNA almost completely disappeared, while the
DNA content did not change (Table XXI).

The only problem with this method is that it is time-consuming.
According to Hess and Thalheimer (1965), the time of extraction can
be reduced by using 16% PCA in place of 10%. Photometry of sections
after various periods of treatment with 16% PCA showed (Table XXI)
that after 48 h under these conditions RNA completely disappeared,
without change of the DNA content. However it was still necessary

TABLE XXI

Effect of perchloric acid on the RNA and DNA content
of neurones of sympathetic ganglia of the cat
(Brumberg and Pevzner, 1966)

Conditions of treatment	Length of treatment period (in hours)	Cytoplasmic RNA content (from photometry of sections stained by Einarson's method)	Nuclear DNA content (from photometry of sections stained by Feulgen's method)
10% perchloric acid at 23-26°C	0	0.61 ± 0.01	0.60 ± 0.02
	4	0.23 ± 0.01	0.79 ± 0.02
	5	0.17 ± 0.01	0.74 ± 0.02
	6	0.16 ± 0.01	
10% perchloric acid at 4°C	0	0.57 ± 0.01	0.44 ± 0.01
	48	0.23 ± 0.01	0.48 ± 0.02
	72	0.07 ± 0.01	0.45 ± 0.02
16% perchloric acid at 4°C	0	0.51 ± 0.01	0.27 ± 0.01
	18	0.24 ± 0.01	0.29 ± 0.01
	48	0.05 ± 0.01	0.25 ± 0.01

The content of DNA and RNA is expressed in the corresponding extinction values.

TABLE XXII

Effect of Corazol convulsions on nucleic acid content
of neurones and glia of rat spinal cord (determined by
different cytospectrophotometric methods) (Saudargene
and Pevzner, 1969)

Type of cell, and method of cytospec-trophotometry	Control, optical density	Convulsions	
		optical density	difference from control
Neurones			
UV-cytospectrophoto-metry	0.56 ± 0.03	0.45 ± 0.02	-20%
Visible cytospectro-photometry	0.80 ± 0.04	0.63 ± 0.02	-21%
Neuroglia			
UV-cytospectrophoto-metry	0.45 ± 0.02	0.39 ± 0.02	-13%
Visible cytospectro-photometry	0.68 ± 0.02	0.61 ± 0.03	-10%

Determinations in the visible region of the spectrum were made on
sections stained with chrome-gallocyanin according to Einarson.

to show that the procedure did not influence the cell protein. The
protein content was estimated by absorption at 280 nm in sections
previously treated with hot TCA. That TCA (5% solution; 90°C;
7-8 min) completely removes both types of nucleic acid without
influencing protein content has been shown in a series of methodological studies (for reviews, see: Brodskii, 1966).

The absorption at 280 nm was found to be almost identical after
removal of RNA with hot TCA as with cold PCA (Brumberg and Pevzner,
1966). Thus, within the limits of accuracy and sensitivity of the
cytospectrophotometric method, we can say that 16% PCA for 48 h at
4°C completely removes the RNA from cells without affecting the DNA
and protein content. This method is clearly not universal, in
that conditions required with other tissues or fixation methods may
well be different.

This method makes a comparison of the properties of the

ribonucleoprotein complex of neurones and glia from spinal cord possible (Venkov and Pevzner, 1975). Ribosomal RNA presumably requires a long extraction time because it is bound to cell protein in the ribonucleoprotein complex. Clearly the extraction process is a gradual one, and it may be expected that different fractions of RNA, differing in their binding to the polypeptide chain of RNP, will be released at different times during the treatment. To test this, we prepared sections of both cervical and lumbar thickening of spinal cord, which were stained in gallocyanin chrome alum, and then incubated in PCA for 2, 4, 8, 18, and 48 h. A fraction washed out with 2 h of acid hydrolysis was found both in the cytoplasm of anterior horn motoneurones and in the cell bodies of the surrounding neuroglia (Fig. 25). This fraction made up two thirds of neuronal RNA, but only half of glial RNA. The remaining RNA required more than 18 h treatment before significant extraction occurred.

Similar experiments have been carried out on superior cervical sympathetic ganglia, and it is interesting that in this case the ratio of the two types of RNA was practically equal in neurones and glia.

Comparison of Cytophotometry in Visible Light and UV-cytophotometry

UV-cytospectrophotometry is based on the absorption of UV-light by nucleic acids which has a maximum in the region of 260 nm. In the nucleic acid molecule, the chromophore is the conjugated double

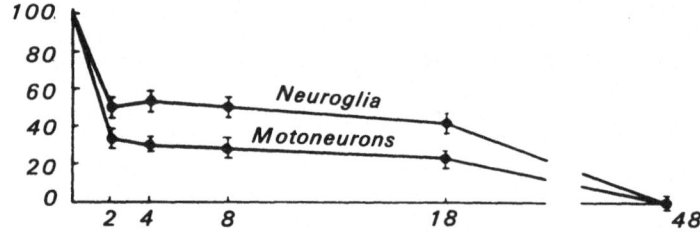

Fig. 25. Effect of extraction of RNA from slices of rabbit spinal cord with cold perchloric acid (Venkov and Pevzner, 1975). Ordinate- percentage of initial quantity of RNA per cell; Abscissa- duration of extraction in hours.

bond in the purine and pyrimidine rings (Cavalieri et al., 1948; Cavalieri and Bendich, 1950). Therefore, UV absorption is a property not only of nucleic acids, but of oligonucleotides. In fact, the UV absorption of the nucleotides formed as a result of total hydrolysis of nucleic acids is substantially higher (approximately 1/3) than the absorption of the acids themselves (Spirin et al., 1959; Shalina et al., 1967). This effect, called the "hyperchromic effect" may be a source of error in UV-cytospectrophotometry. If the influence under study can cause depolymerization of nucleic acids in the cells, the subsequent elevated UV-absorption may be erroneously attributed to an increase in nucleic acid content.

Therefore it is necessary to compare the results of UV-cytospectrophotometry with the results of some other quantitative determination of nucleic acids in cells - a determination which cannot be influenced by the hyperchromic effect. A suitable method is cytophotometry in visible light, in particular photometry of the intensity of staining with gallocyanin chrome alum (Einarson, 1935, 1951). Gallocyanin, being a basophilic stain, binds to the phosphate radical in the nucleic acid molecule. This binding is stoichiometric (Sandritter et al., 1954, 1963; Einarson, 1957; Pakkenberg, 1958, 1962; Kiseli, 1962; Ovchinnikova and Selivanova, 1964). Doubts have been expressed about whether gallocyanin staining of nucleic acids is subject to the Bouguer-Lambert-Beer law (Oram, 1955; Terner and Clark, 1960). However, under the actual conditions of photometry of sections stained by Einarson's method these apprehensions are not borne out. The results of a detailed study, carried out by Sandritter et al. (1966) conclusively showed that photometry of the intensity of the Einarson reaction quantitatively estimates phosphate groups in the nucleic acid molecules.

We compared the results of determination of the concentration of cytoplasmic RNA in neurones using UV-cytospectrophotometry and photometry after gallocyanin chrome alum staining. In this series of determinations, male rats were injected subcutaneously with Corazole (pentylenetetrazole) at a dose of 45-55 mg/kg. After 15-25 min when a state of clear convulsive activity was attained, the animals were killed by decapitation.

The RNA content in the cytoplasm of motoneurones of anterior horn of spinal cord of these rats turned out to be markedly lowered as compared to controls. As is shown in Table XXII, this reduction ·occurred both with cytophotometry in visible light and with UV-cytospectrophotometry, the degree of this lowering being practically identical in both cases.

The data obtained seem to confirm that the hyperchromic effect is not an important source of error in cytospectrophotometric determinations. This is not surprising when it is considered that even a complete breakdown of the nucleic acid molecule into nucleotides would raise the total UV-absorption by no more than one third. But such released individual nucleotides would be completely washed out of the section in the fixation process, and the subsequent histological treatment. Thus any likely reduction in the degree of polymerization of the nucleic acid would not be expected to result in a hyperchromic effect exceeding the general error of the cytospectrophotometric determination.

Comparison of UV-cytospectrophotometry
and Cytointerferometry

As was shown in Chapter 2, cytointerferometry has been successfully used for the determination of total protein in fixed cells. We have used this method in parallel with UV-cytospectrophotometry.

For cytointerferometry the unstained sections, freed from paraffin and taken through to water, are placed under the interference-polarization microscope MBIN-3, with objective 40 x 0.65, and magnitude of bifurcation 150 μm. The optical scheme of this microscope is shown in Fig. 10. The measuring device is a Senarmont compensator (Zakhar'evskii and Kuznetsova, 1961a). The maximal darkening is first determined for the background (a blank region of the preparation) and then for the structure being studied. This is achieved by turning the rotable analyzor, which has a disk with a scale expressed in degrees. The angle of turn of the analyzer is a measure of the dry mass (in fixed preparations, this is practically total protein).

As was stated in Chapter 2, the quantity of dry mass per cell (P) in the sections is related to the optical retardation δ by the equation

$$P = \frac{\delta S}{100\alpha}$$

where S = area of the structure being measured (in cm^2),
       α = the specific refractive increment of the object,
           which for protein solutions equals 0.0018.

The magnitude of the optical retardation (or phase difference) δ is related to the angle of turn ϕ by the equation

$$\delta = \frac{2\lambda}{360} \; \phi$$

where λ is the wavelength of the polarised light (in cm).

Using interference filters, monochromatic light was obtained at a wavelength of 546 nm.

Hence the dry mass is given by

$$P = \frac{2\lambda\phi S}{360 \times 100\alpha} = \frac{2 \times 5.46 \times 10^{-5} \times}{360 \times 100 \times 0.0018}\phi S = 0.017\phi S$$

Here S is expressed in cm^2, the concentration in g/cm^3, and the resulting quantity of protein in g. However, to express the quantity of protein in one cell in terms of grams is gardly convenient. Considering that g/cm^3 is equivalent to $pg/\mu m^3$, we considered it expedient to use the same formula, but to express the area in square microns, and the quantity of protein in picograms.

Cytointerferometry was carried out on sections of superior cervical sympathetic ganglion, subjected to 3 h electro-stimulation. By comparison with control ganglia a marked accumulation of protein was noted in the stimulated ganglia (54%).

Neighboring sections of the same ganglion were mounted on quartz microscope slides and then treated with 5% trichloroacetic acid (90°C; 7-8 min) for removal of nucleic acids, after which the slides were subjected to cytospectrophotometry. The optical density at 280 nm, which corresponds to the maximal absorption of protein, was considered as a measure of the quantity of cell protein. In this way, an increase in the content of cell protein of 43% was demonstrated.

Thus the interferometric data showed a rather greater increase in protein content than UV-cytospectrophotometry. We consider this due to three reasons.

First, interferometry measures total dry mass in which the nucleic acid content cannot exceed a few percent. However this percentage may increase as a consequence of a substantial increase of cell RNA. As will be described in the next chapter, exactly such an increase (more than 40%) in the RNA content occurs in neurones of superior sympathetic ganglia after electrostimulation.

Secondly, in the protein molecule it is only the aromatic amino acids which are determined by UV absorption. It is possible that the proportion of these amino acids in the protein molecule might change as a result of electrostimulation of the ganglion.

Thirdly, the sensitivity of measurement of protein content on the basis of photometry at 280 nm is rather low. According to our data in neurones, the mean optical density of nucleic acids, for example, at 265 nm (that is not even at the very maximum absorption) is rather higher than the optical density of proteins at 280 nm. However the concentration of protein in the cytoplasm of fixed cells is an order of magnitude higher than the concentration of nucleic acids.

Taking into account the limitations of the two methods we consider that the agreement found between the results of UV-cytospectrophotometry and cytointerferometry of proteins is sufficiently good for our purpose. Rather similar comparisons of these two methods were obtained by Brodskii and Kuznetsova (1961) who determined nuclear protein of neurones of frog retina after light stimulation.

* * * *

The data described in the present chapter confirm that variations in the thickness of the histological section may be reduced to a level which allows sufficiently reliable estimation of differences in mean magnitudes of optical densities of nucleic acids between groups of cells.

For nervous tissue fixed in acid formalin ethanol mixtures with subsequent embedding in paraffin, the conditions arrived at for selective extraction with perchloric acid (16% solution; 0-4°C; 48 h) allowed complete extraction of RNA while conserving the initial quantity of DNA and protein.

Comparison of the results of photometry of nerve cells in UV and in visible light using the intensity of staining with gallocyanin chrome alum showed good agreement. This argues against the occurrence of hyperchromic errors in the use of UV-cytospectrophotometry. A small hyperchromic effect may have taken place in the conditions of our determinations, but its magnitude cannot exceed that of the general error of the method.

The method of UV-cytospectrophotometry has several advantages, both for spectrophotometric analysis and for morphological appraisal of the undisrupted tissue structure. However this quantitative cytochemical method suffers from several disadvantages, including the possibility of considerable errors in the measurements and low chemical selectivity. The results of our experiments confirm that the theoretical possibilities of errors may not in practice turn out to be a significant factor in the general error of the method and that the chemical selectivity of UV-cytophotometry may be increased using specially developed methods of extraction.

Cytointerferometry is a useful addition to UV- and visible cytospectrophotometry. With its help it is possible to estimate the content of total protein in individual cells with reasonable accuracy.

CHAPTER 6

NUCLEIC ACIDS IN THE NEURONE NEUROGLIA

UNIT WITH DIFFERING FUNCTIONAL STATES

OF THE NERVOUS SYSTEM

The most important biological changes in the functional state of the nervous system involve specific physiological excitation and inhibition of groups of neurones (but excluding exhaustion). However normal changes in the functional state are only likely to lead to very small changes in RNA metabolism, and these will probably be confined to particular RNA fractions. The method for RNA determination described above can only detect rather large changes in total RNA (in practice, mainly ribosomal RNA). Thus we have had to create conditions in which the experimental animal undergoes abrupt changes in functional state which involve the majority, or at any rate a large number of nerve cells in the region of the CNS being studied. Such an approach is widely used in work on functional neurochemistry; with care, extrapolation of data obtained by the study of sharp changes in the state of the neurones can be used to judge corresponding characteristics in normal physiological processes. It is, after all, unlikely that the underlying biochemical processes for physiological changes and for the more severe changes in functional state (approaching to reactions of the stress type) in the nervous system (in experimental and pathological condition) are completely different.

Starting from this assumption, we carried out a series of experiments in which we subjected animals (mice, rats or cats) to various influences which profoundly change the functional state of the nervous system.

Circadian Variation of RNA and Protein Content

Experimental manipulations which change the functional state

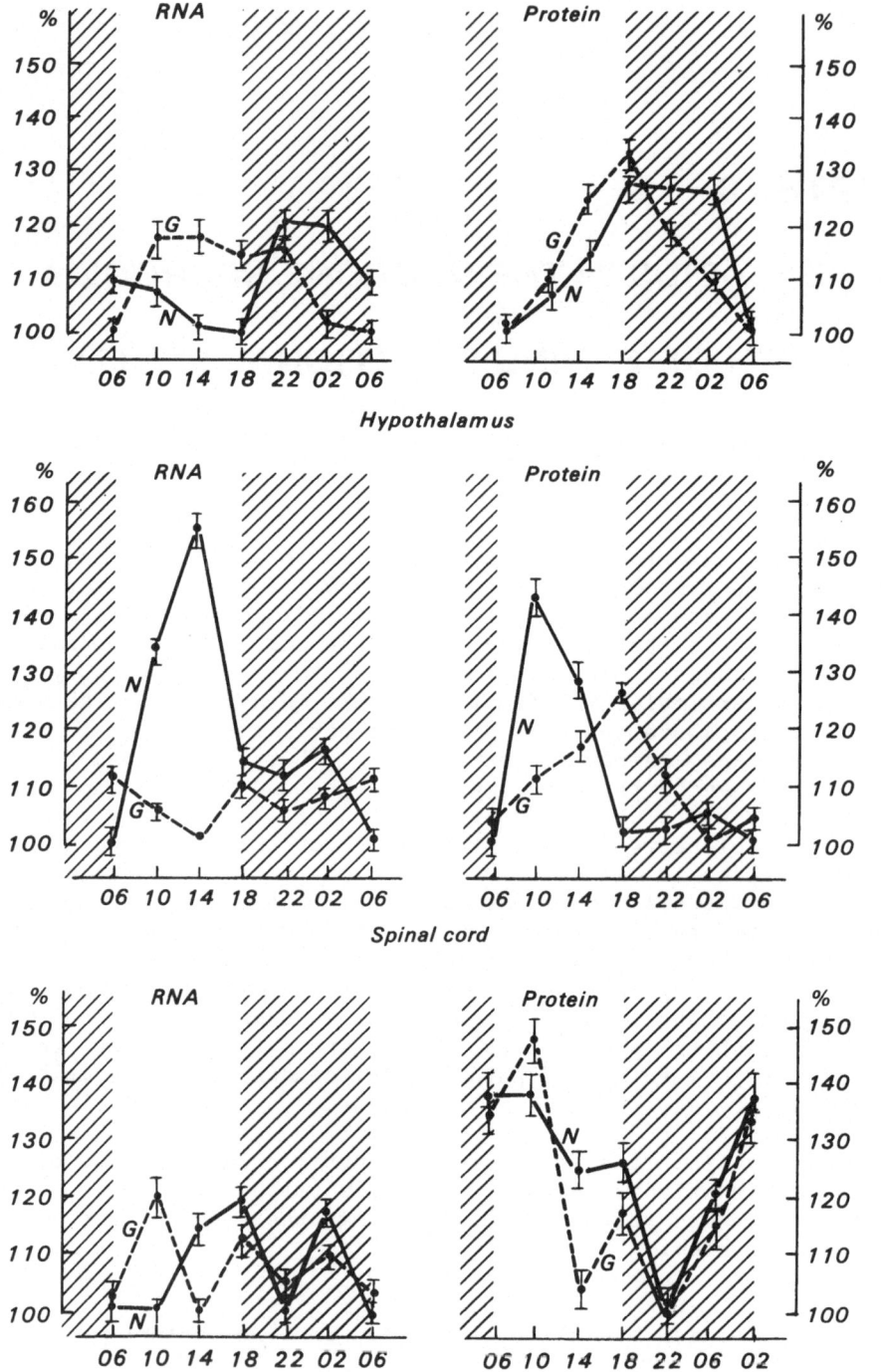

Fig. 26. Circadian rhythm of RNA and protein content in neurones and in perineuronal glia of various regions of rat central nervous system   (Litinskaya et al., 1976). Ordinate- variation of the mean protein or RNA content per cell expressed as a percentage of the minimum value during the 24 h cycle; Solid line- neurones; Dashed line- neuroglia; Abscissa- time of day (h). The shaded area corresponds to the hours of darkness.

of the nervous system are superimposed on an initial state, which is conventionally assumed to be relative rest. But there already exists an influence which is comparable to those used in our experiments. The nervous systems of all kinds of animals living on our planet are subject to changes associated with the alternation of day and night, and of the seasons. External temperature, sensory stimulation, feeding regimes and motor activity are among the factors which show such changes.

Relatively little work has been done on circadian rhythms in RNA and protein metabolism in the nervous system. Merritt and Sulkowski (1970) showed that the concentration of RNA and the amount of RNA polymerase in homogenates of rat brain increased at the beginning of the dark period, and reached a maximum at midnight. Richardson and Rose (1971) studied incorporation of ^{3}H-lysine into rat brain protein in vivo at different times of day. They showed that there was a maximum rate of incorporation between 1200 and 1600 h, and a minimum between 0000 and 0400 h. However, both these studies were carried out on whole brain. We can find no report of quantitative studies of macromolecule metabolism at the cellular level, in particular comparing neurones and glia.

In our work on this problem, male Wistar rats of body weight 180-210 g were used. The animals were kept for several days in the laboratory in order to habituate them to the new environment. Experiments were carried out in natural lighting between 20-22nd March. At this time in Leningrad, day and night are of equal duration. Rats were killed at 4 h intervals by rapid decapitation

without anesthesia. Sections were then prepared from various
regions of the CNS, and stained in gallocyanin chrome alum for
nucleic acids, or in amido black 10B for protein. Although gallocyanin stains both RNA and DNA, it can be assumed that cell DNA will
remain constant, and that circadian fluctuations in nucleic acid
content are entirely due to RNA.

We found that in the supraoptic nucleus of the hypothalamus,
the RNA and protein content of the neurones shows a circadian rhythm,
that is there is one maximum and one minimum in a 24 h period (Fig.
26). An analogous rhythm was also shown for the perineuronal glia
of this nucleus. In the neurones the peak levels occurred in the
dark, while in the glia it was during the light period. Thus the
changes in the two cell types were 12 h out of phase. The total
protein, on the other hand, showed very similar changes in both
neurones and glia, with a peak at 1800 h and a minimum around
0600 h (Fig. 26).

Cymborowski and Dutkowski (1969) found a high rate of incorporation of ^{3}H-uridine into RNA of secretory neurones of the cricket
during the day, and a much lower rate during the night. This is in
the opposite direction to our data. Whether the source of this
discrepancy is in the difficulty of comparing autoradiographic and
cytophotometric data, or in differences between rat and cricket
neurosecretory systems is not clear.

In anterior horn of spinal cord, the picture was different.
In this case, the maximum quantity of RNA in the motoneurones was
found in the daytime (1400 h) while at this time the glial cells
showed a minimum (Fig. 26). In the dark (0200 h) the RNA content
of the glia increased, while that of the neurones decreased. The
overall pattern for changes in protein content of the motoneurones
was rather similar to that for RNA - a maximum during the day and a
minimum during the night - although the exact positions of the peaks
were somewhat different (Fig. 26). This pattern of changes is not
at variance with the histological data of Konecki and Kozubska
(1961) who showed a high density of Tigroid (Nissl substance) in
cytoplasm of mouse spinal motoneurones between 0000 and 0300 h, and
a much lower density around 1500 h. However quite apart from
possible species differences, the interpretation of this work is
made difficult by the fact that Tigroid does not give a quantitative
estimate of nucleic acid. The density of Tigroid could also be
affected by shrinkage of the cell; we have in fact demonstrated a
circadian rhythm of cytoplasmic volume of spinal motoneurones
(Pevzner et al., 1974).

More complex circadian changes occur in Purkyně cells of the

cerebellum. As in spinal cord, the RNA content increases during the
day, although to a lesser extent, and reaches a maximum at 1800 h.
In the dark, the RNA content first falls to a minimum, and then
rises to a second maximum, and then after 0200 h it decreases once
more to a minimum (Fig. 26). In the glial cells, an analogous
rhythm occurs, but shifted by 12 h. Protein content in the neurones
declines somewhat during the day, then shows a further marked
decrease at the onset of darkness. Peak protein content occurs at
dawn. Rather similar changes were found in the surrounding glia,
although the decrease during daylight was more marked (Fig. 26).

The mechanisms of circadian rhythms of cell metabolism have
not been much studied, and the existing data are not always in
agreement. For example, while Richardson and Rose (1971) found a
maximum incorporation of ^{3}H-lysine into rat brain protein during the
day, and a minimum at night, ter Haar and MacKinnon (1972) showed
that a peak of incorporation of ^{35}S-methionine into rat brain
occurred at 2100 h and a minimum at 0900 h. It is interesting that
the circadian rhythm of incorporation of label into proteins of
thalamus is exactly opposite in males and females (Burnet et al.,
1974). Our studies are not detailed enough to clarify the mechanisms
of these circadian rhythms. Such conclusions as may be drawn have
been discussed by us elsewhere (Litinskaya, Pevzner and Khrust, 1976).
One point is directly relevant to the theme of the present book.
From Fig. 26 it can be seen that the period of oscillation (that is
the number of maxima and minima per 24 h) is in general the same for
neurones and the corresponding glial cell-satellites. The phase of
oscillation of protein content in general coincides in neurones and
glia, while that of RNA content is 12 h out of phase. This is the
case whether the neuronal peak occurs during the day or during the
night.

Thus even without any experimental manipulations, circadian
changes in the functioning of the nervous system are accompanied
by oscillations of macromolecule content in the cells of the nervous
system, both in neurones and in the perineuronal glia. This
confirms Hydén's suggestion, that the glia cannot be considered to
be merely a metabolically supportive cell population. The neuroglia,
as well as the neurones, make an active contribution to the metabolic restructuring that accompanies changes in the functional state
of the nervous system. Fig. 26 makes it clear that this restructuring in the glia may reflect that in the neurones, or differ
markedly from it. In order to study the similarities and differences
in such responses of neurones and glia in more detail, it is necessary to carry out experiments in which controlled changes in the
functional state of the neurones occur. Such material forms the
remainder of this chapter.

Motor Activity

Muscular exercise (e.g. prolonged running or swimming) when sufficiently intense or prolonged to lead to a stress reaction in the organism makes demands on many parts of the organism, including the nervous system (Yakovlev, 1955, 1956, 1963; Pogodaev, 1966, 1970; Pogodaev et al., 1968). When the muscular load was taken to the point of exhaustion, there was a depletion of the synthetic apparatus of the nervous system (see reviews by Pevzner, 1966a, b; Brumberg and Pevzner, 1971). Therefore we have been interested in following the changes in RNA content of the neurone-neuroglia unit during intensive but not exhausting motor activity.

Experiments were carried out on male white mice of body weight 28-32 g. Animals were made to swim for 2, 3, or 4 h in water at 35-36°C. At the end of this period, animals were killed by decapitation, and a section of spinal cord in the region of the lumbar enlargement removed, as well as the neighboring spinal ganglia. Other groups of mice were killed 2, 4, 6, 18, or 24 h after the termination of a 3 h swimming period. Histological sections were examined photometrically before and after selective extraction of RNA with cold PCA as described in the previous chapter.

In another series of experiments, male Wistar rats were made to run for one hour in a motor-driven wheel at 2.4 m/min. Sections of motor cortex (zone IV) and of the lumbar thickening of the spinal cord were stained in gallocyanin chrome alum, and RNA content determined at 585 nm in the MVF-5 microspectrophotometer (Bakharev et al., 1964). As with our investigation of circadian rhythms, all changes in nucleic acid staining were attributed to RNA.

Swimming for 2 h does not produce a reliable change in RNA content in the cytoplasm of spinal motoneurones. However, at the end of the third hour a sharp increase occurs in the quantity of cytoplasmic RNA. This was even more pronounced after 4 h uninterrupted swimming (Fig.27) In the neuroglia of anterior horn a small increase in RNA content was noted after 2 and 4 h. This increase of RNA content in the glial cells was considerably less marked than in the neurones.

The quantity of RNA in neurones of sensory spinal ganglia was markedly reduced after 2 h muscular exercise. After 3 h the level of cytoplasmic RNA returned to normal and remained practically unchanged during subsequent swimming; in the glial cell-satellites of the ganglia, by contrast, the RNA content remained unchanged over a period of 3 h and only after 4 h of uninterrupted swimming was the quantity of RNA reduced (Fig. 27).

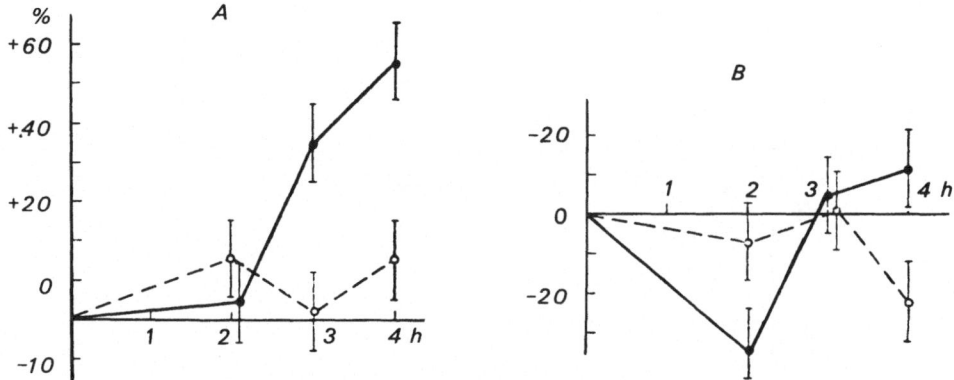

Fig. 27. Changes in RNA content in neurones and neuroglia of mouse spinal cord after different durations of swimming (Brumberg, 1968a). Ordinate- change from resting level (in %); Abscissa- duration of swimming (in h); A- anterior horn of spinal cord; B- spinal ganglia; Solid line- neurones; Dashed line- neuroglia.

After cessation of 3 h muscular activity, the quantity of RNA in the cytoplasm of motoneurones swiftly returned to normal. After 4 h the quantity of cytoplasmic RNA had already stabilized at the normal level. This reestablishment of the initial level of RNA in the neurones is accompanied by a lowering of its quantity in the glial cell satellites of the anterior horn of spinal cord (Fig. 28). At the same time, as the level of neuronal RNA becomes normal the quantity of glial RNA remained lowered for almost 24 h after cessation of swimming (Fig. 28).

The cytoplasmic RNA content in neurones of spinal ganglion remains basically unchanged both at the cessation of 3 h swimming, and at different periods after termination of exercise (Figs. 27, 28). The changes in RNA content in these neurones apparent in Fig. 28 did not exceed the normal individual variability and were not statistically significant. At the same time, however, the content of glial RNA in the ganglia was sharply increased in the course of a 2 h period and only returned to normal after 4 h. Muscular activity might well produce different changes of RNA metabolism in spinal cord neurones. As a rule, in experiments of this type reduction of cytoplasmic RNA content occurred in spinal motoneurones in animals which were made to run to exhaustion (Hydén, 1943, 1947; Hochberg, 1958) or to swim with a load (Aleksandrovskaya et al., 1967). Apparently all these cases involve a sharp excitation followed by exhaustion of the nervous system, which as a rule was accompanied

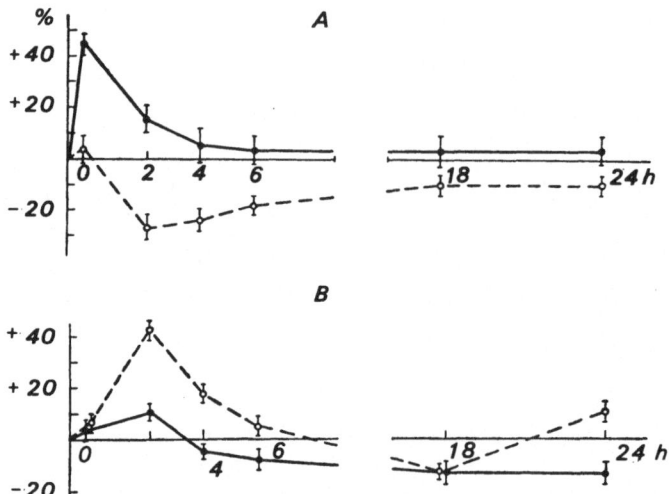

Fig. 28. Changes in RNA levels in neurones and neuroglia
of spinal cord of mouse during recovery after 3 h forced
swimming (Brumberg, 1968b). Ordinate- change from
resting level (in %); Abscissa- period of rest (in h)
after the end of a 3 h swimming period; A- anterior horn
of spinal cord; B- spinal ganglia; Solid lines- neurones;
Dashed lines- neuroglia.

by a lowering of RNA content in the neurones (for reviews, see:
Hydén, 1960, 1962; Pevzner, 1963a, 1966a; Brodskii, 1966). In the
early stages of muscular exercise entirely different biochemical
changes may occur. The dependence of the metabolic changes of the
nervous tissue on the changing phases of muscular activity was
convincingly shown by Yakovlev (1955, 1956, 1963) and Pogodaev
(1966, 1970; Pogodaev et al., 1968).

The increase in RNA content in cytoplasm of spinal motoneurones
during swimming confirms that this form of exercise does not exceed
the normal capacity of the experimental animal. Swimming in water
close to body temperature, and without a load, corresponds more
closely to normal physiological stressors than such commonly used
experimental influences as electroshock, or pharmacologically
induced convulsions. Mice and rats are naturally mobile animals and
even several hours of such muscular exercise does not seem to exceed
what might occur under natural conditions. Moreover in this type
of situation the pace of the muscular activity is to a large extent
set by the animal itself: the rhythmical functioning of various
muscle groups and their corresponding motor centers led to well
defined periods of raised activity, and corresponding rest periods.

It is striking that while there is an increase in RNA content
in motoneurones under these conditions, this is not accompanied by
any marked changes in RNA of their glial cell-satellites. However,
in the period following the end of swimming, there are marked
changes in RNA content in the neuroglia both of spinal cord anterior
horn, and spinal ganglia. This is indicative of the important role
played by glial cells in the period of recuperation after the
cessation of load. This question will be discussed in more detail
below.

Running rats in the wheel for one hour had almost no effect on
the RNA content of motoneurones in the lumbar thickening of the
spinal cord. Such changes as could be observed did not exceed the
normal variability, and were not statistically significant. Nor
were significant changes found in RNA content of the perineuronal
glia of the spinal cord anterior horn. All the same, it is sugges-
tive that these changes are in the same direction as in the data
from the previous experiment with swimming in mice. A significant
increase in RNA content was found, however, in the motoneurones of
cerebral cortex (Table XXIII), without changes being observed in
the corresponding glia. It is possible that the neurones of motor
cortex are more labile with respect to their macromolecule metabolism
than those of spinal cord, although much may depend on the particular type of exercise being used. For example, in the experiments
of Watson (1972), running rats in a treadmill led to an accumulation
of protein (estimated by cytophotometric determination of dry mass)
in neuroglia of the cervical region of the spinal cord. It is
interesting that these changes are found only in astrocytes (as well
as in ependymal cells of the central canal) but not in oligodendroglia. Neurones were not studied in this work.

Some of the most important studies in the field of functional
neurochemistry are those of Brodskii (1961, 1966). He has analyzed
the effects of stimulation with a flashing light on RNA content of
retinal ganglion cells. The RNA content was shown to increase after
the specific activation of the retinal neurones.

This led Brodskii to the conclusion that neuronal RNA does not
take part in the first biochemical reactions, leading to the excitation of the nerve cells, but in the subsequent secondary processes
most probably of a compensatory character. Knowing the role of
RNA in protein synthesis, it may be thought a priori that these
compensatory processes involve the synthesis of protein in the
neuronal cell body. This was further shown by Brodskii et al.
(1966) in an elegant experiment, the results of which will be
discussed in the next chapter (pp. 208). Our data support
Brodskii's views on the position of RNA in the chain of biochemical
events in neuronal activation. For example, in one of our experiments, mice performed intensive work for 2 h and for all this time
the mean RNA content in the cytoplasm of the excited neurones

TABLE XXIII

RNA content of motoneurones and perineuronal glia
from brain and spinal cord of rat at rest and
after a one hour period of motor activity
(Tiplady et al., 1974)

	Neuronal Cytoplasm	Neuronal Nucleus	Neuroglial Cells
(i) Motor area of cerebral cortex			
Control	77.2 ± 3.9*	29.4 ± 2.1*	6.66 ± 0.36
Active	88.8 ± 2.5*	36.2 ± 1.1*	7.38 ± 0.76
(ii) Anterior horn of spinal cord			
Control	1183 ± 56	121 ± 6.7	16.0 ± 1.1
Active	1366 ± 87	138 ± 8.0	14.8 ± 0.73

RNA content is expressed in arbitrary units (the product of the optical density and the volume of the cell, or the relevant part of it).

*Statistically significant difference from control ($p < 0.05$).

remained unchanged (see Fig. 27); only after 3 h of swimming was an accumulation of cytoplasmic RNA noted. Evidently this does not exclude the possibility either of changes in the quantity of nuclear RNA in the cells, or of changes in the rate of synthesis of RNA which are compensated for by equivalent changes in breakdown. It may also be suggested that changes in RNA content might have occurred which were quantitatively less than the error of the cytospectrophotometric method, or confined to a small percentage of neurones. However it is undoubtedly true that the massive increase in quantity of cytoplasmic RNA in the majority of neurones occurs only after 3 h swimming. It follows that the occurrence of such substantial changes in RNA metabolism is not a necessary condition of the early stages of activation of specific neuronal functions.

Thus the data obtained confirm the secondary role of RNA metabolism in the activation of nerve cell function, and demonstrate the well-defined autonomy of neuronal metabolic processes in the neurone-neuroglia unit - changes in neuronal metabolism need not necessarily be accompanied by analogous changes in the glia.

Electrostimulation

Electrical stimulation of isolated nerves of preparations, and also electrical skin stimulation of the whole animal are widely used in functional neurochemistry (see reviews by Pevzner, 1966a, 1969b, 1972b; Pogodaev, 1966, 1970). Although such a stimulus cannot be considered physiologically normal, it is of value on account of its accurately quantifiable parameters (voltage applied, frequency, length of influence, etc.) which aids reproducibility of results and facilitates the comparison of data obtained by different investigators. We have thus used this method of producing acute changes in the functional state of the nervous system.

In these experiments, female Wistar rats were placed in a plexiglass chamber which was a simplified version of that described by Vladimirova (1956). An electric current of 20-40 V, at a frequency of 40 pulses/min was applied to the metal grid floor. During the first 5-10 min of uninterrupted stimulation the rats were observed to be in a state of considerable motor excitation: they jumped, rushed about the chamber, or gnawed the metal grid. Subsequently, this excitement subsided, the amplitude of the movements became less, signs of exhaustion became apparent, and after 55-60 min the rats lay on their bellies and almost ceased to respond to the electrical stimulus. This condition lasted for several minutes after the cessation of electrostimulation. The ability to move the extremities was restored only after a period of 5-10 min. The animals then began to move about the cage, chose a comfortable place, and after several minutes fell into a deep sleep. Control animals at this time of day were normally awake.

Animals were killed 5, 10, 20, or 60 min after the beginning of stimulation, and also at 1, 2, 4, or 18 h after cessation of 60 min electrostimulation. The spinal cord in the region of the lumbar enlargement together with the adjacent spinal ganglia were removed and fixed by Brodskii's method.

Experiments were also carried out with electrostimulation of the superior cervical sympathetic ganglia of cats. The cervical sympathetic chain of the cat was exposed under urethane anesthesia, and using contractions of the third eyelid as a criterion of sympathetic activation, the preganglionic stump of the superior cervical ganglion was stimulated with an alternating current of supra-maximal intensity, and a frequency of 300 Hz for 3 h. Some of the ganglia were subjected to fatiguing antidromic stimulation for 30-45 min. In this case, the degree of fatigue was judged by the amplitude of the overall biopotential of the ganglion arising in response to orthodromic excitation (the electrophysiological part of this work was carried out

at our request by Yu.P. Pushkarev). Control animals were
subjected to anesthesia for the same period as the experimental animals.

In superior cervical ganglion of cat, 30-45 min stimulation
with antidromic electric current did not cause any change of nucleic
acid content either in neurones or glia. Only after a more prolonged period (3 h) of orthodromic stimulation of the preganglionic
trunk was a marked change noted - the quantity of RNA increased in
the cytoplasm of sympathetic neurones, and fell in their glial
cell-satellites (Fig. 29).

An analogous relationship between the quantity of nucleic acid
in neurones and glia was demonstrated in spinal cord. Five minutes
after the beginning of electrical stimulation of rats the content

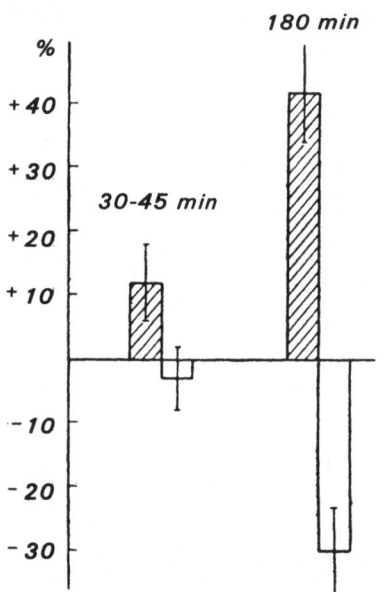

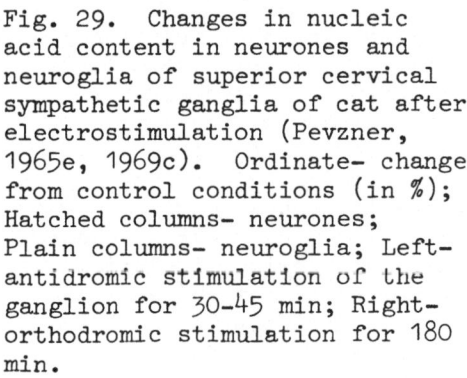

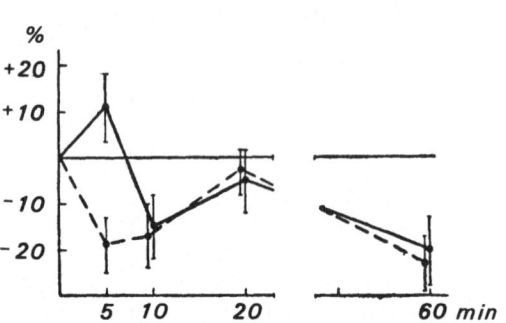

Fig. 29. Changes in nucleic
acid content in neurones and
neuroglia of superior cervical
sympathetic ganglia of cat after
electrostimulation (Pevzner,
1965e, 1969c). Ordinate- change
from control conditions (in %);
Hatched columns- neurones;
Plain columns- neuroglia; Left-
antidromic stimulation of the
ganglion for 30-45 min; Right-
orthodromic stimulation for 180
min.

Fig. 30. Changes in nucleic acid
content in neurones and neuroglia of spinal cord of rat during electrical skin stimulation
(Pevzner and Khaidarliu, 1967).
Ordinate- difference from control level, taken as 100%; Ab-
scissa- period of stimulation
(in min); Solid line- motoneurones of spinal cord anterior
horn; Dashed line- neuroglia.

of nucleic acid was increased in motoneurones of anterior horn of
spinal cord, and decreased in the glia (Fig. 30). Subsequently it
was lowered in both types of cell of anterior horn, but already
after 20 min of uninterrupted electrical excitation it had normalized
(the phenomenon of pseudonormalization). After 60 min of electrical
skin stimulation a reduction in RNA content was again observed in
both types of cell (Fig 30).

   At the end of stimulation the RNA content in these cells
returns to normal considerably more slowly (Fig. 31). Normalization
of the amount of RNA in motoneurones and their glia does not begin
until 2 h after the end of electrostimulation and was completed in
the glial cells of anterior horn considerably sooner than in the
motoneurones. Hypercompensation was also noted: the content of
nucleic acid in the glia rose to the normal level; however it did
not then stabilize but continued to increase, although at a reduced
rate (Fig. 31).

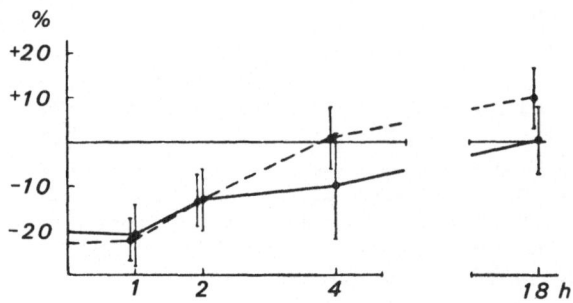

Fig. 31. Changes in nucleic acid content in neurones and neuroglia
of spinal cord of rat during recovery after 1 h of electrical skin
stimulation (Pevzner and Khaidarliu, 1967). Ordinate- difference
from control level, taken as 100%; Abscissa- period of rest (in h)
after cessation of a 1 h period of stimulation; Solid line- motoneurones of spinal cord anterior horn; Dashed line- neuroglia.

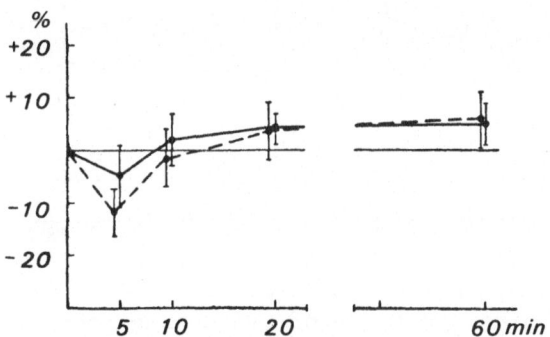

Fig. 32. Changes in nucleic acid content in neurones and glia of
spinal ganglia of rat during the course of electrical skin stimulation (Pevzner and Khaidarliu, 1967). Solid line- neurones of spinal
ganglia. Other designations as in Fig. 30.

In the spinal ganglion, the nucliec acid content showed
considerable stability (Fig. 32). Only 4 h after the cessation of
60 min electrical skin stimulation of rats was an increase of
nucleic acid in the glial cells of the ganglion noted, which disappeared after 18 h rest (Fig 33). As was stated above, an
analogous increase in RNA content in glial (but not in neuronal)
cells of spinal ganglion was observed in the period of rest after
swimming and it also quickly disappeared during recovery.

A biochemical analysis of homogenates of corresponding regions
of the nervous system was also carried out. Gray matter was taken
from spinal cord in the region of the upper part of the lumbar
enlargement and 2-4 mg samples were submitted to analysis of DNA and
RNA. Differential extraction of these acids was carried out
according to Ogur and Rosen (1950) with several modifications
(Baranov, 1962; Hess and Thalheimer, 1965). Two-wavelength spectrophotometry was then carried out according to Tsanev and Markov
(1960). For calculation of RNA content, we used the coefficient
703 calculated by Broun and Goncharova (1962) for nervous tissue,
in place of that of 561 suggested by Tsanev and Markov for yeast.

Similar determinations were carried out on superior cervical
sympathetic ganglia by M.N. Baranov. Tissue was thoroughly freed
from the attached capsule, frozen, and divided lengthwise. One half
was used for fixation according to Brodskii followed by cytophotometric determination of RNA, the other for homogenization and subsequent microchemical determination of RNA and DNA according to
Ogur and Rosen as modified by Baranov (1962).

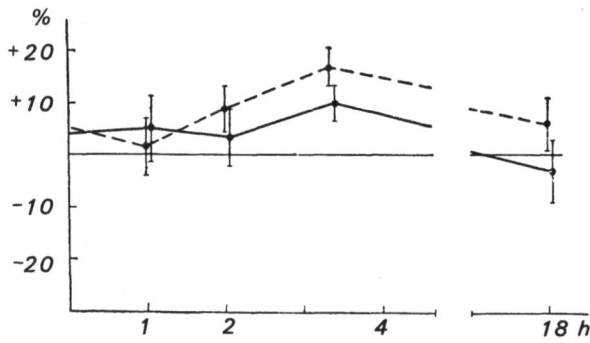

Fig. 33. Changes in nucleic acid content in neurones and
neuroglia of spinal ganglion of rat during recovery after
1 h of electrical skin stimulation (Pevzner and Khaidarliu,
1967). Solid line- neurones of spinal ganglion. Other
designations as in Fig. 31.

TABLE XXIV

Concentration of DNA and RNA in homogenates of
spinal cord anterior horn of the rat after 5 min
electrical skin stimulation (Khaidarliu, 1967b)

Experimental Conditions	DNA Concentration (in mg% P)	RNA Concentration (in mg% P)
Rest (control)	11.1 ± 0.5	11.3 ± 0.5
Stimulation	11.2 ± 0.4	10.1 ± 0.2
	(+1%; p > 0.9)	(−11%; p < 0.05)

In brackets - difference between experimental and control values.

The concentration of DNA in the homogenate of spinal cord
gray matter (Table XXIV) and of cervical sympathetic ganglion
(Table XXV) showed practically no change. We naturally did not
expect that such a short experimental period (3 h) could produce a
change in the metabolism of such a stable biochemical component of

TABLE XXV

DNA and RNA concentration in homogenate of superior
cervical sympathetic ganglion of the cat after 3 h
electrical stimulation (Baranov and Pevzner, 1963b)

Experimental Conditions	DNA Concentration (in mg% P)	RNA Concentration (in mg% P)
Rest (control)	34.1 ± 4.1	37.5 ± 1.7
Stimulation	29.2 ± 3.3	34.3 ± 2.1
	(−14%; p > 0.6)	(−9%; p > 0.7)

In brackets - difference between experimental and control values.

the nervous system as DNA. The absence of changes in the DNA content
in the homogenates confirmed that changes in the water content of
the tissue were not occurring, thus permitting subsequent calculations of RNA concentration to be made in terms of wet weight without
correction.

As can be seen from Table XXIV, the concentration of RNA in
the homogenate of anterior horn of spinal cord is only slightly
lowered after electrical skin stimulation for 5 min. However in
the tissue of sympathetic ganglia under the influence of 3 h
electrostimulation no reliable changes in the concentration of RNA
were noted (Table XXV).

The dynamics of the shift in RNA content observed in the motoneurones (initially an increase followed by a decrease) with electrical skin stimulation corresponds well with the suggestion that
has been made by a number of neurochemists (for reviews, see:
Hydén, 1960, 1962; Palladin, 1963, 1965; Pevzner, 1963a, 1966a;
Brodskii, 1966) that excitation of the neurone is accompanied by
an increase in RNA synthesis while over-excitation and exhaustion
lead to a lowering of the cell's RNA supply.

In the present series of experiments, as in the experiments on
swimming, we were dealing with a situation involving motor excitation of the animal. However in the case of swimming, the magnitude
and rhythm of the muscular load is largely determined by the animal
itself. On the other hand the aversive stimulation of the paw of
the animal by the electric current, being both powerful and uninterrupted is an unnatural and very stressful influence which rapidly
leads to exhaustion of the animal's motor apparatus.

This seems to be the cause of the differences in the biochemical
changes in the spinal cord. With swimming the content of RNA
changes only in the motor neurones themselves, while electrical
skin stimulation results in changes both in the neurones and in
their surrounding glia. Further, the quantity of RNA in the cytoplasm of motoneurones with swimming remained elevated after 4 h of
uninterrupted load, whereas with electrostimulation an hour was
sufficient to reveal a sharp and stable reduction in RNA in neurones
and glia of the anterior horn.

At the same time, in the reparative period marked similarity
of the two types of influence on the organism may be seen. At their
cessation very active biochemical changes proceed in the glial cells.
The impression is created that the stable normal level of RNA in the
neurones, which are passing from an active to a resting state, is
connected with the reduction in glial RNA content.

Cessation of electrical skin stimulation leads to a normalization of RNA levels more quickly in the glia than in the neurones

(Fig. 31).

        Antidromic stimulation with an electrical current for 30-45 min
did not lead to changes in RNA content in sympathetic neurones or
their glia (Fig. 29) although the total amplitude of biopotentials
in the sympathetic ganglia was markedly changed. However, orthodromic stimulation for 3 h led to noticeable changes in RNA content
in neurones and their glia in the ganglion. This divergence was
first considered to be the result of different durations of electrostimulation (Pevzner, 1969b). Later, however, the very interesting
data of Geinisman et al. (1970a) were published showing that with
equal duration of orthodromic and antidromic stimulation of motoneurones of spinal cord only synaptic, orthodromic, influences
produced clear changes in the content of cytoplasmic RNA in these
neurones. This presumably explains the fact that electrical skin
stimulation whose effect should be produced initially in the sensory
neurones, does not influence the RNA content in the neurones or glia
of the sensory ganglia (Fig. 32). Subsequent interneurone-mediated
stimulation of the motoneurones which clearly involves a polysynaptic
path, leads to large and complex biochemical changes in the post-
synaptic motoneurones (Fig. 30).

        One of the important aspects of the present series of experi-
ments is the relationship between cytophotometric analysis of
individual cells, and chemical analysis of the same regions of the
nervous system. During the period when the RNA content of neurones
is raised, and that of the surrounding glia is lowered, the content
of RNA in homogenate of whole nerve tissue either does not change
(sympathetic ganglia) or is very slightly lowered (anterior horn of
spinal cord). Thus, the opposed changes in RNA content in the two
types of cells of the neurone-neuroglia unit are not accompanied by
an overall change in the whole tissue in which it is localized.
Considering the short time in which this process occurs, there is
every reason to suggest a spatial redistribution of RNA within the
neurone-neuroglia unit. Such a migration of RNA from the neuroglia
to the nerve cells was first postulated by Hydén (1959a, b, 1960,
1962, 1964). Although there has still been no direct demonstration
of its occurrence, the indirect data taken together make it an
attractive suggestion. Hydén's hypothesis will be discussed in
detail in the next chapter.

        Thus, the data show that exhaustion of the nervous system is
accompanied by a single type of change in nucleic acid content in
nerve and glial cells. From the onset of exhaustion or at its
cessation the nucleic acid metabolism of the neurones and glia
differs. In particular the reparative synthetic processes in glial
cells proceed more intensively than in the neurones.

        The fact that changes in opposite directions in nucleic acid
content of neurones and glia are not accompanied by overall changes

in RNA content in whole tissue homogenates may be considered to be indirect evidence in favor of the transfer of RNA from glial cells to neurones.

Convulsive States

The study of convulsions seems appropriate when it is necessary to study the effect of generalized excitation in the nervous system, involving many neurones and showing a characteristic behavioral pattern.

There is a number of well studied analeptics available, including Corazol, strychnine, and picrotoxin, which may be used to produce well defined convulsions in an experimental animal.

The close interdependence of the metabolism of nucleic acids and protein in the nerve cell, suggests that the basic regularity of functionally conditioned changes of the metabolism of nucleic acids in the neurone-neuroglia unit would occur also in protein metabolism. Therefore, although the immediate problems of the present work were the study of RNA metabolism in this system, it seemed appropriate in various cases to carry out determinations of protein content as well.

Experiments were carried out on male Wistar rats, which were injected either subcutaneously or intraperitoneally with Corazol (pentamethylene tetrazole, Metrazol) at a dose of 45-55 mg/kg.

After 1-5 min (depending on the method of injection) well defined clonic convulsions began, lasting around 40 min, after the cessation of which the animals were in a state of complete exhaustion (drowsiness, complete immobility, marked weakness of muscular tonus) for 1 h. After 2 h the animals began now and again to change their body position, and the muscular tonus increased. Sleep lasted 5-6 h after cessation of convulsions, after which the rats began to actively move about. However their activity still seemed somewhat sluggish. Only after 18 h of rest was the behavioral pattern indistinguishable from controls.

Animals were killed during clonic convulsions (15-20 min after injection of Corazole) and also at various times from 3 to 18 h after cessation of convulsions.

The administration of Corazole to rats led to a marked reduction of the RNA content in cytoplasm of both spinal cord motoneurones and their glial cell-satellites (Fig. 34). In spinal ganglia a similar reduction was seen in the neurones, while in the glia there was practically no change (Fig. 34). With cessation of convulsions

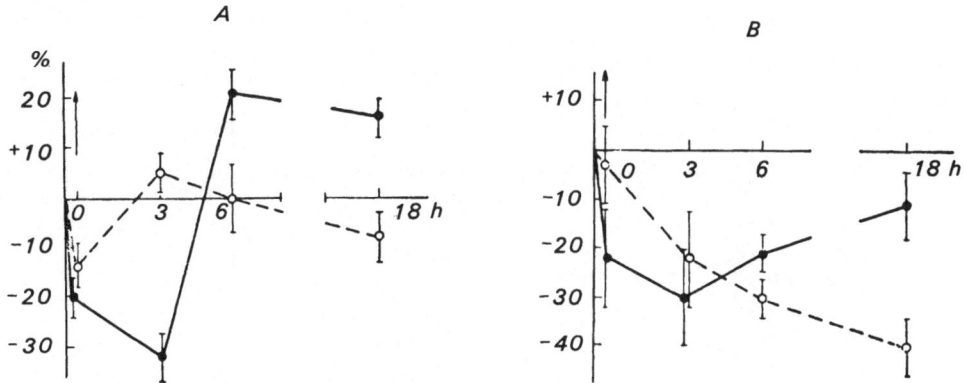

Fig. 34. Changes in nucleic acid content in neurones and neuroglia of spinal cord of rat during corazol convulsions and subsequent recovery (Saudargene, 1969a). Ordinate- difference from control level, taken as 100%; Abscissa- period of rest (in h) after cessation of convulsions; A- spinal cord anterior horn; B- spinal ganglia; Solid lines- neurones; Dashed lines- neuroglia; Arrow- end of convulsions.

the RNA content in the glial cells of anterior horn of spinal cord rapidly returned to normal reaching the resting level after 3 h and showing practically no further change. In the motoneurones, however, a longer lasting reduction of RNA content was noted, and only after 3 h of rest did a rapid normalization occur, leading in this case to an elevation above the resting level (Fig. 34).

In neurones of the spinal ganglia the content of cytoplasmic RNA remained at a reduced level for several hours after the cessa- tion of convulsions. Only after 18 h of rest did the RNA content return to the control level. In the glia of spinal ganglia a pro- gressive lowering of nucleic acid content was observed (Fig. 34).

The dynamics of the changes in amino and imino groups (total protein) and in protein sulphydryl groups was also investigated (Figs. 35 and 36). In the spinal cord, a similar picture emerged for both total protein and for sulphydryl groups. An initial reduction occurred in both neurones and glia, with a subsequent rapid normalization and hypercompensation in the glia, and a much slower recovery in the neurones.

A different picture emerged in the spinal ganglia. Here the lowering of content of total protein and of protein sulphydryl groups at the height of convulsions turned out to be more marked in the neurones than in the glia; after cessation of convulsions the

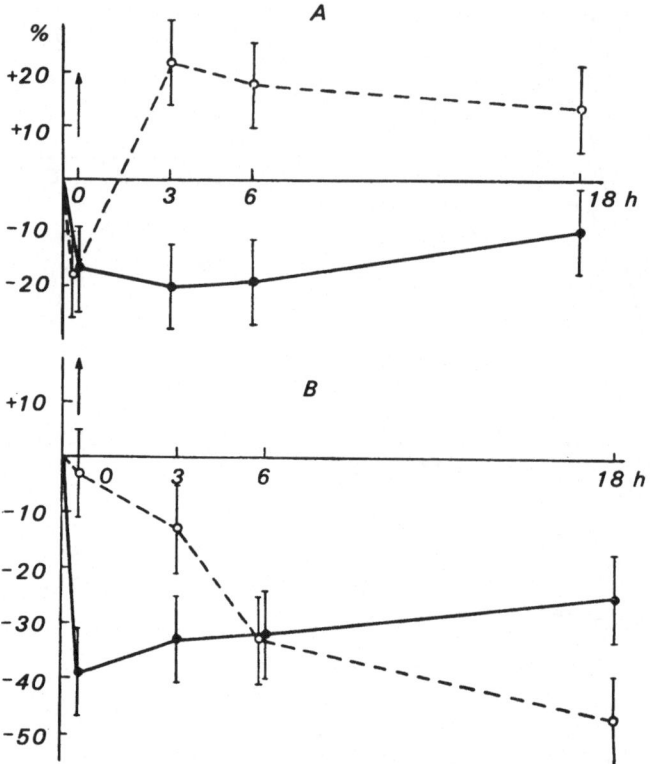

Fig. 35. Changes in total protein content in neurones and neuroglia of spinal cord of rat as a result of corazol con-vulsions and subsequent recovery (Saudargene, 1969b). Same designations as in Fig. 34.

content of sulphydryl groups of protein progressively returned to the resting level both in neurones and in glia (in 18 h) while the content of total protein in the neurones of the ganglia increased considerably more slowly. In the glia of the ganglia a progressive lowering of total protein content was observed during the period of recovery after the cessation of convulsions (Fig. 36).

Convulsions produced by Corazol and similar drugs appear on a number of grounds to be useful models for functional biochemical studies of the CNS. In particular, the excitation of the nervous system produced by the analeptic is of a generalized character. This process apparently involves such a great number of nervous structures that an analysis of whole brain homogenates allows the demonstration of regular changes in content and turnover of nucleic acids (Baranov, 1963; Talwar et al., 1961, 1966; Chitre et al.,

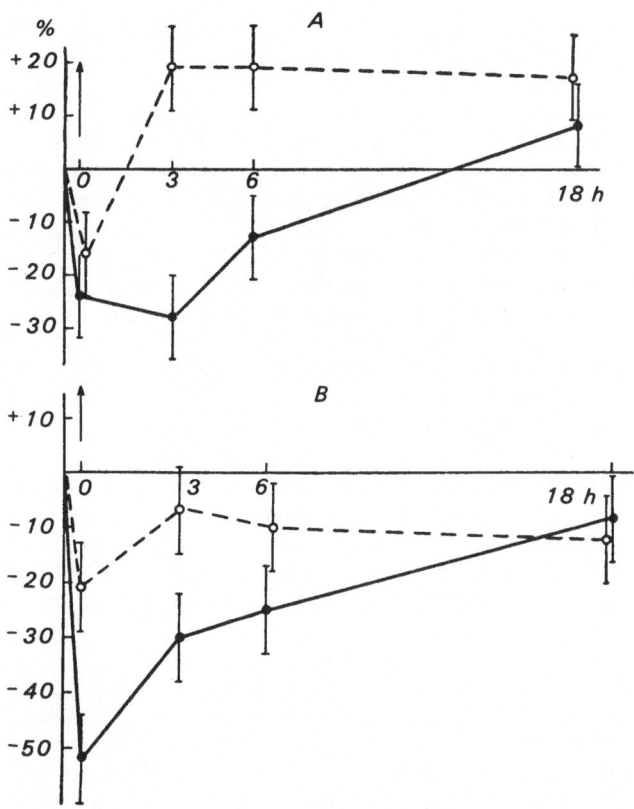

Fig. 36.   Changes in protein sulphydryl group content in
neurones and glia of spinal cord of rat as a result of corazol
convulsions and subsequent recovery   (Saudargene, 1969b).
Same designations as in Fig. 34.

1964).

     Our results illustrate this at the level of the neurone-
neuroglia unit. Cytophotometric determinations in both visible and
UV-light (see Chapter 5) confirmed the published reports that a
reduction occurs in RNA content of cytoplasm of spinal motoneurones,
and also showed a marked reduction in the glial cell-satellites
(Table XXII).

     The character of the changes induced evidently depends both
upon the nature of Corazol convulsions, and on the type of neurone-
neuroglia unit being studied. Two hours after electroconvulsive
shock, administered via transcorneal electrodes, an accumulation of
RNA and protein was found in the cytoplasm of hippocampal neurones,

while no changes were found in the perineuronal glia (Tencheva and Pevzner, 1974). Application of mescaline (3,4,5-trimethoxyphenethyl-amine) to the surface of the cerebral cortex of cat leads to profound epileptogenic activity, as shown by extracellularly recorded evoked potentials. This activity was accompanied by a marked accumulation of RNA in the cytoplasm of neurones - of layer II of the cerebral cortex in the region of the epileptogenic focus. The cytoplasmic RNA content later normalized, but the total protein in the cyto-plasm of these cells remained lowered. No changes were observed in the surrounding glial cells (Tencheva and Pevzner, 1973). In these experiments, the neuronal excitation in paleo- and neocortex seems similar to that resulting from muscular load, in that only the neuronal metabolism is affected (see Fig 27a and Table XXIII). In the case of Corazol convulsions, the metabolic changes are more profound, involving both neurones and glia.

Analysis of the reparative changes in the neurone-neuroglia unit after the cessation of convulsions showed that after substantial changes in RNA content in both types of cell, the reparative, compensatory processes, leading to normalization of RNA content, proceed more quickly in the glial cell satellites than in the neurones (Fig. 34).

Simultaneous determinations of RNA and total protein content in cells of spinal motor nuclei allows a comparison to be made of the time course of biosynthesis of these two types of macromolecule at the cellular level. As can be seen from a comparison of Figs. 34 and 35, convulsions lead to a reduction in both RNA and protein in spinal motoneurones. When the convulsions cease, there is an increase in cytoplasmic RNA which reaches the normal level and then exceeds it (Fig. 34). The gradual normalization of protein levels in the motoneurone cytoplasm then begins (Fig. 35). This is consis-tent with the view of RNA as a template for protein synthesis. Thus 6 h after the end of convulsions the content of RNA is asympto-tically returning to its steady state level, while the protein content is still depressed. Such a dynamic picture corresponds well with our understanding of the biosynthetic scheme of protein synthesis as well as with the epigenetic system of delayed negative feedback. This has been discussed in detail by Saudargene and Pevzner (1969).

For cytochemical analysis of proteins, the combination of color reactions for amino- and imino-groups and the reaction for SH-groups is generally suitable. The Sakaguchi reaction, a staining method of the first type, is specific for α-amino acids, and may thus be used to estimate total cell protein. However sulphydryl groups are considerably more labile, and their number can change with with changes in the conformation of the protein macromolecule (Ungar and Romans, 1958; Sandritter and Krygier, 1959; Gayevskaya et al., 1966). Therefore differential changes in the amino- and imino groups on the one hand, and the sulphydryl groups on the other may be an indication that changes are occurring in the higher structure of the protein molecule as well as in protein content. Such a pattern is seen in the relationship of neuronal and glial protein in spinal ganglia after Corazol administration (Fig 37).

However, in the motor neurones of anterior horn of spinal cord and their glial cell satellites, both during convulsions, and in the subsequent rest period after their cessation, such a divergence does not occur (Fig. 37). The dynamics of the changes of both amino- and sulphydryl-groups of protein turned out to be practically identical. Thus acute changes may occur in the functional state of the motor structures of the spinal cord which result in changes in total protein content in both neurones and glia without the occurrence of conformational changes (in any case, of the kind that affect the number of SH groups on the surface of the protein molecule).

Acute changes in the functional state of the nervous system can thus lead to changes in nucleic acid and protein metabolism not only in the neurones but also in the glial cell-satellites. During the

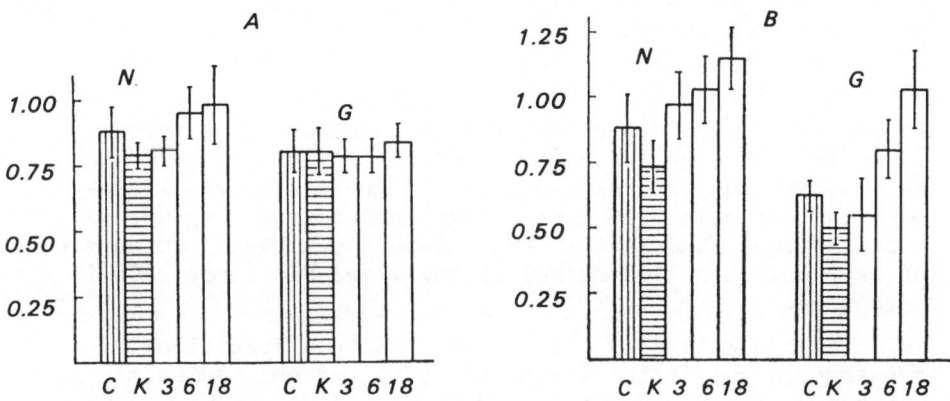

Fig. 37. Relative sulphydryl group content in neuronal (N) and neuroglial (G) protein of spinal cord of rat during corazol convulsions and subsequent recovery (Pevzner and Saudargene, 1971). Ordinate- ratio of optical density of protein SH-groups to optical density of total protein; A- anterior horn of spinal cord; B- spinal ganglia; C- control; K- corazol convulsions; 3, 6, & 18- period of rest (in h) after cessation of convulsions.

reparative processes, metabolic changes occur first in the neuroglia, which appear to assist the recovery of the initial metabolic state in the neurone.

Deprivation of Normal Sleep

The alternation of wakefulness and sleep is a universal phenomenon in the animal world, which in recent years has attracted much attention not only from physiologists, but also biochemists. The biochemical processes associated with the commencement, continuance, and termination of sleep have been very little studied (for reviews, see: Laborit and Laborit, 1965; Jouvet, 1967). Our studies do not pretend to be an analysis of the mechanism of sleep. However, we have considered it possible to use the model of disturbances in normal sleep developed from neurophysiological work to compare their influence on neuronal and glial metabolism.

Male Wistar rats were deprived of sleep for 24, 48, or 96 h. Two situations were compared: "full insomnia" and selective deprivation of the paradoxical, rapid eye movement (REM) phase of sleep. The former was achieved by subcutaneous injection of an aqueous solution of amphetamine (6 mg/kg) each 3 h (Svorad and Novakova, 1960). The latter used the method of Jouvet et al. (1964), placing the rat in a basin of water on a square wooden platform 5 x 5 cm, standing 3-4 cm above the water level; at the onset of REM sleep the weakening of the muscular tone of the body and particularly of the neck led to the animal dipping its muzzle into the water, or even falling completely in, resulting in its awakening. Every 4 h the rat was removed from its perch for 10-20 min for feeding.

At the end of the experiment the animals were transferred to individual cages where they fell asleep within 5-10 min. We also studied the influence of different degrees of amobarbital narcosis - 40 min of light narcosis (70 mg/kg, subcutaneous) (Kudryavina, 1954; Dement'eva, 1961) or 60 min of deep narcosis (100 mg/kg) (Arbuzov and Nikiforov, 1967) were used.

Animals were killed by decapitation using a specifically constructed guillotine (Voronka, 1971) and only animals which did not wake up were used.

Animals of the same background and weight kept in a waking condition were used as controls. Moreover, for the series of experiments with phenamine-induced insomnia, rats given subcutaneous injections of 0.2 ml distilled water were used as additional controls.

Content of total and basic proteins was determined by cytophotometry of sections stained with amido black 10B on the microspectrophotometer MUF-5 (Bakharev et al., 1964).

When neurones (and their corresponding glial cell-satellites) of either the supraoptic nucleus of the hypothalamus **or** of red nucleus of midbrain were compared, a marked depletion of protein in neurones and glia was observed after both forms of insomnia (Fig. 38). Deprivation of REM sleep led to a greater change in the neurones than deprivation of both phases of sleep. In the neuroglia on the other hand, full insomnia (produced by amphetamine) was accompanied by a more marked reduction in total protein (Fig. 38).

Under the influence of amobarbital narcosis, the protein content in neurones and glia of both nuclei was depressed. In supraoptic nucleus this reduction was greater in neurones than in glia while in red nucleus the degree of reduction of total protein (Fig. 39) as well as that of basic protein (Fig. 40) was, as a rule, the same in neurones and glia. With light narcosis a more marked change was usually noted than with deep narcosis.

In the last few years, particular attention has been paid to the study of the REM phase of sleep, which is accompanied by desynchronization of the EEG, fast, low-amplitude rhythms of the biopotentials, rapid eye movements and considerable weakening of the muscular tonus. The active character of sleep, of which Pavlov wrote half a century ago, seems to be particularly associated with this phase of sleep. However, the biochemistry of the phases of sleep has been little studied and few data are available. It has been shown, however, that an increased demand for oxygen by brain tissue, a strengthening of blood supply, and an intensifying of brain protein synthesis occur during REM sleep (see reviews by Glushchenko and Demin, 1971; Voronka et al, 1971,1972; Rubinskaya, 1973; Demin and Rubinskaya, 1974).

Laborit and Laborit (1965) linked this phase of sleep with the neuroglial cells. A rather speculative argument led these authors to the suggestion that sodium ions pass out of the neurones and are accumulated in the neuroglia during the slow wave phase of sleep, while a reverse transfer from glia to neurones leads to the development of the REM phase of sleep. Jouvet (1967) also writes on the possible role of the neuroglia in REM sleep - also without any real basis in the factual evidence.

Although our data are connected with the selective deprivation of REM sleep, rather than with a direct analysis of it, and are thus to a considerable degree indirect, they create rather the opposite impression. From Fig. 38 it can be seen that amphetamine insomnia (which involves the deprivation of both phases of sleep) led to a quantitatively greater reduction in protein in the neuroglia of the two nuclei studied than deprivation of REM sleep only. Thus the effect of insomnia here shows a gradual character: deprivation of one phase of sleep leads to a relatively small change, while deprivation of both phases of sleep acts in the same direction,

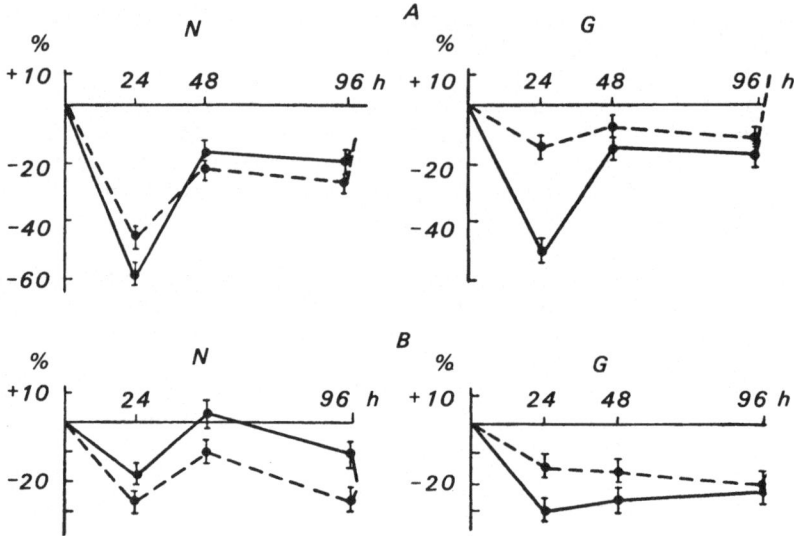

Fig. 38. Changes in the content of total protein in
neurones (N) and neuroglia (G) of supraoptic nucleus of
hypothalamus (A) and of red nucleus of the midbrain (B)
of rat during various forms of insomnia   (Voronka, 1971;
Voronka et al., 1971).   Ordinate- difference from corresponding level for waking animals, taken as 100%; Abscissa-
period of insomnia (in h), changes after 96 h correspond to
15-20 min sleep after cessation of 96 h forced insomnia.
Solid line- deprivation of both phases of sleep; Dashed
line- selective deprivation of rapid-eye-movement (paradoxical) sleep.

but more strongly. A quite different picture emerges for the
neurones (Fig. 38). Here deprivation only of the REM phase of sleep
led to a considerable reduction in protein content, which is the same
as, or even greater than that with deprivation of both types of
sleep. This seems to indicate that neuronal protein metabolism is
specifically involved in REM sleep while in the glia both phases
have a comparable importance. If this assumption is justified,
then REM sleep - i.e. the phase of active reconstruction of brain
metabolism - may be considered to be associated predominantly with
specific neuronal rather than glial functions. This supports the
comment made in connection with the swimming experiments - that of
the two cell types in the neurone-neuroglia unit, the role of the
neurones is the more specific for brain function.

A depression of protein content in neurones and glia was also
noted during amobarbital narcosis. In Figs. 39 and 40 the changes
in cell protein produced with natural sleep are presented for
comparison. It is perfectly clear that both for total protein and

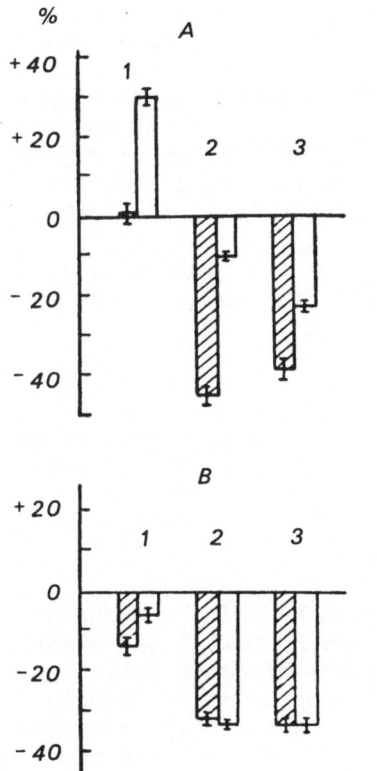

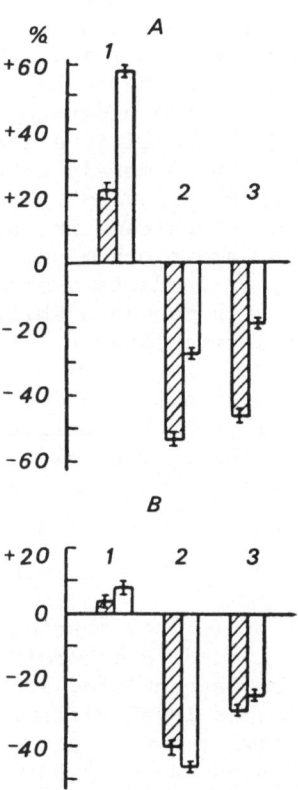

Fig. 39. Changes in content of total protein in neurones and neuroglia of supraoptic nucleus of hypothalamus (A) and red nucleus of midbrain (B) of rat during natural sleep and barbiturate narcosis (Voronka and Pevzner, 1972). Ordinate- difference from corresponding waking level, taken as 100%; 1- natural sleep for 15- 20 min.; 2- light barbiturate narcosis (40 min after administration of 70 mg barbiturate/ kg body weight); 3- deep barbiturate narcosis (60 min after administration of 100 mg barbiturate/kg body weight; Hatched columns- neurones; Plain columns- neuroglia.

Fig. 40. Changes in content of basic proteins in neurones and neuroglia of supraoptic nucleus of hypothalamus (A) and red nucleus (B) of rat during natural sleep, and barbiturate narcosis (Voronka and Pevzer, 1972). Same designations as in Fig. 39.

for basic protein in neurones and glia the changes in natural sleep
and for narcosis are quite different. Narcotic sleep seems to be
very unsuitable as a model for natural sleep. Leaving to one side
the question of the mechanism of the depression of protein in cells
of nervous system under conditions of narcosis (this question was
discussed in detail in a separate review of Voronka and Pevzner,
1972) we wish to merely underline that in this series of experiments
the same regularity was observed as in the preceding ones: an acute,
in some cases extreme influence on the nervous system either produces the same biochemical changes in the glial cells as in the
neurones, or leads to a selective change of macromolecules in the
neurones with marked stability of RNA and protein levels in the
glial cell-satellites.

The data obtained confirm the role of the neurone as the
functionally and metabolically more active partner in the neurone-
neuroglia unit. They also confirm the relatively greater stability
of neuroglial metabolism under the influence of extreme factors.

Forced Hypokinesia

The biological characteristics of man and of the majority of
animals (including laboratory animals) are such that the organism's
metabolism is regulated at a constant level of motor activity
although this level differs depending on the animal. Reduction of
motor activity for a defined period (e.g. deep sleep or hyber-
nation) in no way contradicts this concept. Any substantial
restriction of mobility for prolonged periods will thus lead to
various functional and biochemical changes in the organism, including
its nervous system (Brumberg and Pevzner, 1968; Portugalov et al.,
1968).

We have compared the biochemical changes in the spinal neurones
and their glia resulting from hypokinesia.

Most of the experiments in this series were carried out on
male white mice. Each animal was placed in an individual
cage measuring 7.0 x 2.5 x 2.5 cm. The chambers considerably
limited the movement of the animals, but did not cause
complete immobilization. Food and water were provided ad
libitum. In the course of the first few days after con-
finement, the mice showed restlessness, but gradually
approached the hypokinetic condition, sitting immobile in
the chamber. At the end of the third week the usual
symptoms - loss of body weight of 15-20%, paresis of the
rear extremity and disturbance of motor coordination - were
noted. Only 2-3 days after the cessation of 3 weeks
hypokinesia did the mice show a return of normal motor
activity.

Animals were killed by decapitation after 3 weeks hypokinesia, and also 2, 6, 24, or 72 h after their removal
from the chamber. Spinal cord in the region of the lumbar
enlargement and the associated spinal ganglia were studied.

Male rats were also used. These were placed for 3 weeks in
individual plexiglass chambers, limiting the animals' movements. In this case only the anterior horn of the spinal
cord was taken (without the spinal ganglia) in the region of
the cervical enlargement.

The nucleic acid content was determined using UV-cytospectrophotometry.

At the end of 3 weeks hypokinesia, when the animals seemed to
have completely adapted, the cytoplasmic RNA content was normal in
both the motoneurones of anterior horn of spinal cord and in the
neurones of the spinal ganglia. No changes in nucleic acid content
were found in the glial cells of rat or mouse anterior horn, while
in the neuroglia of mouse spinal ganglia RNA content was somewhat
decreased (Fig. 41). During the first 2-6 h of normal motor activity
a marked depression of RNA content was noted in neurones and glia
of anterior horn and spinal ganglia. In both cases, this effect
was greater in the glial cells than in the neurones, and the subsequent restoration of normal RNA content occurred more rapidly in
the glia. At the end of the third day of unrestricted motor

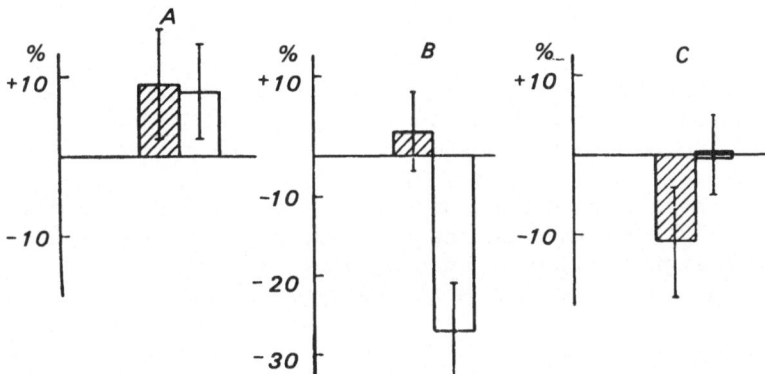

Fig. 41. Changes in RNA content in neurones and neuroglia
of spinal cord of mouse and rat after 3 weeks forced hypokinesia (Brumberg and Pevzner, 1968). Ordinate- difference
from corresponding control level, taken as 100%; A- anterior
horn of spinal cord of mouse; B- spinal ganglia of mouse;
C- anterior horn of spinal cord of rat at the end of 3 weeks'
restriction of activity; Hatched columns- neurones; Plain
columns- neuroglia.

activity the RNA content of cytoplasm of both types of neurone was
practically normal. However, a reduction in RNA content was again
found in neuroglia, both in spinal cord anterior horn, and in
spinal ganglia.

Essman (1971) isolated 23 day old mice for 4 weeks, and then
prepared enriched glial fractions from cerebellum. A 2.5 fold
increase in RNA content was found. If mice were subjected to
mocerate sensory stimulation immediately after the period of isolation, then RNA content was considerably reduced. The levels of
sensory stimulation used were insufficient to cause changes in glial
RNA levels in normal mice. This is in general agreement with our
data, although the mechanisms of hypokinesis and sensory deprivation
must differ. A detailed comparison can not be made because Essman
separated only glial fractions, and did not compare them to neurones.

Comparison of the reparative processes in neurones and neuroglia makes it clear that at the beginning of the post-hypokinetic
period, the biochemical changes in the glial cells were at least as
great if not greater than those in the neurones (see Fig. 42).
However, the rate of the subsequent normalization of RNA content was
higher in the glia than in the neurones. This is particularly
clearly seen in anterior horn of spinal cord, where the depression
of RNA in the cytoplasm of the motoneurones returned to normal only
after 15-16 h, while in the cell bodies of their glia the normal
level was already reached 6 h after the removal of mice from the
chamber (Fig. 42).

At the end of the third day of free motor activity, the RNA
content of glial cells of anterior horn of spinal cord and spinal
ganglia was again lowered. At this time the behavior of the
animal suggested a normal function of the nervous system. RNA
levels were also normal in both motoneurones and sensory neurones.

Thus, longer term changes have again been demonstrated which
were selectively localized in the glial cells. The possible
physiological significance of such subsequent changes in the neuroglia will be discussed in the next chapter.

Acute and Chronic Hypoxia

Few experimental influences on the nervous system have been
used so frequently or studied so carefully as hypoxia. This is
mainly due to the general biological importance of hypoxia with its
ecological and clinical significance.

The study of hypoxia is also important in investigations on the
functional biochemistry of the nervous system. The properties of
energy metabolism of the nervous system, and the changes in the

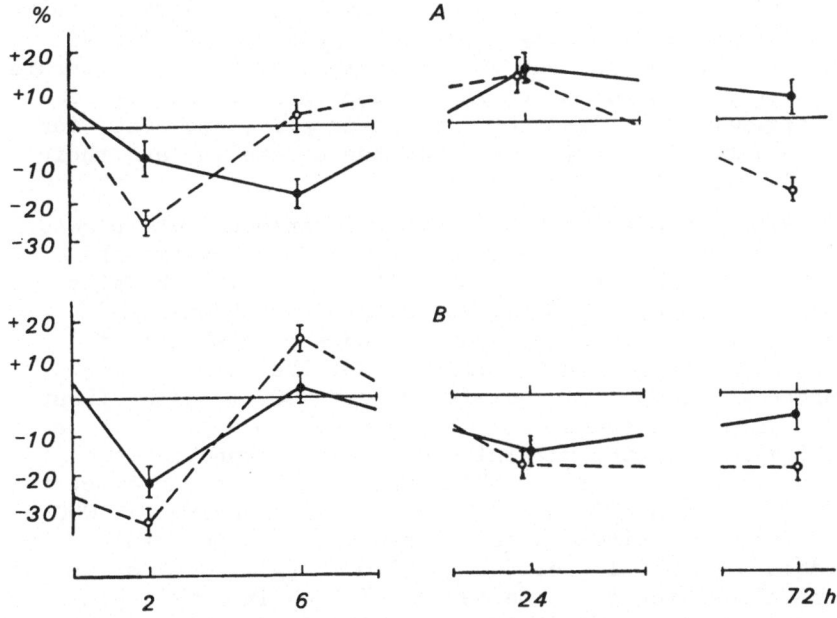

Fig. 42. Changes in RNA content of neurones and neuroglia of spinal cord of mouse after 3 weeks' hypokinesia (Brumberg and Pevzner, 1968). Ordinate- difference from control level, taken as 100%; Abscissa- length of period of free motor activity (in h) after cessation of 3 weeks' restriction of movement; A- spinal cord anterior horn; B- spinal ganglia; Solid line- neurones; Dashed line- glia.

cellular metabolism of the brain under conditions of increased neuronal functioning leads to the suggestion that activation of the nervous system may lead to intracellular hypoxia (for reviews, see: Brodskii, 1966; Pevzner, 1966a,b; Mchedlishvili, 1968, 1969).

Therefore in another series of experiments, animals were subjected to hypobaric hypoxia.

Male Wistar rats were placed for 2 h in a barochamber at a simulated altitude of 8,700 m (240 mm Hg pressure; partial pressure of O_2 55 mm Hg). The "ascent" of the animal to the given altitude was accomplished as follows (Chetverikov, 1966, 1967). A barochamber with a capacity of 36 litres was used. The pressure was lowered over a period of 16 min, with pauses every 100 mm Hg. As a rule, no mortalities occurred under these conditions. Control animals were kept for 2 h at normal atmospheric pressure. In the middle of the experimental period, the chamber was ventilated, without raising the pressure, to avoid hypercapnia. Animals were killed immediately after 2 h hypoxia, and also at intervals of 6, 18, and 48 h. Nucleic acids were determined by cytospectrophotometry in motoneurones of the cervical enlargement of spinal cord and their glia, and also, for comparison, in Purkyně cells of the cerebellum and their glial cell-satellites.

In experiments with chronic hypoxia, animals were placed in a barochamber of the type described by Gazenko and Malkin (Gazenko et al., 1968) either for 5.5-6 h daily for 14 h, or continuously for 21 days (with only a 1 h decompression for cage cleaning and replenishment of food and water). In the first case, the animals were subjected to a simulated altitude of 3000 m on the first day, with an increase of 500 m per day to a final value of 7000 m for all the remaining days, corresponding to a barometric pressure of 290 mm Hg. In the second case, the "altitude" was 3000 m the first day, increasing by 1000 m, and from the third day to the end of the experiment remained at a value of 5500 m (350 mm Hg). Control animals were placed for the same period of time in normal atmospheric pressure and standard condition of food and drink.

Rats kept for 2 h in the barochamber at a simulated altitude of 8,700 m showed a marked increase of RNA content in both types of neurones studied: spinal cord anterior horn motoneurones and cerebellar Purkyně cells (Fig. 43). However in the case of the cerebellum, this increase was also found in the glial cells.

After hypoxia, the increased RNA content in cerebellum gradually normalized. This occurred earlier in the glia than in the neurones. 18 h after the end of hypoxia, the RNA level became lower than the control level in the whole neurone-neuroglia unit, and only after 48 h were normal values reached in both Purkyně cells and their glial cell-satellites (Fig. 44). In spinal cord motoneurones, where the initial effects of hypoxia were not seen in the surrounding glia, the normalization was more complex. The original level of RNA was attained more quickly than in the Purkyně cells, but was then followed by a lowering of RNA levels. There was then an increase in the amount of cytoplasmic RNA, but in this case the glia were also involved, although to a lesser extent (Fig. 44).

Prolonged hypobaric hypoxia (2-3 weeks) led to different changes in nucleic acid content in cerebellum and spinal cord, depending on the regime of hypoxia (Fig. 43). However the same characteristic appears here as in acute hypoxia: in cerebellum the RNA content is increased both in the Purkyně cells and in their glial cell-satellites, while in spinal cord the accumulation of RNA was found only in the motoneurones, and did not reach statistical

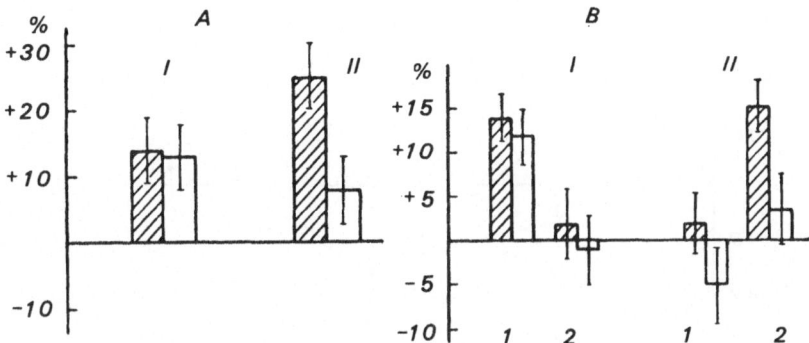

Fig. 43. Changes in nucleic acid content in neurones and
neuroglia of cerebellum and spinal cord of rat after acute
and chronic hypoxic hypoxia   (Pevzner, 1968; Gazenko, Demin,
Malkin and Pevzner, 1968).   Ordinate- difference from corresponding control level, taken as 100%; A- acute hypoxia; B-
chronic hypoxia; I- Purkyně cell layer of cerebellum; II-
anterior horn of spinal cord; 1- intermittent acclimatization
(daily "ascent" in the barochamber to 3000-7000 m simulated
altitude for 5-6 h/day during 2 weeks); 2- uninterrupted
acclimatization (continuous exposure to an "altitude" of
5500 m for 3 weeks); Hatched columns- neurones; Plain columns-
neuroglia.

significance in the neuroglia (Fig. 44).

     Nucleic acid synthesis is closely linked to the aerobic
processes of carbohydrate catabolism and terminal oxidation, and
therefore the synthesis of nucleic acids may be expected to be
disturbed under hypoxic conditions.

     However, compensatory mechanisms exist in the organism, which
tend to counter the tissue hypoxia. Chetverikov and co-workers
(Gasteva et al.,1960; Chetverikov, 1966, 1967; Chetverikov et al.,
1970) showed that the inhibition of biosynthetic processes in
hypoxia is not the result of disturbance of the biochemical processes
in the tissue (in particular exhaustion of the supply of high-energy
phosphate in the cells) but on the contrary is an adaptive reaction,
which leads to a reduction in the sensitivity of the cell to oxygen
deficiency.

     We have shown acute hypoxia to lead to an increase in the
nucleic acid content in cells of the nervous system. This increase
was particularly marked in spinal cord motoneurones. This includes
motoneurones innervating the respiratory musculature, the activation

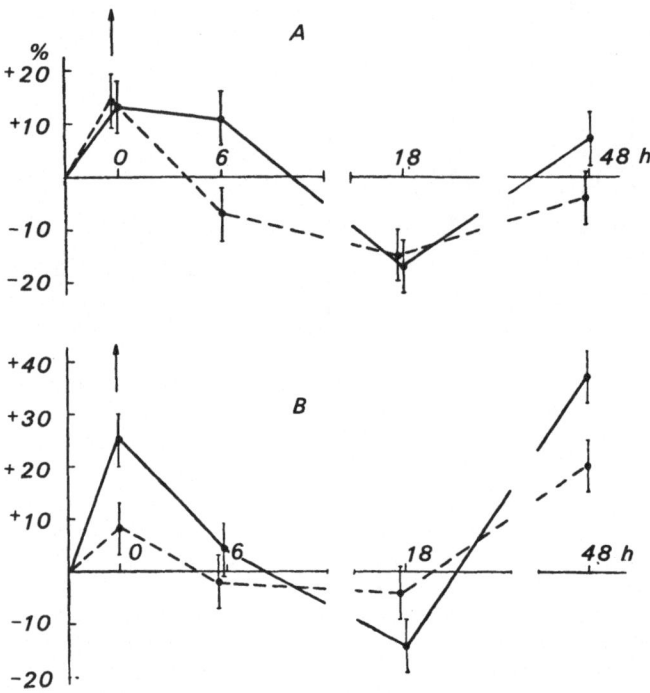

Fig. 44. Changes in nucleic acid content in neurones and
neuroglia of cerebellum and spinal cord at the end of 2 h
of hypoxic hypoxia (Pevzner, 1968). Ordinate- difference
from control level, taken as 100%; Abscissa- duration of
hypoxia (to the left of zero point) and post-hypoxic period
(to the right of zero) in h; A- Purkyně cell layer of the
cerebellum; B- spinal cord anterior horn; Solid lines-
neurones; Dashed lines- neuroglia.

of which is an important part of the compensatory reaction of the
organism to oxygen deficiency. Activation of neurones is generally
accompanied by an increase in RNA content (for reviews, see: Hydén,
1960, 1962; Pevzner, 1963a; Brodskii, 1966). Therefore the accumulation of RNA in motoneurones appears to be a biochemical aspect of
the complex compensatory mechanisms which maintain the vital
functions of the organism during oxygen deprivation.

The survival time for a Purkyně cell after complete cessation
of the cerebral blood supply is about 8 min, while that for a spinal
motoneurone is 13 min (see VanLiere and Stickney, 1963). The higher
resistance of motoneurones to hypoxia may be related to the greater
stability of the surrounding glial cells under such conditions
(Fig. 43).

   The Polish authors Albrecht and Smialek (1975) have obtained
results which are complementary to ours. In their work, rats
breathed a hypoxic atmosphere (4% O_2, or half the partial pressure
used in our experiments). The incorporation of ^{35}S-methionine
(intravenous injection) into enriched fractions was studied, and was
found to be lower than control values in both cell types both
immediately after hypoxia, and 2 h later. This effect, however,
was greater in neurones than in glia. Normalization of the rate of
incorporation during the 24 h after hypoxia occurred more rapidly
in the glia than in neurones.

   Hydén and his co-workers have obtained quantitative data on
changes in the metabolism of perineuronal glial cells during hypoxia
(Hydén and Lange, 1961; Hamberger and Hydén, 1963). They subjected
rabbits to an atmosphere containing 8% O_2 and 92% N_2 for 12-15 h
and found that the succinate and cytochrome oxidases were
activated in nerve cells of the vestibular nucleus, while in
the glia the activities of these enzyme systems remained unchanged.
Thus the metabolism of the neuroglia again seems to be more stable
than that of the neurones under conditions of oxygen deficiency,
which is in good agreement with our studies on spinal cord (Fig.
44) in which the RNA content in the glia showed practically no
change either in the period of acute hypoxia or in the subsequent
24 h. In the cerebellum on the other hand, the metabolism of RNA
in glial cells changes as markedly during hypoxia as that in the
Purkyně cells (Fig. 44). It is clear that not all neurones show
the same properties as Deiters' cells, and this illustrates once
more the limitations of data obtained from only one kind of nerve
cell. In Hydén's laboratory, enriched fractions have been used to
study the influence of 5-30 min of anoxia on RNA synthesis of
neurones and glia from rabbit cerebrum. RNA synthesis was shown to
be inhibited in both cell types under these conditions (Yanagihara,
1974a). The depth of hypoxia achieved in brain structures under
in vivo conditions (by placing animals in a barochamber, or even by
tying off the blood vessels supplying the brain) is, of course, much
less than can be attained in vitro. In the first case, the compensatory mechanisms of the organism remain intact, including the
circulatory and respiratory systems, while in the second case only
endogenous sources of energy (ATP, creatine phosphate, glucose,
glycogen) are available, and these are very limited. Placing a rat
in a barochamber leads very rapidly to a twofold increase in blood
glucose levels (Borgström et al., 1976) and this will evidently find
its way initially to the glial cells rather than to the neuronal
cell body.

   Three properties of the recuperation period may be observed
from this group of experiments which have already been noted in
previous series.

   First, after the end of the period of hypoxia, the rate of

normalization of the RNA content in the neuroglia was markedly
higher than in neurones (Fig. 44). A number of studies have shown
no changes in content of various substances or enzyme activities
after hypoxia. Clearly this is not due to metabolic inertia, as
some authors conclude, but to the high reparative capacity of glial
cell metabolism.

Secondly, reparative changes in the neuronal and glial cells
are not completed when the RNA content reaches the normal level.
Subsequent changes may occur in RNA content, and may be in the
opposite direction (Fig. 44). Such metabolic hypercompensation
has been noted in the period after the cessation of various stresses.
Without doubt, stressful influences trigger a series of compensatory
reactions, the actions of which continue after their cessation.
Similar biochemical changes, according to the data of N.N. Yakovlev
(1955), are said to be at the basis of training, or more broadly, of
adaption (Yakovlev, 1955). This is further discussed below.

Thirdly, as can be seen from Fig. 44, while the hypoxia itself
has no effect on the nucleic acid content of glia of spinal cord,
24 h after cessation of hypoxia such changes do occur. This again
emphasizes the active role played by neuroglial metabolism in
compensatory processes proceeding in the nervous system.

Our studies of the effect of intermittent and uninterrupted
acclimatization to hypoxia were carried out in conjunction with
O.G. Gazenko and V.B. Malkin who determined the blood erythrocyte
content and hemoglobin levels in the rats subjected to acclimatization. As can be seen from Table XXVI, the adaptive changes in
the oxidative capacity of blood were markedly greater with uninterrupted than with intermittent acclimatization. According to our
data uninterrupted acclimatization led to a normalization of the
level of nucleic acid only in the cerebellum, while in motoneurones
under these conditions changes in the nucleic acid content were basically the same as with acute hypoxia (Fig. 43). In cerebellum the
same picture was also found with intermittent hypoxia, while with
uninterrupted acclimatization the RNA content in neurones and glia
of cerebellum did not change (Fig. 43).

These data hardly permit any definitive conclusions to be drawn
concerning the mechanism of the adaptive processes in the nervous
system (which was not dealt with in the present study either). We
wished only to demonstrate that both chronic and acute hypoxia
showed the same regularities as other functional conditions of the
nervous system that we have studied, i.e. that changes in RNA
content in the glia are either absent, or generally in the same
direction as in the neurones.

Thus our data indicate the greater stability of neuroglia as
compared to neurones under the influence of hypoxia, and confirm

TABLE XXVI

Changes in red blood cell count and hemoglobin content
in blood of the rat after various forms of acclimatization
to hypoxia (Gazenko et al., 1968)

Type of acclimatization	Red blood cell content (x 10^6)		Hemoglobin content (in g%)	
	before acclimati-zation	after acclimati-zation	before acclimati-zation	after acclimati-zation
Stepwise (from 3000 to 7000 m)	6.45	7.25	15.3	20.3
Uninterrupted (5500 m)	5.40	9.05	14.2	21.5

the higher reparative capacity of glial cell-satellite metabolism.

Cold Adaptation

Cold, like hypoxia, is a natural stimulus, and the mechanisms
involved are equally diverse, including cardiovascular, pulmonary,
endocrine and muscular changes. Changes in the metabolism of the
nervous system no doubt also occur both in response to the cooling
itself, and during the subsequent adaptation. Nayeemunnissa and
Rao (1975) have shown that such effects are not merely secondary
ones. They investigated the acclimatization of earthworms to cold,
as well as the administration to intact worms of nervous system
extracts from worms previously acclimatized to cold. Identical
changes in a number of parameters of lipid metabolism, as well as
accumulation of RNA were found in the nervous systems of both groups.

One form of adaptation which is commonly found in higher animals
is hibernation. This may be described as a reversible anabiosis,
accompanied by a lowering of body temperature, which occurs in
homoiothermic animals. It is accompanied by changes in various
regions of the CNS, in particular the hypothalamus (Mroskovsky,
1971). We decided to use this natural model before embarking on
artificial experiments on cold adaptation in homoiothermic animals.

We studied the ground squirrel Citellus erythrogenys. Animals

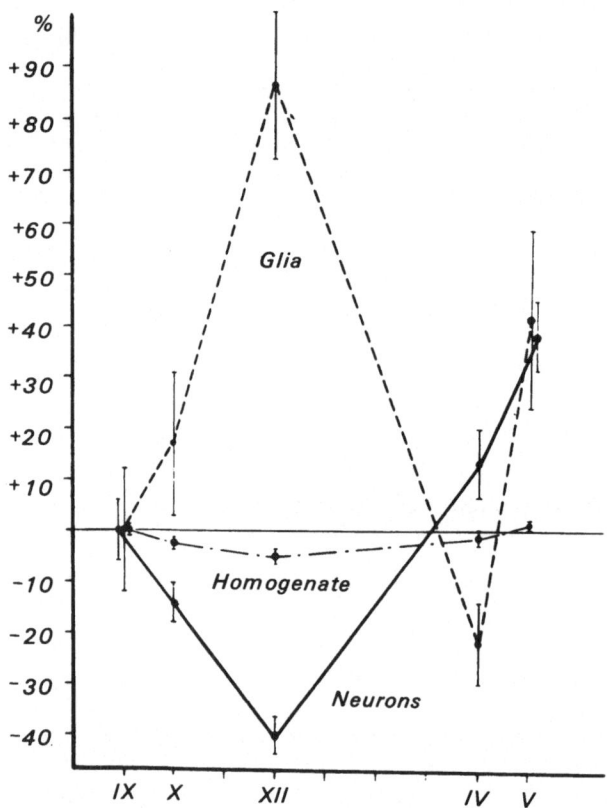

Fig. 45. Changes in RNA content per cell in neurones
and perineuronal glia, and RNA concentrations in homogenate,
of the supraoptic nucleus of the hypothalamus of ground
squirrel during hibernation, and subsequent awakening
(Pevzner and Semeshina, 1976). Ordinate- percentage change
from the RNA content in September; Abscissa- month of the
year. Hibernation lasted from October to April.

were captured in June in the steppes around Novosibirsk. While active, they were housed in a summer vivarium, and early in September were transferred to a dark environment at 8-10°C, where they went into hibernation after a week. At the beginning of May they were transferred to a warm environment, where they emerged from hibernation after 1.5-2 h.

We compared five groups of animals: in September (active condition, before hibernation); in October (start of hibernation); in December (deep hibernation); in April (end of hibernation); and in May (active condition, after end of hibernation). All animals were killed by rapid decapitation, care being taken not to awaken hibernating animals.

Sections of hypothalamus were stained in gallocyanin chrome alum, and scanning cytophotometry carried out at 546 nm using the MUF-5 microspectrophotometer.

Fig. 45 shows that the cytoplasmic RNA content of neurones of the supraoptic nucleus of hypothalamus became progressively lower as the depth of hibernation increased, while RNA content of the surrounding glia increased. Before emergence from hibernation, the neuronal RNA content increased while that of the glia decreased, in both cases "overshooting" the level found in September (i.e. before hibernation). RNA concentrations in whole hypothalamus homogenates are also shown in Fig. 45. On the whole, the changes in total RNA resemble those occurring in the neurones rather than those in the glia. The problem with using whole hypothalamus (about 60 mg dry weight) is of course that it contains a number of regions besides the supraoptic nucleus, which in fact make up the major part of the sample.

From the point of view of neurone-glia interrelationships, the period of hibernation may be divided into two stages. In the first stage, the changes in RNA content in neurones and glia are reciprocal. In the second stage, when the animals return to their active condition after hibernation, the RNA content of both neurones and glia is markedly increased. There is evidence from physiological data (Shtark, 1970; Mrosovsky, 1971) that considerable activation of the neurones occurs during this period of emergence from hibernation. This seems to be a generalized activation that involves the whole neurone-neuroglia unit.

We have also studied cold adaptation in rats. Male albino laboratory rats were placed in a cold room at 2-4°C. In the first few days, the animals showed little activity, and lost weight. At the end of the second week their behavior became more normal, and they regained weight. The rectal temperature remained unchanged throughout the experiment. As controls, we used rats from the same litter, which were kept under normal conditions at 19-21°C.

Sections of hypothalamus were stained for histones with fast green FCF and two-wavelength cytophotometry carried out at 595 and 650 nm.

Fig. 46 shows that towards the end of the first 24 h of cold exposure the content of nuclear histones in neurones of supraoptic nucleus of hypothalamus is increased, while towards the end of the third day it is lower than the initial level. It returns to the initial level after 15 days in the cold room. The pattern of changes in the glial cells is diametrically opposite.

The medial preoptic area was also studied. This area of the hypothalamus is more directly concerned with temperature regulation. The changes observed in the neurones were closely similar to those in the supraoptic nucleus. However, a different pattern was seen in the glial cells. After 24 h of cold exposure, the histone content was unchanged. After 3 days it was reduced, and by 15 days it had returned to normal (Fig. 46).

The first 24 h may represent a period during which the heat-sensitive neurones of the medial preoptic area receive a moderate

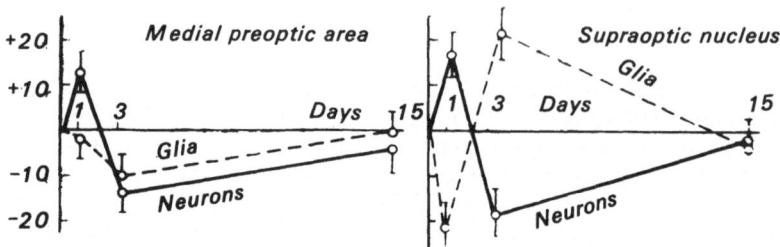

Fig. 46. Changes in nuclear histone content per cell in neurones and perineuronal glia of rat hypothalamus during uninterrupted cooling (Krichevskaya et al., 1976). Ordinate- percentage change from control level; Abscissa- duration of exposure of animal to 4oC (in days); Solid line- neurones; Dashed line- glia.

input which is not sufficient to affect the glial cells. In the
subsequent period the prolonged cold exposure leads to a metabolic
reorganization which involves the whole neurone-neuroglia unit.
This may be a similar process to that occurring during muscular
effort, where a state of fatigue seems to be necessary to initiate
the process of adaptation.

While the neurones of the medial preoptic area are directly
involved in temperature regulation, the neurosecretory neurones of
the supraoptic nucleus of the hypothalamus are part of the hypo-
thalamo-hypophyseo-adrenal axis. This system responds primarily
to the non-specific component of stress, which undoubtedly occurs
in the early stages of cold exposure. (It should not be forgotten
that we are talking of animals that have been raised in a thermo-
statically controlled animal house.) The process of adaptation
begins later, and the subsequent pattern of changes (Fig. 46) is
similar to that already seen in hibernation (Fig. 45). Whether this
similarity is a chance one, or whether it reflects a common mechanism
of adaptation in the two situations is still not clear.

Slonim and Shvetsova (1973) suggest that an acute, severe
challenge to the homeostatic mechanisms of a homoiothermic organism
resulted in a more marked adaptive response than a mild, chronic
stimulus. They drew this conclusion from a comparison of prolonged
exposure of rats to the cold, under conditions similar to those
described above, with a short-lasting, intermittent exposure to
extreme cold.

We have used the scheme suggested by Slonim and Shvetsova.
Male Wistar rats in individual cages were placed for 2 min
in a cold room at -20°C, then kept for 5 min at 25°C, and
then once again transferred to -20°C for 2 min. This cycle
was repeated 15 times. Experiments thus lasted about 1.5 h,
which included a total of 30 min cooling. No effect was
observed on the rectal temperature of these animals. The
control group of animals was kept in individual cages, and
subjected to similar manipulations as the experimental group,
but kept at a constant 25°C throughout. At the end of the
experimental period, animals were returned to standard
animal house conditions, before being killed at intervals
from 1 h to 30 days after the end of the cooling period.

Hypothalamic sections were stained in gallocyanin chrome
alum, and photometry carried out on an aperture-photoelectric
cytophotometer (Carl Zeiss, Jena).

One hour, 2 days and 15 days after the end of the stimulus, the
RNA content of the neurone-neuroglia unit of supraoptic nucleus
remained essentially unchanged. In this particular series, no
resting controls were used - our reference group had been subjected

Fig. 47. Changes in RNA content per cell in neurones
and perineuronal glia of the medial preoptic region and
the mamillary bodies of rat brain after intermittent severe
cooling (Filipchenko et al., 1975). Ordinate- percentage
change from control level; Abscissa- duration of exposure
of the animal to room temperature after the cessation of
intermittent severe cooling; Solid line- neurones; Dashed
line- neuroglia. The open circles denote significant
differences from control (p< 0.05); Closed circles indicate no statistical significance.

to the repeated transfer from one room to another. This stimulus
itself may act as a stressor, to which the effect of cooling would
add relatively little. Such a "masking" effect might be relatively
less marked in a region concerned specifically with thermoregulation
rather than with stress.

When the medial preoptic area was studied, significant
changes were indeed found. One hour after the cessation of the
intermittent cold exposure the RNA content was greater in experimental animals than in controls in both neurones and glia. This
elevation remains for 2 days, while at 5 days the RNA content had
returned to the control level (Fig. 47).

In the mamillary bodies of hypothalamus changes in RNA content
were only found 2 days after cessation of the stimulus. However,
the accumulation of RNA in the whole neurone-neuroglia unit is
similar to that in the medial preoptic region, but occurring rather
later, at 5-15 days after the return of the animals to the animal
house. Only after 30 days did the RNA content of neurones in this
region return to normal, while that in the glia dropped somewhat
below the control value (Fig. 47).

Physiological data (see: Hensel, 1973) indicates that the
anterior hypothalamus (which includes the medial preoptic area) is
responsible for the reduction in heat generation, and the increase
in heat loss by the organism. The posterior hypothalamus (including
the mamillary bodies) control increases in heat production, and
conservation of body heat.

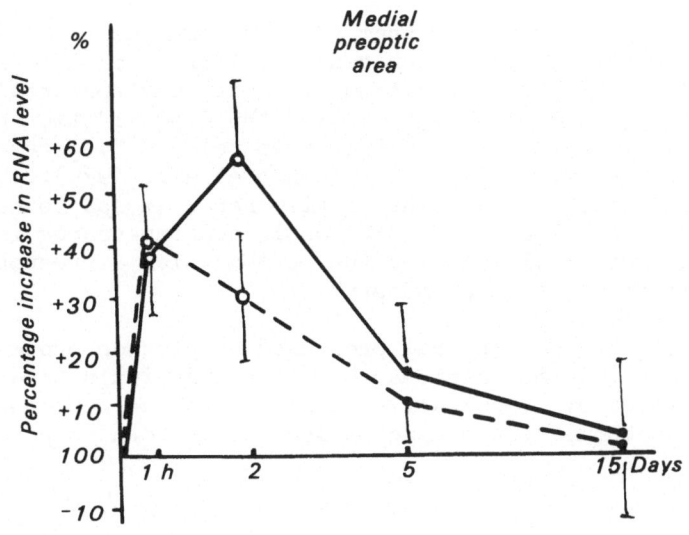

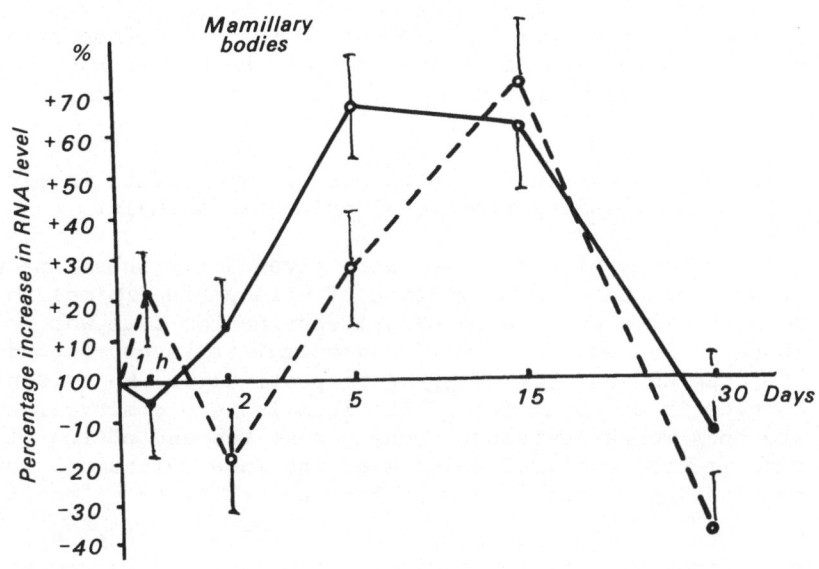

The time courses of the changes in these two regions do not coincide (Fig. 47), the more persistent changes occurring in the mamillary bodies.

These experiments show that adaptive changes to an acute, intermittent cold exposure can continue long after the cessation of the stimulus, and the restoration of normal environmental temperature. This phenomenon brings to mind the consolidation of memory, a consideration which has led Slonim to talk of "vegetative memory" (Slonim and Svetsova, 1973; Filipchenko et al., 1975). We cannot examine this interesting point of view here, except to note that an accumulation of RNA occurs in the whole neurone-neuroglia unit of the hypothalamic nuclei during the period which corresponds to the consolidation of this "memory".

Thus it can be seen that prolonged adaptation processes, which are important for the maintenance of the organism's homeostasis, are accompanied by a generalized response in the neurone-neuroglia unit - the accumulation of macromolecules in both components of the system.

Effects of Epinephrine

Hypoxia, electrostimulation, convulsions and the other stressors which we have studied undoubtedly involve the sympathetic nervous system, and greatly increase the secretion of epinephrine into the circulation. It is possible that one of the mechanisms of action of the preceding series of experiments on changes in RNA content in the neurone-neuroglia unit is the influence of epinephrine on these structures.

Therefore we carried out a series of experiments involving administration of epinephrine to experimental animals.

Cats and rats of both sexes were given daily subcutaneous injections of a 0.01% solution of epinephrine in saline for 14 days. The daily dose of epinephrine for cats was 60 µg/animal and for rats 30 µg/animal (this corresponds to 25-30 µg/kg and 150-170 µg/kg respectively). No evident changes in the behavior of the animals were observed, and the body weight remained unchanged at the end of the experiments. Control animals of the same weight did not receive injections, and were kept under the same conditions.

The superior cervical sympathetic ganglion of the cat, and the spinal cord in the region of the lumbar enlargement in the rat were studied 3 h after the last injection.

Administration of epinephrine led to an accumulation of nucleic

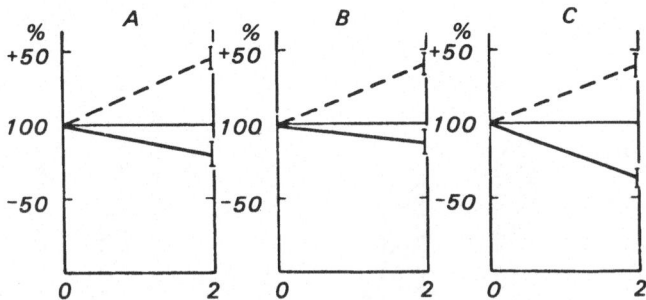

Fig. 48. Changes in nucleic acid content in neurones and
neuroglia of various regions of the nervous system resulting
from 2 weeks daily administration of adrenaline   (Pevzner,
1964, 1967). Ordinate- difference from corresponding
control level, taken as 100%; Abscissa- duration of course
of injections (in weeks); A- superior cervical sympathetic
ganglion of cat; B- lateral horn of spinal cord; C- anterior
horn of spinal cord of rat; Solid line- neurones; Dashed
line- neuroglia.

acids in glial cells of the anterior and lateral horns of the spinal
cord of rat, and of cervical sympathetic ganglion of the cat. At
the same time, there were no changes in RNA content of cytoplasm of
neurones of the cympathetic ganglion and lateral horns of spinal
cord, while in motoneurones of anterior horns the content of RNA
in the cytoplasm diminished (Fig. 48).

Studies of the superior cervical sympathetic ganglia under
these conditions also showed an increase (of 42%) in the number of
glial cells attached to the external membrane of the nerve cell
(Pevzner, 1969a). Thus the mass of glial nucleic acid, based on
unit mass of neuronal nucleic acid, represented an even greater
increase.

For complete removal of cellular RNA in fixed sections of
nervous tissue, treatment with cold 16% perchloric acid for 48 h is
necessary (Table XXI). Obviously the RNA more loosely connected to
cell protein will be leached out sooner than the more tightly
bound. Therefore some of the sections were extracted only for
18 h. Comparison of the amount of RNA removed in 18 and 48 h then
made it possible to determine the ratio between loosely bound and
tightly bound RNA. Both in the neurones and glia of superior
cervical sympathetic ganglion of control animals, the loosely bound
RNA predominated (Fig. 49). Under the influence of epinephrine,
however, both in neurones and in glia of the ganglia the proportion

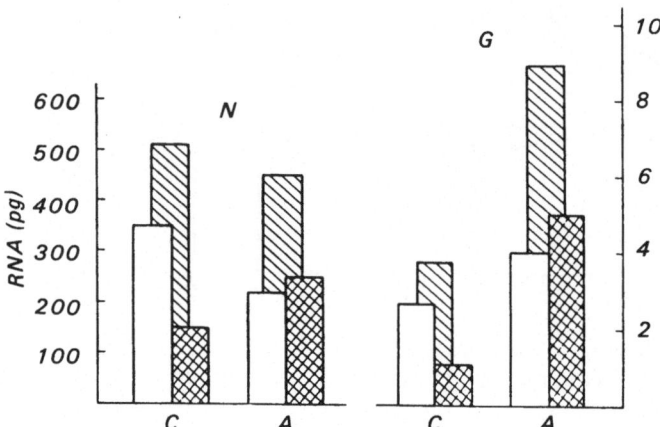

Fig. 49. Contents of different RNA fractions in neurones (N) and neuroglia (G) of superior cervical sympathetic ganglion of cat under normal conditions, and under the influence of 2 weeks daily administration of adrenaline (Pevzner, 1969c). Ordinate- content of RNA (in pg) per cell; C- control; A- with adrenaline; Plain columns- loosely bound RNA; Cross hatched columns- tightly bound RNA; Single hatched columns- total RNA.

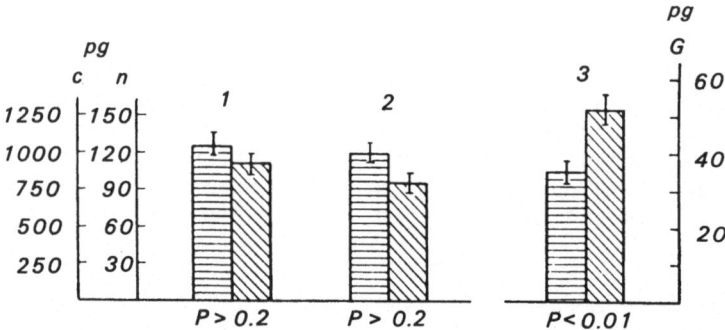

Fig. 50. Dry mass (in pg) in neurones and neuroglia of superior cervical sympathetic ganglion of cat under normal conditions, and under the influence of 2 weeks daily administration of adrenaline (Pevzner, 1965e). Ordinate- dry mass (in pg) per cell; Left- scales for neuronal cytoplasm (c) and nuclei (n); Right- scale for glial cell satellites (G); 1-neuronal cytoplasm; 2-neuroal nuclei; 3-neuroglial cells; Horizontal hatching- control; Diagonal hatching-adrenaline.

between the two fractions became 1:1. The absolute level of RNA
(the total of the two fractions) in the neurones was practically
unchanged, while in the glia it was considerably increased (Fig. 49).

Using cytophotometry, the dry mass of nuclei and cytoplasm of
neurones of cervical sympathetic ganglia and of glial cell bodies
was determined. It was shown that administration of epinephrine to
cats led to an accumulation of dry mass (in practice, protein) in
the cytoplasm of neurones and in the bodies of glial cells of
sympathetic ganglia (Fig. 50). In the neuronal nuclei, however,
no significant changes in protein content occurred.

The physiological significance of the selective influence of
epinephrine on accumulation of nucleic acid in the neuroglia became
clearer with the study of Orbeli (1938) on the trophic role of the
sympathetic nervous system. He first postulated this role from
studying the influence of direct stimulation of a sympathetic nerve
on the work capacity of mice (the Orbeli-Ginetsinskii phenomenon).
Subsequently Orbeli's co-workers also demonstrated the neurotrophic
influence of the sympathetic nervous system, this influence being
clearly shown in experiments with parenteral administration of
epinephrine (Aleksanyan, 1964; Tonkikh, 1964).

We suggest that regular administration of epinephrine may mimic
the trophic influence of the sympathetic nervous system. The
accumulation of nucleic acid in neuroglia as we have shown, is more
significant that that in the neurones. Activation of the functioning
of any cell is accompanied by an increase in RNA content - while
inhibition of the cell exhausts the supply of RNA. If the suggestion
of Hydén concerning transfer of RNA from glia to neurones is confirmed, then it is logical to suppose that the increase in RNA
stores in the glia might prolong adequate functioning of the neurone.
No increase in the RNA content of neurones occurs under the influence
of epinephrine. Thus epinephrine, while not acting as a specific
activator of these neurones, assists their function, increasing their
work capacity, and postponing exhaustion. This is exactly the kind
of influence that may be considered trophic.

As can be seen from Fig. 48, the cytoplasmic RNA content of
sympathetic neurones of cervical ganglia, and lateral horn of spinal
cord remained practically unchanged under the influence of epinephrine injections, but it was somewhat reduced in the cytoplasm
of motoneurones of anterior horn. Data from electrophysiological
studies have shown that injection of adrenaline inhibits the
electrical activity of the majority of motoneurones (Engberg and
Ryall, 1966; Weight and Salmoiraghi, 1967). Such repeated inhibition apparently leads to a reduction in cytoplasmic RNA content.

Thus the selective accumulation of RNA and protein in glial
cells under the influence of repeated administration of epinephrine

(which may be considered to imitate effects of sympathetic activation) indicates that the target of the neurotrophic influence of the sympathetic nervous system is neuroglial metabolism.

It is worth mentioning here the comparison we have made of histone content in various types of neurones which differ in their biogenic amine content. Sections of brain tissue from male Wistar rats were stained with ammoniacal silver for arginine and lysine histones (Black and Ansley, 1966). The specificity of this method is demonstrated by applying it to sections previously treated with hydrochloric acid to remove the histones, or with sodium nitrite, which causes them to be deaminated. In either case, the color reaction does not occur. To assess the method, aqueous solutions of arginine-rich (fraction f_3) and lysine-rich (fraction f_1) histones from rat brain, and an equal mixture of the two, were loaded onto polyacrylamide gels. Sections of these gels were mounted on microscope slides, and stained with ammoniacal silver. The maximum absorption of f_3 occurred at 418 nm, while that of f_1 was at 452 nm.

Sections of subcortical nuclei of rat brain were stained in the same way. Table XXVII shows that for the three brain nuclei the maximum absorption was in between the figures for pure lyophylized preparations of histone fractions f_1 and f_3. Thus practically identical absorption maxima were found for brain regions in which different neurotransmitters predominated. According to Dahlstrom and Fuxe (1964) and Anden et al. (1965), the lateral nucleus of the reticular formation (group A1) contains mainly noradrenergic neurones, the paragigantocellular nucleus of the reticular formation (group B1) primarily serotonin, while the substantia nigra contains mainly dopaminergic neurones. The ratio of arginine-rich to lysine-rich histones in the three types of neurone was equal, however, definite differences were found between the spectral curves of their glial cell-satellites (Table XXVII). We may suppose that the other monoamines resemble adrenaline, and have an influence on the glial compartment of the neurone-neuroglia unit.

Adrenalectomy and Substitution Therapy with Hydrocortisone

Studies of stress reactions have demonstrated the important regulatory role of the corticosteroids, the secretion of which may be raised several fold above normal during stress (Selye, 1952, 1957). In the kinds of biochemical changes which we have observed in the neurone-neuroglia unit the corticosteroids may play an important role.

Therefore it seemed of interest to compare the sensitivity of neuronal and glial metabolism to adrenal cortex hormones.

TABLE XXVII

Ratios of arginine- and lysine-rich histones in neurones
and perineuronal glia from various subcortical nuclei
of rat brain (Raigorodskaya et al., 1973)

Brain Region	Predominant Neurotransmitter	Absorption Maximum of the Ammoniacal silver stained preparation (nm)	
		neurones	neuroglia
Nucleus reticularis lateralis	norepinephrine	432 ± 1	452 ± 1
Nucleus reticularis paragigantocellularis	serotonin	431 ± 1	441 ± 1
Substantia nigra	dopamine	432 ± 1	433 ± 1

Lyophylized and ammoniacal silver stained
preparations of:
    arginine-rich histones, f_3 418 nm
    lysine-rich histones f_1 452 nm

Male Wistar rats were adrenalectomized under sterile
conditions using ether anesthesia. After the operation,
they received normal food ad libitum; 1% NaCl was added
to drinking water. Some of the adrenalectomized animals
were given daily subcutaneous injections of hydrocortisone-
acetate (Richter, Hungary) at a dose of 20 mg/kg. On the
fourth day (half an hour before killing) the last injection
was given intraperitoneally at a dose of 50 mg/kg. For
controls, both normal, intact animals, and sham-operated
animals were used.

All groups of animals were killed on the fourth day after
the operation by decapitation without anesthetic. Cerebellum,
spinal cord (cervical enlargement) and a sample of brain
tissue in the region of the optic chiasm were taken for
analysis. Purkyně cells of the cerebellum, motoneurones of
the anterior horn, and neurones of the supraoptic nucleus
of the hypothalamus were subjected to photometry, together
with their corresponding glia.

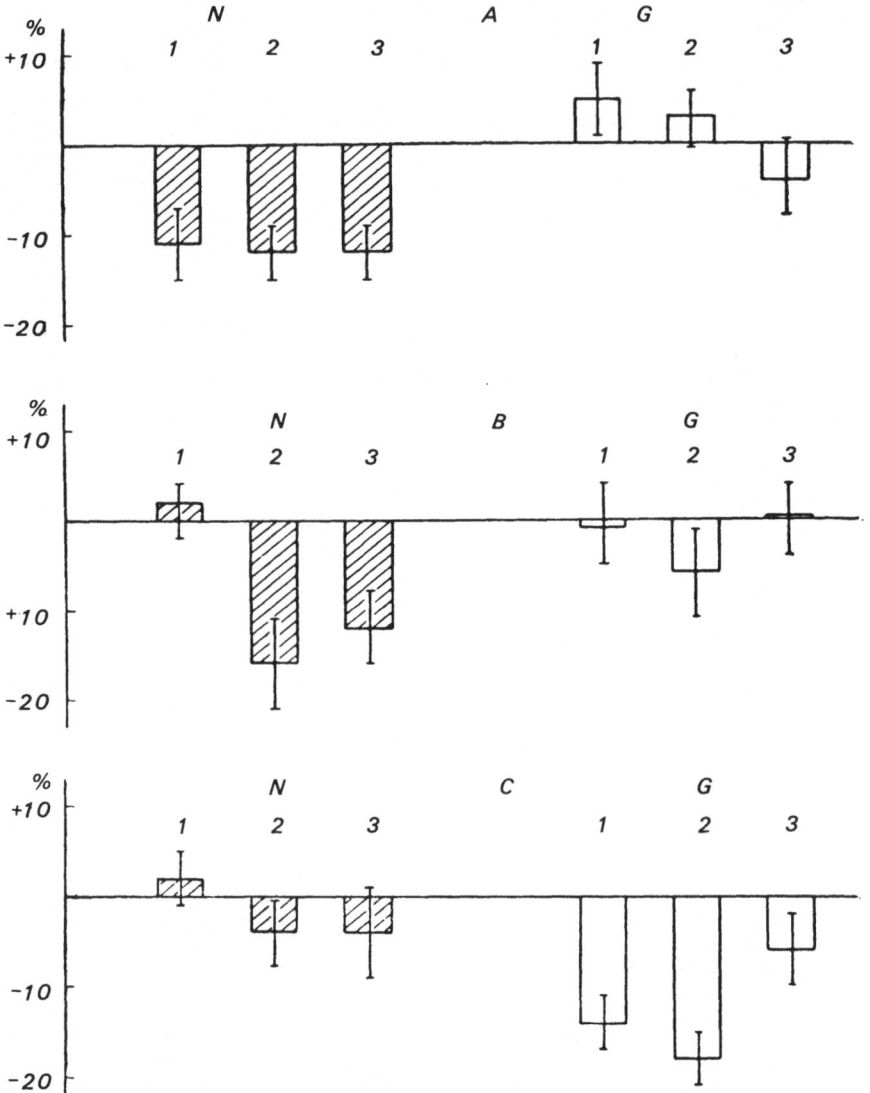

Fig. 51. Changes in nucleic acid content in neurones and
neuroglia of various regions of the nervous system of rat
after adrenalectomy   (Antonov and Pevzner, 1968; Pevzner,
1969c). Ordinate- difference from control level, taken as
100%; A- Purkyně cell layer of the cerebellum; B- spinal
cord anterior horn; C- supraoptic nucleus of hypothalamus;
N- neurones; G- neuroglia; 1- sham operated rats; 2- adrenalectomized rats (4th day after operation); 3- adrenalectomized
rats, which received daily injections of hydrocortisone.

Before killing, the corticosteroid levels in blood plasma of all four groups was determined according to Sweat (1954) as modified by Van der Vies et al., (1960). Determinations were carried out for us by L.M. Antonov and V.G. Shalyapina.

Blood corticosteroid levels did not differ significantly between intact and sham-operated animals. No corticosteroids were found in the blood of adrenalectomized animals. Administration of hydrocortisone to adrenalectomized rats led to a corticosteroid concentration in blood considerably over the norm (Antonov and Pevzner, 1968). However it is necessary to make the point that this magnitude does not represent the mean corticosteroid concentration in blood of this group of animals during the experimental period: the determinations were carried out half an hour after an intraperitoneal injection of hydrocortisone which was more than twice the dose administered daily after the operation.

After 4 days, a lowering of RNA content was noted in the cytoplasm of cerebellar Purkyně cells in the sham-operated and adrenalectomized animals. Substitution therapy with hydrocortisone did not affect this reduction (Fig. 51). No changes were found in glial cells surrounding the Purkyně cells in any of the experimental groups.

Cytoplasmic RNA content of anterior horn motoneurones was also lowered by adrenalectomy, and again this reduction was not abolished by hydrocortisone. No statistically significant changes in the glial cells occurred (Fig. 51).

A different picture emerged in the hypothalamus - a region of the CNS particularly sensitive to the corticosteroid balance. Here, by contrast, changes were found only in the neuroglia; in the sham-operated and adrenalectomized animals the content of RNA in glial cells of supraoptic nucleus was markedly reduced. Administration of hydrocortisone in this case caused a normalization of nucleic acid levels. In nerve cells of supraoptic nucleus the RNA content was practically unchanged in all of the experimental groups (Fig. 51).

The hypothalamus has a high permeability to circulating corticosteroids (Eik-Nes and Brizzee, 1967; Peterson and Chaikoff, 1963). While in the cerebellum and spinal cord the changes in nucleic acid content were non-specific (i.e. not compensated for by hydrocortisone substitution therapy) in the hypothalamus the lowering of nucleic acid content was evidently connected to the corticosteroids deficiency in the organism, as indicated by the full normalization of nucleic acid content which followed administration of hydrocortisone to adrenalectomized rats. The difference in the localization of the changes produces is also striking - the non-specific changes in nucleic acid metabolism occur in the neurones, while the specific changes occur in the neuroglia.

We have already referred to the markedly greater stability of neuroglial metabolism to the influence of extreme, non-specific factors in the discussion of previous results. The high sensitivity of glial cells to hormonal influence is also apparent in the preceding section (Figs. 48 and 49). Thus the glial cells may be the basic site of action of hormonal influences on nervous system metabolism. It is possible that the metabolism of neuroglia - the cellular barrier standing between the other structures of the nervous system and the capillaries - is the general target of humoral factors.

The data obtained allow us to make the suggestion of the role of neuroglia as the basic site of action of hormonal (and possibly more generally of **humoral**) influences on nervous system metabolism.

Effects of Inhibitors of Nucleic Acid Biosynthesis

One of the convenient methods for the study of intracellular nucleic acid metabolism is the use of antimetabolites and inhibitors of its biosynthesis. The possibilities of using such inhibitors for resolving the problems of functional biochemistry of the nervous system have been discussed in detail in a review by Squire and Barondes (1972).

We have studied the effect of two inhibitors - 6-mercaptopurine and aurantin (a mixture of actinomycin C_2, C_3 and D). The first interrupts the nucleic acid biosynthesis in not less than four places, the second selectively inhibits DNA dependent RNA-polymerase (for literature, see: Pevzner, 1969d).

Experiments were carried out on young rats of both sexes, at a body weight of 100-120 g. Animals were injected subcutaneously with 6-mercaptopurine daily at a dose of 4-5 mg/kg for 4 days; the last injection was made 1-2 h before killing. Since mercaptopurine was injected as a 0.05% solution in 1% $NaHCO_3$, control animals of the same weight received similar daily injections of 1% $NaHCO_3$. Aurantin was injected subcutaneously in small (0.7 mg/kg) or large (3 mg/kg) doses. Rats were killed 18 h later, except for some animals which received the smaller dose, and were killed after 48 h.

The Purkyně cell layer of the cerebellum and the anterior horn of the cervical enlargement of the spinal cord were studied by cytophotometry.

In cerebellar Purkyně cells, injection of either aurantin or 6-mercaptopurine resulted in a lowering of cytoplasmic RNA levels. With aurantin, the content of nucleic acid in the perineuronal glia was even more sharply reduced than in the Purkyně cells, while with

6-mercaptopurine, no significant changes were noted in the glia.

In the anterior horn of spinal cord the influence of aurantin was also found in both neurones and glial cells, while the influence of 6-mercaptopurine was confined to the neurones. But here, by contrast, both the inhibitors led to an increase rather than a decrease in RNA content (Fig. 52).

The influence of aurantin on cerebellum and spinal cord also differed in that in cerebellum this inhibitor was effective only in a small dose, and had no effect in a large dose, while in spinal cord, it was effective, on the contrary, only in a large dose (Fig. 52).

Interpretations of the changes produced by an influence in vivo are always complex. While 6-mercaptopurine is a small molecular weight compound, aurantin is a mixture of complex cyclic decapeptides. Therefore while it is likely that mercaptopurine can penetrate the blood-brain barrier, it is improbable that the influence of aurantin is a direct one. This inhibitor is administered into the blood stream and must act indirectly on the nervous system - through

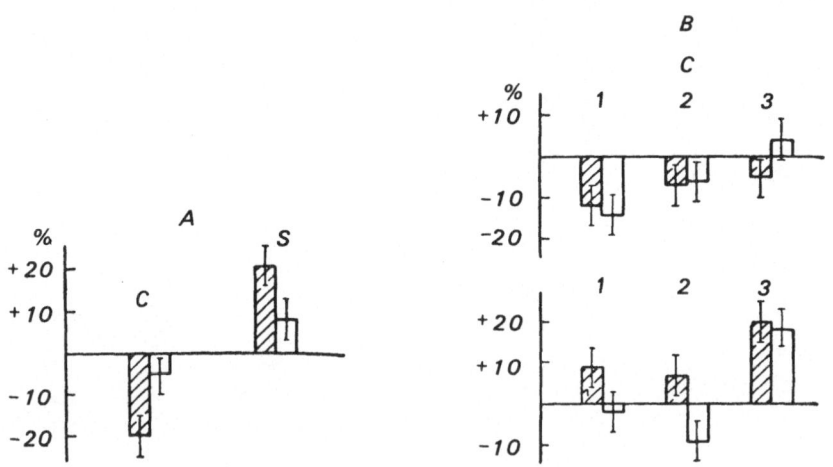

Fig. 52. Changes in nucleic acid content in neurones and neuroglia of cerebellum and spinal cord of rat under the influence of 6-mercaptopurine (A) and aurantin (B) (Pevzner, 1969d). Ordinate- difference from corresponding control level, taken as 100%; C- cerebellum; S- spinal cord; 1- 18 h after injection of aurantin at a dose of 0.7 mg/kg; 2- 48 h after injection of aurantin at a dose of 0.7 mg/kg; 3- 18 h after injection of aurantin at a dose of 3.0 mg/kg; Hatched columns- neurones; Plain columns- neuroglia.

inhibition of RNA and protein synthesis in blood cells, brain
capillary walls, and liver endocrine glands and other peripheral
organs. According to Alipova (1952) intramuscular injection of
aurantin in mice leads to glial hyperplasia in brain tissue. In a
simpler system - isolated nerve ganglia of molluscs - addition of
aurantin to an incubation medium containing labelled precursors of
RNA inhibited incorporation into the giant identified neurones to
a greater degree than in the perineuronal glia (Kuz'min et al.,
1975).

Thus with inhibition of nucleic acid biosynthesis, as with the
other traumatic conditions described above (hypoxia, hypokinesia,
convulsions), the nucleic acid content of the glial cells either
does not change, or changes in the same direction as the neuronal
RNA content.

CHAPTER 7

THE ROLE OF THE NEUROGLIA IN NEURONAL

FUNCTION — A GENERAL SUMMARY OF

THE EXPERIMENTAL EVIDENCE

Nerve cells are constantly undergoing alterations of their metabolism which are necessary for the maintenance of their changing functional state. These alterations may be transient and limited, consisting mainly of changes in ionic distribution, transformations of labile phosphorus compounds and of ammonia and related compounds, conformational changes in protein molecules, or alterations in the permeability of cell membranes. More profound metabolic shifts may follow, resulting in changes in the amounts of macromolecules: nucleic acids, proteins, lipids. Only changes of the second kind-more profound and prolonged - can be detected by the quantitative cytochemical method at our disposal. Thus it is these aspects of t the role of the glial cells in neuronal function that we are trying to elucidate.

It is apparent from the data presented that not only neuronal, but also glial cell metabolism is involved (Fig. 53). Thus the various stages of cold-adaptation are accompanied by parallel changes in RNA content in the whole neurone-neuroglia unit (Figs. 45 and 47). Similarly the protein content of neurones in the spinal cord (Fig. 35), hypothalamus (Figs. 38 and 46), and red nucleus (Figs. 39 and 40) is reduced by a number of experimental conditions to the same degree as in their glial cell-satellites. Thus it is impossible to regard the neuroglia as merely a support, as meta-bolically inert structural cells in nervous tissue. On the other hand, neuroglial metabolism can clearly be distinguished from that of the neurones. Under a number of conditions the RNA content of the neuroglia is more stable than that of the neurones (Fig. 54). Examples are the accumulation of RNA in cortical motoneurones (Table XXIII) and of RNA and protein in neurones of hippocampus (Tencheva and Pevzner, 1974) and of cerebral cortex (Tencheva and

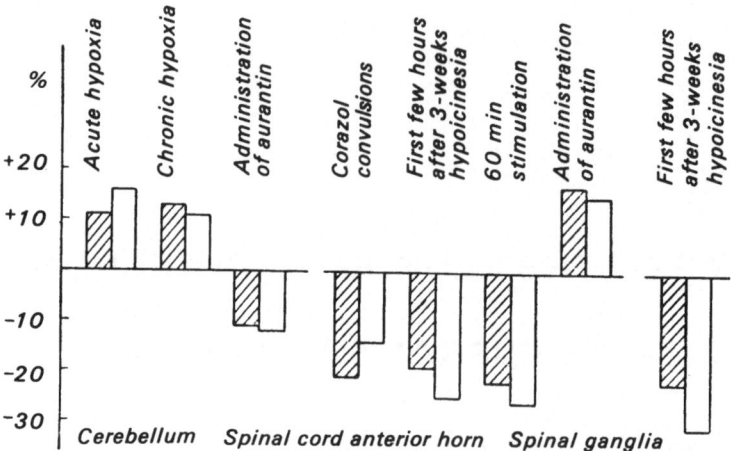

Fig. 53. Overall changes of nucleic acid content in the
neurone-neuroglia unit of various regions of the nervous
system   (Pevzner, 1968). Ordinate- difference from control
level, taken as 100%; Hatched columns- neurones; Plain
columns- neuroglia.

Pevzner, 1973) where no corresponding changes occur in the glial
cell-satellites.The reduction in protein content of neurones in the
spinal ganglia (Fig. 35) and the supraoptic nucleus (Figs. 38, 39,
and 40) is accompanied by a significantly smaller decrease in
protein content in neuroglial cells.

    The neuroglia, then, as far as RNA and protein content are
concerned, are characterized by a greater resistance than the
neurones to factors altering the functional conditions of the nervous
system. This resistance might be associated with a lower sensitivity
of glial cell metabolism, but might also be explained by a higher
capacity of the neuroglia for regeneration. Under conditions where
distinct changes of RNA content occurred both in neurones and glia
the rate of return to normal demonstrates that the restorative
processes in the glial cells do indeed occur faster than in the
neurones. In fact, in the Purkyně cells of the cerebellum (Fig.
44), in neurones of the spinal ganglia (Fig. 42), and in spinal
motoneurones (Figs. 31, 34, and 42) normalization of RNA levels
occurs considerably later than in their glial cell-satellites.

    The role of the neuroglia in restorative processes is also
suggested by the fact that even when the glial RNA content was not
substantially altered under the experimental conditions, marked
changes often occurred later. Thus three hours swimming leads to
an increase in the RNA content in rat spinal motoneurones while that

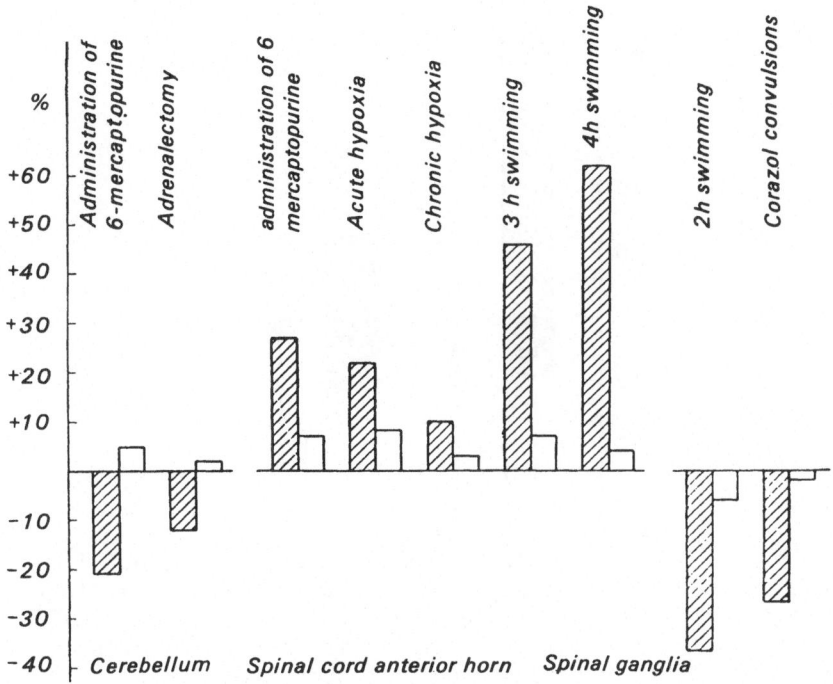

Fig. 54. Relative stability of nucleic acid content in
glial cell-satellites (by comparison with their corresponding neurones) in various functional conditions of the
nervous system   (Pevzner, 1968). Ordinate- difference
from control level, taken as 100%; Hatched columns-
neurones; Plain columns- neuroglia.

of their glial satellites did not change. When these rats were
removed from the water into quiet cages, significant changes in glial
RNA content occurred (Fig. 28). The normalization of RNA levels in
the motoneurones took place against this background of changes in
the neuroglia, being completed after four hours of quiescence; as is
clear from Figure 28, changes in the neuroglia were apparent for
several hours longer. Consequently, if these animals are made to
swim again on the next day, all the biochemical changes in response
to the repeated load would occur in a system already altered in its
neurone-neuroglia relations. It is characteristic of this situation
that the stable changes in RNA metabolism are localized not in the
motoneurones, but in their glia. As it is well known that adaptive
changes in the organism (including the intracellular biochemical
reorganizations) occur as a result of repeated physical load, it
may be suggested that the neuroglia are the basic morphological
substrate of the process of training, or adaptation, in the nervous
system.

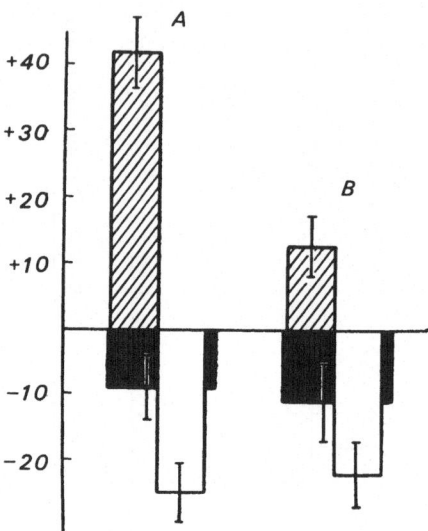

Fig. 55. Effect of electrostimulation on RNA content in
single cells and in homogenates of whole tissue of different regions of the nervous system   (Pevzner, 1970a, 1971).
Ordinate- difference from corresponding control levels,
taken as 100%; A- superior cervical sympathetic ganglion of
cat (electrostimulation of the preganglionic trunk for
180 min; B- anterior horn of spinal cord of rat (electrical skin stimulation for 5 min); Hatched columns- neurones;
Plain columns- neuroglia; Black columns- homogenate of
corresponding region of the nervous system.

        In a number of situations (electric skin stimulation, recovery
after three weeks' hypokinesia) a full functional-biochemical resti-
tution could be observed in the neurones, while marked changes in
the glia continued to be apparent. These lasting changes take place
a considerable time (of the order of 1 - 3 days) after the cessation
of the influence (Figs. 31 and 42).

        The early stages of nerve cell excitation by electrostimulation
are of particular interest for an understanding of the role of the
neuroglia. An increase in the RNA content of neurones is in this
case accompanied by a decrease in the neuroglia (Fig. 55). From the
accumulated data on the greater resistance of glial metabolism to
stress factors, it is hardly possible that the reduction of glial
RNA content is due to fatigue. In fact, as is clear from Figure 30,
the initially lowered level of RNA returns to normal while the
electrostimulation continues. Moreover, an overall reduction in RNA
content of both the neuronal and glial cells, continuing for several
hours, and indicating true fatigue of the neurone-neuroglia unit,
begins only after more prolonged electrostimulation. Therefore we
suggest that such initial reciprocal changes in RNA levels of the

neurones and glia as a result of neuronal excitation may be explained by the direct transfer of RNA from the glia to the neurone. The proposal that such a transfer might occur was put forward first by Hydén. This question will be discussed in more detail below.

The neuroglia, although the more resistant part of the neurone-neuroglia unit to the action of extreme factors, is at the same time more sensitive to trophic influences on the nervous system, for example to the daily moderate stimulation of the sympathetic nervous system by the administration of small doses of adrenalin. This treatment typically leads to accumulation of RNA in the glial cell-satellites (Fig. 48) as if to increase the reserves of RNA which may subsequently be necessary for maintenance of prolonged neuronal excitation.

The hypothalamus is the most sensitive region of the nervous system to the influences of adrenal cortex hormones. A lack of corticosteroids and subsequent replacement therapy with these hormones produces corresponding changes which are again confined to the glia (Fig. 51). Possibly the effect of all hormonal factors which influence the metabolism of the nervous system is localized primarily at the level of the glial cells.

Scheme of Inter-relationships of Metabolism of Nucleic Acid and Protein in the Neurone-Neuroglia Unit

The general conclusion arrived at above is rather schematic and reflects only the most commonly met regularities of the biochemical processes system of neurone-neuroglia unit. In each particular case the metabolic changes in this system are determined by complex relations of their properties and the character of the influences acting upon it. One and the same influence on different neurone-neuroglia units may lead to divergent metabolic changes and one and the same neurone-neuroglia unit may react differently to two apparently similar experimental influences. This conclusion may be repeatedly demonstrated from our data, which are summarized in Figs. 53 and 54.

Therefore we may expect that individual results of functional-biochemical studies may not fit in with such a scheme. Thus, Geinisman et al. (1970b) and Rubinskaya (1971) showed changes in RNA levels in glial cell-satellites which were more marked than the changes in motoneurones of anterior horn. Discussing their data, these authors suggest the possible compensatory role of the lowering of RNA in glia to support the maintenance of normal levels of neuronal RNA.

Therefore the scheme we have suggested (Fig. 56) must be considered only as a hypothetical attempt to generalize the most frequently met interrelationship of nucleic acid (and protein)

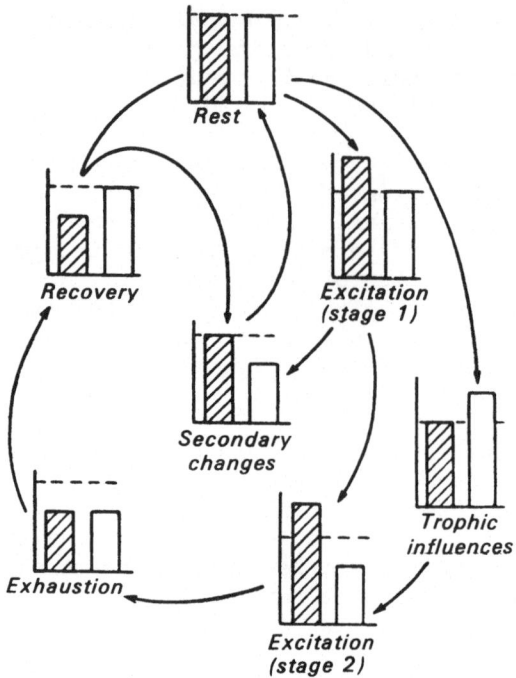

Fig. 56. Scheme of metabolic interrelationships for nucleic
acids (and proteins) in neurones and their glial cell-satellites in different functional states of the nervous system
(Pevzner, 1970a, 1971). Height of columns- nucleic acid
(or protein) content in neurones (hatched columns) and
neuroglial cells (plain columns); Dashed line- corresponding resting level. Explanation in text.

metabolism in the neurone-neuroglia unit. Such a generalization
was first suggested by us (Pevzner, 1965e, 1966b) on the basis of
the early experiments. Subsequently, as the accumulation of factual
data has continued the scheme has by degrees changed and become
more comprehensive (Pevzner, 1967, 1968, 1970a) and at the present
time is as it appears in Fig. 56. The following sections are
devoted to detailed discussion of the consecutive steps of this
scheme.

Neuroglia during Adequate Excitation
of the Neurone

In the state of relative rest of the nervous system, the
catabolism and anabolism of nucleic acid in cell structures of
nervous tissue are well balanced one with the other. Therefore

the quantity of nucleic acid remains at a defined relatively stable level.

Moderate, non-exhaustive, excitation of the neurone leads eventually to an accumulation of RNA in its cytoplasm. Apparently in the early stages of excitation no changes in glial RNA content occur. Thus, for example, swimming of mice in our experiments led to a marked increase in content of cytoplasmic RNA in motoneurones of spinal cord, while changes in RNA content of the surrounding glia either did not occur, or showed very slight increases (see Fig. 26). Other results showing marked changes in RNA and protein content which do not involve changes in the glia are summarized in Fig. 54 and discussed at the beginning of this chapter. Analagous relationships have been described by Rose et al. (1973) for experiments involving ^{3}H-lysine incorporation after intraventricular injection as a function of light deprivation. In rats reared in the dark from birth, exposure to uninterrupted light for 0.5-2 h led to an increase in incorporation of label into neuronal protein, while no such changes were shown in the glia. Using autoradiography after injection of ^{3}H-UMP into the lateral ventricle, Pohle and Matthies (1974) showed that visual stimulation of rats in a Y-maze led to increased incorporation of label into RNA of hippocampal neurones, while again there was no corresponding change in the glial cells. Electrophysiological data indicate that excitation leads to weaker and slower acting changes in membrane potentials in the glia than in the neurones (Hild et al., 1958; Birks et al., 1960; Sokolov, 1962; Aladzhalova and Kol'tsova, 1964; Roitbak, 1965, 1968). This correlates with the greater biochemical lability of neurones, for instance in the response to hypoxia (Hamberger and Hydén, 1963; Hydén, 1964). The much higher specificity of the metabolic response in neurones should also be noted. This has already been mentioned in dealing with the response to deprivation of slow wave and REM sleep in rats. It is also seen in the response to inhibitors such as GABA and K$^+$. While GABA inhibits neuronal oxidative processes (Aleksidze and Blomstrand, 1968) K$^+$ causes an activation (Bradford and Rose, 1967; Haljamäe and Hamberger, 1971; Hultborn and Hydén, 1974). However in glial cells, both GABA and K$^+$ lead to a similar, non-specific influence - an increase in oxidation of substrates (Blomstrand, 1968, 1969).

This difference between neuronal and glial metabolism makes sense if we bear in mind that the role of the neurone is to react to diverse, rapidly changing situations. The functional differentiation of neurones is paralleled by their striking morphological differentiation in the course of development. Nerve cells achieve maturity very rapidly, and neither divide nor grow during the life of the animal (Glees, 1955; Kuhlenbeck, 1970). Neuroglia however arise from a highly cambial element of the neural tube; it is not by chance that inflammation, regeneration, or tumor formation in nervous tissue involve the glial cells. Such a functional

"division of labor" between neurones and glia leads to a biochemical
divergence: the metabolism of the neurone, being concerned with the
immediate function of nervous tissue, reacts with much more sensitivity and rapidity to changes in the condition of the nervous system,
while the metabolism of the glial cell (the auxilliary element in
the neurone-neuroglia system) is more stable.

Neuroglia during Prolonged or Sharp Excitation
of the Neurone

The observation of an elevation of RNA content in the cytoplasm
of an excited neurone, first made in 1943 by Hydén, has been repeated
many times and may be considered as firmly established. However the
mechanism of this phenomenon is still not established.

A major contribution to this problem was made by the studies
of V. Ya. Brodskii and his co-workers (Brodskii, 1966; Brodskii et
al., 1966). It was shown that while under normal conditions light
stimulation of retina led to a marked accumulation of protein in
retinal ganglion cells, in animals which had previously been
injected with the RNA synthesis inhibitor aurantin, such stimulation
not only failed to increase the protein content in these neurones,
but in fact considerably lowered it. Probably initial excitation of
the neurone is accompanied by an increased breakdown of protein.
This could only be demonstrated in such experiments with aurantin,
since normally such breakdown is obscured by the active biosynthesis of protein in the neurone. Subsequently, analogous data were
obtained by Hydén (1967b) with respect to RNA: injection of rabbits
with amphetamine led to accumulation of RNA in Deiters' cells, while
previous intraveneous injection of 0.3 mg actinomycin D (3-7 h
previous) led to a lowering of RNA content in these neurones. The
mechanism of this breakdown of RNA and protein remains unclear, but
the fact itself corresponds well with the ideas of D.N. Nasonov's
School (Nasonov and Aleksandrov, 1940; Nasonov, 1956, 1963; Troshin,
1956) of an increased alteration of cell protein in the period of
active functioning of the cell. The beginning of the destructive
phase of excitation of the neurone explains the biological significance of the accumulation of RNA in activated cells. It became
obvious that such an accumulation was a compensatory reaction
providing for the subsequent increase in protein biosynthesis.
However, the basic problem - the mechanism of the initial lowering
in protein content, and the subsequent corresponding compensatory
biosynthetic processes - still requires detailed investigation.

Intensive activation of the nervous system leads to a generalized
metabolic response: an accumulation of RNA and protein in the whole
neurone-neuroglia unit. This has been shown in thermosensitive
neurones of the hypothalamus (medial preoptic area, mamillary bodies)
under various conditions of cooling of homoiothermic animals (Figs.

45, 46, and 47). An analagous accumulation of RNA occurs in motor neurones of dog cortex and their glial cell-satellites during the development of conditioned avoidance reflexes (Svanidze et al., 1972) as well as in neurones and neuroglia of hippocampus in experiments involving reversal of handedness in rats (Kazakhashvili, 1974). A "model learning" situation involving electrical stimulation of isolated ganglia from a gastropod mollusc was accompanied by increased incorporation of labelled precursors into RNA and protein of both giant identified neurones, and of their surrounding glia, according to autoradiographic data (Adzhimolaeva et al., 1972; Kuz'min et al., 1975). In our experiments involving single and repeated preganglionic stimulation of sympathetic ganglia of the cat, accumulation of RNA was also observed to an equal degree in neurones and in their glial cell-satellites (Pevzner et al., 1973). In all this work, the activation of the neurones is accompanied by learning, or by adaptation. One may consider that excitation of the nervous system which has an active character, and is connected to the complete mechanisms of adaptation, learning and memory, is accompanied by an increase in the biosynthetic processes of the whole neurone-neuroglia unit (a suggestion not incorporated in the scheme in Fig. 56). This is based mainly on metabolic data for RNA and protein. However Abdel-Latif et al. (1974) have showed that neurotransmitters (noradrenaline, dopamine, and, in particular, acetylcholine) markedly increased the biosynthesis of various types of phospholipid in enriched fractions of both neurones and glia from rat cerebral cortex in vitro.

These biosynthetic processes depend on the breaking of high energy bonds, formed in the processes of oxidative and glycolytic phosphorylation. They also depend on the presence of a pool of low molecular weight precursors, in particular nucleotides and amino acids. If the excitation is too strong, or too prolonged, the biosynthetic apparatus of the neurone will cease to cope with the load.

The Question of Transfer of RNA
from Glia to Neurones

We suggested that it is exactly at this stage of neuronal functioning that the transfer of glial RNA to neuronal cell bodies by pinocytosis, postulated by Hyden (1960, 1962), may occur. Direct proof of such a transfer has not yet been obtained. However there are several indirect indications that this occurs.

Firstly, pinocytosis is a widespread phenomenon in cell function (for reviews, see: Holter, 1959; Chapman-Andressen, 1965; Steinman et al., 1976). Numerous observations indicate that pinocytosis of macromolecules by neuronal membranes (Tobias, 1960; Rosenbluth and Wissing, 1964) and by oligodendroglia may occur

(Pomerat, 1952; Klatzo and Miquel, 1960; Geiger, 1963; Mizuno and Okamoto, 1964; Brightman, 1965b; Nicholls and Wolfe, 1967). The capacity for pinocytosis is possessed by ependymal cells, (Brightman, 1965a) and also by Schwann cells (Kaye et al., 1963; Rosenberg, 1970). The occurrence of pinocytosis has also been shown between glial cells, and capillary endothelial cells (King and Schwyn, 1970). Moreover, studies of labelled proteins have shown the transfer of a protein fraction (with a molecular weight between 12,000 and 20,000) from axons to Schwann cell glia (Droz et al., 1974) and from Schwann cells to axoplasm (Lasek et al., 1974). The transport of protein, and in particular ATPase, may be facilitated by trans-glial canals in the regions of intercellular contacts between the neurones and the perineuronal glia (Shivers, 1976; Mashanskii et al., 1974). Rose and Sinha (1974) have also suggested the possibility of a transfer of proteins between neurones and glia from incorporation studies in which intraperitoneal injections of rats with ^{3}H-lysine and ^{14}C-phenylalanine were used, followed by the separation of enriched fractions. In the first hour, the radioactivity of the proteins was higher in the neurones than in the glia, but this ratio was subsequently reversed. The synaptosomes did not accumulate any of the rapidly labelled material so the involvement of axonal flow in this process could be ruled out. Thus direct transfer from the neuronal perikaryon to the glial cell body may be occurring. Thus there is considerable support for the suggestion that not only ions and low molecular weight substances, but also macromolecules may pass through the membranes of the neurone-neuroglia unit.

Secondly, states of activation of the nervous system character-ized by increased RNA content in the neurones and a reduction in the glia have been repeatedly shown in different types of nerve cell, in various animals, and using methods differing in principle (Hydén and Egyhazi, 1963; Hydén, 1964; see also our Figs. 29 and 30). In our experiments with electrical stimulation, this relationship was already clear 5 min after the beginning of the stimulus. It is difficult to see how these changes could occur in such a short time simply by alterations in rate of synthesis and breakdown of neuronal and glial RNA, since even the most rapidly turning over macro-molecule fractions have a half-life of the order of several hours (Lajtha et al., 1976).

Thirdly, biochemical determination of RNA in homogenates of samples of the same regions of brain where marked reciprocal changes in RNA content in neurones and glia have been shown by cytophotometry indicated the absence of significant changes of RNA in whole nerve tissue (Fig. 55). This argues in favour of redistribution of RNA within the tissue, i.e. a transfer of RNA from glial cells to neurones.

Finally, the migration of RNA from glia to neurones is suggested by determinations of the nucleotide composition of RNA carried out by Hydén and Lange (1966).   These data deserve a detailed discussion.

    The authors injected rabbits with triap (tricyanoaminopropene)
which stimulates protein synthesis and activates oxidative enzymes
in Deiters' cells of the lateral vestibular nucleus.   The base
composition of RNA in Deiters' neurones and their glia in control
animals differed somewhat (Table XXVIII).   RNA content was increased
by 570 pg per neurone one hour after triap administration while in
the glia a reduction of 55 pg occurred in a roughly equal volume of
capsule.   From morphological data, Hydén and Lange conclude that the
total volume of glial capsule surrounding one Deiters' cell is about
ten times the volume of the neurone.   It follows that for the whole
neurone-neuroglia unit there is an increase in mass of neuronal RNA
of 570 pg, and a decrease in mass of glial RNA of 55 x 10 = 550 pg
(Fig 57).   This close correspondence does not favor an explanation
in terms of changes of rates of synthesis and breakdown.   But most
important, in our view, are the data obtained concerning the base
composition of RNA extracted from neurones and glia of rabbits

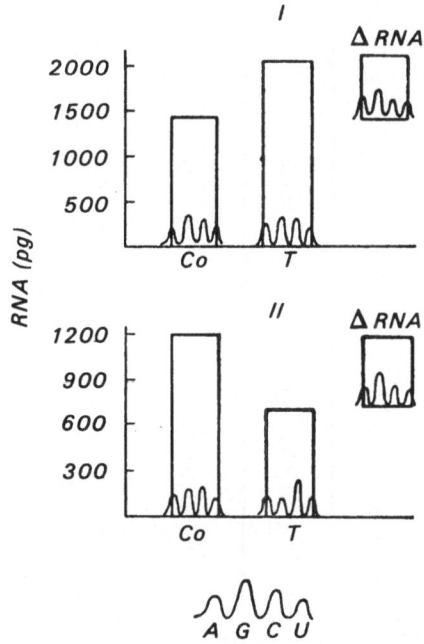

Fig. 57. Quantitative and qualitative changes in RNA in
Deiters' neurones and their glial capsule of lateral vestibular nucleus of rabbit 1 h after administration of
tricyanoaminopropene (triap) at a dose of 20 mg/kg (Hydén
and Lange, 1966). Ordinate- quantity of RNA (in pg) per
cell (I) or per tenfold volume of glial capsule (II);
Base composition of RNA: A- adenine; G- guanine; C- cytosine;
U- uracil; Co- control; T- triap. Explanation in text.

TABLE XXVIII

Nucleotide composition of RNA from neurones and glia of Deiters' nucleus of lateral vestibular nucleus of the rabbit one hour after administration of tricyanoaminopropene (triap)   (Hydén and Lange, 1966)

Base	Neurones		Glia		Δ RNA	
	Control	Triap	Control	Triap	Control	Triap
Adenine	19.7 ± 0.37	20.5 ± 0.31	20.8 ± 0.28	20.1 ± 0.74	22.5 ± 1.61	21.6 ± 1.11
Guanine	33.5 ± 0.39	34.6 ± 0.28	28.8 ± 0.64	21.9 ± 2.15	37.7 ± 1.66	37.8 ± 3.19
Cytosine	28.8 ± 0.36	26.7 ± 0.24	31.8 ± 0.27	38.6 ± 2.40	21.0 ± 2.03	23.0 ± 3.33
Uracil	18.0 ± 0.18	18.2 ± 0.20	18.6 ± 0.55	19.4 ± 0.65	18.8 ± 0.91	17.6 ± 1.49

Base compositions are calculated on the basis of molar proportions (in %).

receiving triap, which turned out to be somewhat different from that
in control animals. Knowing the initial and final amount of RNA
and the ratios of the four nucleotide bases in the two cases, the
authors were able to calculate the nucleotide composition of ΔRNA,
i.e. that fraction of RNA which appeared in the neurones, and that
which disappeared in the glia (Table XXVIII).

     Calculation shows that not only is the mass of ΔRNA in neurones
and glia very similar (Fig. 57), but the nucleotide composition in
neuronal and glial ΔRNA is practically identical (Table XXVIII).
This supports the idea that there is just one ΔRNA fraction which
migrates from the perineuronal glia to the neuronal cell body.

     Autoradiographic data in support of the same suggestion has
been reported by D'yakonova (1972). Visceral ganglia of Limnaea
stagnalis were incubated in 3H-uridine and 1.5-2 h later, radioactivity was found in two regions of the cytoplasm of the gigantic
neurones - in the perinuclear zone, and in the peripheral zone in
areas adjacent to attached neuroglial cells. The glial cell-
satellites themselves were labelled more quickly and intensively
than the neurones. Controls were treated with ribonuclease, and
released 90-95% of label, confirming the involvement of RNA.

     In Hydén's laboratory, Jarlstedt and Hamberger (1971) have
carried out experiments in which slices of cerebral cortex of
rabbits were incubated in a medium containing 3H-uridine. Thirty
or 60 min later, the incorporation of label was virtually identical
in neurones and glia. After 180 min, the neuronal radioactivity
was greatly increased, while much less was found in glial cells.
The authors suggest that at later times, migration of RNA from glia
to neurones may be occurring. Thus a number of studies using very
varied methodology support the idea of transneuronal and transglial
transport of macromolecules.

                Neuroglia during Fatigue of the Neurones.
                Neuroglia as the Site of Trophic Influences
                        on the Nervous System

     Even if we accept the transfer of RNA from glia to neurones as
a fact, the biological significance of this transfer is completely
unclear. In spite of the common origin of neuronal and glial cells,
their RNA composition is significantly different; there is no doubt
that glial RNA migrating to the neuronal cell body is foreign to it.
Whether it becomes a template for the biosynthesis of new neuronal
proteins or supplies the nucleotides used by RNA-polymerase or
whether it has some quite new function (perhaps connected with its
polyanion structure) is a question to which it is at present impossible to give even a very preliminary answer. In one of his reviews
Hydén (1970) offers the suggestion that glial RNA enters the

neuronal cell body and then binds to the histones which act as
repressors of certain genomes. The neuronal DNA would then become
active in transcription of new RNA, and increased biosynthesis of
protein in the neuronal cell body would result. This suggestion
that the glial RNA acts as a derepressor is highly interesting, in
principle completely plausible, but unfortunately hardly based on
any convincing evidence.

The phase in which increased RNA content in the neurones is
coupled with a reduction in the glial cells is found in an actively
functioning neurone, before it passes from excitation to fatigue,
and exhaustion. Evidently the biosynthetic capacity of the glial
cell is the limiting step in the maintenance of intensive or prolonged activity of the nerve cell. When the application of the load
turns out to be excessive for the apparatus of the neuroglia as well,
the source of replenishment of supplies of neuronal RNA ceases to
function, and (if the exhausting influence does not cease to operate
on the nervous system) a marked lowering of RNA content occurs in
the whole neurone-neuroglia unit (Figs. 30, 34, and 42). Judging
both by the behavior of experimental animals, and by the physiological and cytochemical data, we conclude that this lowering of RNA
content corresponds to inhibition and exhaustion of the nerve cell.
As is shown by the cytophotometric results, analogous changes in
protein content occur in the neurone-neuroglia unit (Figs. 35, 38,
39, and 40).

Hydén's group have not observed such an overall reduction in the
macromolecule content of the whole neurone-neuroglia unit. This
difference in biochemical trends is due to the use of different
functional states, and not to a different methodological approach,
as has been confirmed by the studies of Grenell et al. (1968).
These authors, working in the field of space biology, subjected rats
to prolonged rotation, artificially creating an increased gravity
(2.65 g) for these animals, corresponding to that on Jupiter. After
1, 7 or 30 days of uninterrupted rotation, the content of RNA in the
Deiters' cells of the lateral vestibular nucleus and their glial
capsule was determined by the micromanipulation method of Hydén. In
all cases a lowering of RNA content was noted in both neurones and
in the perineuronal glia (Table XXIX). As is clear from Fig. 56,
the initial lowering of RNA content in the glia occurs the later
stages of excitation, preceding exhaustion, during which the RNA
content in the neurones is elevated. Similar relationships are
clearly shown in our experiments, and were repeatedly demonstrated
by Hydén and his co-workers. Other work (Geinisman et al., 1970b;
Rubinskaya, 1971) has dealt with conditions of intensive functioning
of motor neurones (swimming with a load, picrotoxin convulsions),
when the lowering of RNA content in the glial cell satellites
occurred in the context of considerable stability of levels of RNA
in neurones. Geinisman et al. (1970b) made the interesting suggestion that glial cells, surrounding the actively functioning neurones,

TABLE XXIX

Effect of rotation of rats at 2.65 g on RNA content of
isolated Deiters' cells and their glial capsule of
lateral vestibular nucleus   (Grenell et al., 1968)

Duration of period of rotation	Neurones		Glia	
	Quantity of RNA (in pg)	Change (in %)	Quantity of RNA (in pg)	Change (in %)
Control	647.9	-	330.3	-
24 hour	338.9	-48	176.9	-47
7 days	454.8	-30	240.8	-27
30 days	477.1	-26	238.9	-28

are characterized by the alternation of two metabolic cycles:
initially RNA migrates from glia to neurones (which results in a
lowering of its concentration in glial cells), after which there is
an increase in intensity of glial RNA synthesis leading to the replenishment of glial RNA reserves. Subsequently it is possible once
again to arrive at a phase of removal of RNA from the perineuronal
glia to the neuronal cell body. If real inhibition of the nerve
cell occurs only after exhaustion of the reserve of glial RNA which
is capable of being transferred to the neuronal cell body, then it
is logical to suggest that each influence that leads to a selective
accumulation of RNA in glia, will permit a more prolonged functioning
of the neurone-neuroglia unit. Such factors are the injection of
adrenaline and apparently hydrocortisone. The fact that the elevation of RNA levels in the glial cell under the influence of these
agents is not accompanied by analogous increase in the neurones is
in our view of great significance (Figs. 48, 49, and 50).

Thus we are talking of influences on the nervous system which
do not show stimulating effects on the neurones themselves, but in
which the influence on the glia allows more effective functioning of
the nervous system, delaying the onset of exhaustion of the neurone.
It is exactly this aspect of the influences which leads us to
designate them as neurotrophic, and suggests that the neuroglia may
constitute the primary site of trophic influences on the nervous
system. It may be suggested that the information coded in the form
of nerve impulses initially acts on the neurones, which are the

more rapid and specific structures of the nervous system in terms of
functional and metabolic reactions. Slower and less specific humoral
influences, on the other hand, initially affect the neuroglia, which
form the morphological barrier separating the neurone from the surrounding structures of nervous tissues, and at the same time play
the role of connecting link between the blood flow and the neuronal
intracellular space.

Neuroglia during Reparative Processes
in the Nervous System

When a stimulus that has led to exhaustion of the nervous
system ceases, reparative processes are set in motion. The reduced
levels of RNA and protein in the glial cells by degrees return to
normal. It follows that in the reparative period, the anabolism of
nucleic acid and protein markedly exceeds the catabolism. The speed
of reestablishment of the initial RNA content in neuroglia is
markedly higher in the glial cells than in the neurones, as can be
seen from Figs 31, 34, 42, and 44.

The neuroglia, being in immediate contact with the capillaries,
are well situated to increase their macromolecule synthesis from low
molecular weight precursors. It is also of note that the nucleus of
the glial cell, which contains the regulatory mechanisms for protein
biosynthesis, makes up the bulk of the cell volume.

In experiments in vitro Daneholt and Brattgård (1966) incubated
tritium-labelled nucleotides with isolated Deiters' cell bodies and
samples of the glial cell capsule, obtained by Hydén's method. Then
they separated the RNA, and determined its radioactivity. This
allowed the authors to calculate that during a 4 h incubation about
15% of whole RNA turned over in the neurones compared to around 30%
in the glia. Similar results were found by Volpe and Giuditta
(1967) who measured the rate of incorporation of labelled nucleotides into RNA in vivo. One hour after subarachnoid administration
of $6-^{14}C$-orotic acid in rabbits the incorporation into glial RNA was
higher than that into neuronal RNA. However studies using enriched
fractions in vitro have shown higher rates of incorporation of label
into neuronal RNA than into glial RNA. As pointed out in Chapter
Three, the in vivo data do not allow firm conclusions to be drawn
about the ratio of neuronal to glial RNA synthesis. As far as
protein synthesis is concerned, the consensus is that under normal
conditions there is a more rapid rate of synthesis in the neurones
than in the glia, both in vivo and in vitro. What is of more significance to the present discussion, however, is the capacity of
the glial cells to change their rate of biosynthesis during the
reparative period. This has been repeatedly confirmed, and leads
to the suggestion that the glial cells have a much greater capacity
to increase the rate of a number of biochemical and physiological

processes than have the neurones.

Brodskii et al. (1961, 1966) consider that there are two basic
types of cell, having "nuclear" and "cytoplasmic" types of protein
synthesis. They showed that the cytoplasmic type is found in cells
synthesizing large amounts of protein, not only for itself, but
"for export". In the case of neurones, this "export" takes the form
of axonal flow (see Chapter Three), which, naturally, is a great
load on the synthetic apparatus of the cell body. Glial cells, on
the other hand, have the nuclear type of protein synthesis. In this
case the rate of accumulation of cell proteins is not decreased by
any pronounced transport of material away from the cell.

Histochemical data are frequently characterized by considerable
diversity and contradiction. However there is one group of enzymes
for which general agreement has been found, the enzymes of the
hexose-monophosphate shunt. As described in Chapter Three, these
enzymes are found predominantly in the glial cells, a result
confirmed also by biochemical data. The pentose cycle is a highly
effective pathway of glucose oxidation for ensuring the maintenance
of biosynthetic processes, making both pentoses and the reduced form
of NADP available. A great deal of information has been obtained
recently on the compensatory role of the pentose cycle during changes
in the functional state of the nervous system (Khachatryan, 1963,
1967, 1968; Guerra Martinieri, 1967; Jongking, 1967; Zakhar'in, 1968;
Appel and Parrot, 1970; Brue et al., 1972; Domanska-Janik and
Windeman, 1974; Kimura et al., 1974; Hakim et al., 1976).

Hydén (1964) has concluded that when the neurone-neuroglia unit
is continuously active, the Pasteur effect is found in the nerve
cells, while the Crabtree effect occurs in the glia. However, this
suggestion awaits confirmation by other authors. The Pasteur effect
is the inhibition of glycolysis by respiration, and transfers the
energy metabolism of the cell to the more efficient pathway of
aerobic oxidation. The Crabtree effect is the reverse of the Pasteur
effect and is energetically unfavorable. However, the maintenance
of energy metabolism in the glia seems to be less of a problem than
in the neurones, since the former are in direct contact with the
brain capillaries. Moreover glycolysis results in the formation of
phosphorylated intermediate products, which favor increased biosynthesis.

Lastly, the increased neuronal function described above is
accompanied by a rapid breakdown of high-energy compounds, and an
increased utilization of endogenous substrates. When this is
prolonged, a state analogous to intracellular hypoxia arises in the
neurones (Brodskii, 1966; Pevzner, 1963c, 1966b, 1970a; Jakoubek,
1974). The principal function of the reparative processes which
occur after such continuous activity is to alleviate this tissue
hypoxia. The greater resistance to various forms of hypoxia shown

by the glia, as well as its higher reparative capacity can be seen
in Figs. 39 and 40, and in a number of other published reports
(Hydén and Lange, 1961; Hamberger and Hydén, 1963; Yanigahara, 1974a;
Albrecht and Smialek, 1975) and is of relevance here. Other characteristics of glial cells which should be borne in mind are the high
intensity of oxidative phosphorylation (Rapava et al., 1973), the
high ATPase activity (Cummins and Hydén, 1962; Hamberger et al.,
1970; Sellinger et al., 1971; Nagata et al., 1974), the rapid uptake
of amino acids (Rose, 1973; Henn and Hamberger, 1973; Minchin, 1974;
Schon and Kelly, 1974a, 1974b; Hamberger et al., 1975), and the wide
range of non-carbohydrate substrates that can be metabolized, which
include fatty acids and amino acids (Balasz et al., 1973). Given
the lower mass of the glial cell, as well as the absence of energy
demands such as that made by axonal flow, the ability of the glial
cell to overcome hypoxia, and its biosynthetic capacity during
reparative process must be considerably higher than those of the
neurone.

That the neuroglia are important in reparative processes in the
nervous system can be seen even in those cases where the initial
influence on the nervous system does not result in observable changes
in glial RNA content. In such cases marked changes in RNA and
protein content of glial cells occur during the subsequent rest
period (Figs. 28, 33, 34, and 35).

Neuroglia during Secondary Changes
in Nervous System Metabolism

After the termination of a number of influences on the nervous
system, changes may be seen in various types of nerve cell which we
shall term "secondary". (Examples of such changes may be seen in
Figs. 28, 30, 34, and 42.) In general, when a factor affecting the
nervous system ceases to operate, the RNA levels initially normalize,
first in the glia and then in the neurones. The secondary changes
to which we are referring occur later, typically after 1–3 days when
the behavior of the animal has become normal. While the RNA levels
in the neurones remain stable, changes may be observed in glial RNA.

The fact that these changes occur in glia but not in neurones
seems to us to have great physiological significance. During
evolution, neurones developed as the dominant cell in nervous tissue,
and neuronal metabolism reflects the specialist function of the
neurone. The most important aspect of this function is the capacity
for rapid response to changes in the external or internal environment. This appears to be incompatible with the occurrence of long-
lasting changes in neurones in direct response to environmental
change. However the nervous system clearly shows metabolic changes
in response to many stimuli, and this is the basis of such phenomena
as habituation, training and adaptation. We propose that this

reconstruction begins with the secondary changes, which are localized in the neuroglia. Thus the rapid functional biochemical fluctuations can occur in the neurones independently of the secondary, longer-term changes in the glia. If the factor in question is a random one (i.e. in the absence of reinforcement) the effect on the glia will disappear, while regularly occuring, systematic influences will have a cumulative effect on the metabolism of the neurone-neuroglia unit. This will initially be localized in the glia, but reinforcement will lead to adaptive changes in neuronal metabolism as well. Such adaptation at the cellular level may in fact represent the process of consolidation, whereby the secondary, trace changes in the glia are transformed into a permanent metabolic reconstruction which includes the neurone as well.

These secondary changes are particularly marked after total exhaustion of the nervous system. This is in agreement with work which involves physical training (Yakovlev, 1955). The mechanisms of nerve and muscle are of course very different, but changes in the nervous system form one of the components of physical training. Excitation of the nervous system which does not directly affect the neuroglia is thus followed by secondary changes lasting for several days.

The Neurone-Neuroglia Unit

Both neurones and neuroglia have the complete range of basic biochemical mechanisms required for independent functioning (see Chapter Three). Nevertheless neurones and glia show many important metabolic differences, relating to their specialized functions. These become particularly important during stimulation of the nervous system. This section will attempt to summarize the data which has been described under the various aspects of glial metabolism.

The most obvious metabolic characteristic of glial cells as compared to neurones is their higher rate of oxidative phosphorylation (Ranava et al., 1973) and the higher activity of ATPase in the glia (Cummins and Hydén, 1962; Hamberger et al., 1970; Sellinger et al., 1971; Nagata et al., 1974). Thus the glia show a greater capacity to accumulate and release energy in the form of ATP than do the neurones. However, the energy requirements of the neurones must be rather high, in order to sustain nerve impulses, ionic transport, the transport of macromolecules from the nucleus to the cytoplasm, and axonal flow. The glial cells having a smaller mass of cytoplasm, short processes, and slow potentials, should have a considerably lower energy requirement. This provides the basis for the suggestion that the glial cells have a role in the supply of energy to the neurones. This is supported by the data of Hamberger and Hydén (1963) who showed that excitation of the nervous system is accompanied by the activation of respiration and the inhibition of glycolysis (the Pasteur effect) in the neurones, and by the activation of glycolysis and the inhibition of respiration (the Crabtree effect) in the neuroglia. The difference in substrates for tissue metabolism is also of interest. Following the ideas of Waelsch on the compartmentation of oxidative metabolism in nerve tissue (Waelsch, 1959; Waelsch et al., 1964; Van den Berg et al., 1969) Balasz et al. (1973) showed that the large compartment (neurones and nerve processes) uses mainly glucose, while the small compartment (neuroglial cells) uses a wide range of substrates, including fatty acids and amino acids. This gives the neuroglial cells a considerable advantage under conditions of exhaustion of carbohydrate substrates such as occurs during hypoxia, increased functioning, etc. While the synthesis of proteins from amino acids occurs more rapidly in neurones than in glia (Tiplady and Rose, 1971; Hemminki and Holmila, 1971; Haglid and Hamberger, 1973; Lisý and Lodin, 1973; Albrecht and Smialek, 1975; Blomstrand et al., 1975) the accumulation of amino acids from the medium, their active transport and interconversion occur more readily in glial cells (Bradford and Rose, 1967; Rose, 1970, 1973; Hamberger, 1971; Hamberger and Henn, 1973; Schon and Kelly, 1974a, 1974b; Minchin, 1974; Bowery et al., 1975; Hamberger et al., 1975; Sellstrom et al., 1975). This supports the idea that glial amino acid metabolism provides precursors for protein synthesis in both the neurones and the glia, as well as possibly having a regulatory function.

The brain-specific proteins also show compartmentation - one is found primarily in the neurones, the other in the neuroglia (for reviews, see: Moore, 1969, 1972, 1975; Calissano, 1973; Jakonbek, 1974; Palladin et al., 1976; Dahl, 1976).

Glial cells are characterized by a high phospholipid content (Freysz et al., 1967; Hamberger and Svennerholm, 1971; Norton and Poduslo, 1971). But the rate of synthesis of phospholipids from UDP-choline and UDP-ethanolamine is an order of magnitude lower in glial enriched fractions than in the neurones (Freysz et al., 1969; Goracci et al., 1975). This has led Porcellati (1974) to suggest that glial phospholipid may have, in part at least, a neuronal origin.

The compartmentation of transmitter metabolism is of particular interest. Choline acetyltransferase, the enzyme which catalyzes the synthesis of the ubiquitous excitory transmitter acetylcholine (ACh) is either absent from the glia or much less active than in the neurones (Vernadakis and Gibson, 1973; Arbogast and Arsenis, 1974; Nagata et al., 1974). On the other hand, acetylcholinesterase, which catalyzes the hydrolysis of ACh is only slightly less active in the glia than in the neurones (Hemminki et al., 1973; Vernadakis and Gibson, 1973; Aleksidze et al., 1974). Butyrylcholinesterase also shows a high activity in the glia. This enzyme is also capable of breaking down acetylcholine, and it is not inhibited by excess

substrate. The compartmentation of those transmitters derived from
amino acids has been described in the discussion of animo acid
metabolism. Hydrolysis of cyclic AMP (phosphodiesterase activity)
is very active in neuroglia (Desmukh et al., 1974; Nagata et al.,
1974) in particular in oligodendroglia (Poduslo and Norton, 1972).
Cyclic GMP synthesis however (guanyl cyclase) occurs mainly in the
neurones (Goridis et al., 1974).

Not all of these data have been confirmed by other authors, but
on the whole the evidence supports the idea of the complementary
metabolic properties of neurones and neuroglia, that is of the
neurone-neuroglia unit as a single metabolic system. The effective
functioning of such a system depends on the existence of a mechanism
regulating the biochemical organization of each compartment. There
is evidence for the existence of such a system.

The role of potassium ions is of great significance. Activation
of all types of neurones leads to the efflux of K^+. This K^+ may be
involved in the metabolic changes which occur in the neuroglia when
the neurones are excited (Aleksidze, 1974). Even under in vitro
conditions, glial cells show an increased tissue respiration in
response to an increased K^+ concentration (Aleksidze and Blomstrand,
1969; Haljamäe and Hamberger, 1971; Hertz et al., 1973). Under these
conditions, an increased incorporation of ^{14}C-glucose into succinate,
glutamate and aspartate is also found (Salem et al., 1975), as well
as increased oxidation of NADH (Orkand et al., 1973), increased
breakdown of ATP (Schousboe et al., 1970) and activation of ATPase
(Sellinger et al., 1971; Hamberger et al., 1970) and pyruvate kinase
(Nagata et al., 1976). Such reactions are either not found in the
neurones, or are much less marked.

Since all these results were obtained in vitro, the authors'
calculations showing that the K^+ concentrations used were within the
expected physiological range are very important. Moreover, both in
the experiments with enriched fractions and those with microdissected
samples, the uptake of K^+ from the medium was much higher for glial
cells than for the neurones (Hamberger and Röckert, 1964; Bradford
and Rose, 1967; Franck, 1970; Haljamäe and Hamberger, 1971). Thus
the K^+ released by stimulation of the neurones in vivo may be
rapidly taken up by the glial cells, thus resulting in the activation
of glial metabolism.

An analogous role may be played by amino acids. As noted above,
they are rapidly formed and released when the neurone is activated,
and can be rapidly accumulated by glial cells. Balazs et al. (1973)
and Aleksidze (1974) consider that the most important of these
possible amino acid triggers is γ-aminobutyric acid (GABA). Administration of GABA was shown by Aleksidze and Blomstrand (1968) to
inhibit the utilization of succinate by Deiters' cells, but to
considerably increase its oxidation by the corresponding perineuronal

glia. GABA seems to be a non-specific activator of glial oxidative
metabolism, showing an effect similar to that of K^+ ions. However,
the influence of GABA on neuronal oxidative metabolism is the
opposite of that of K^+ ions. Thus the neurone again appears to be
more specific, more sensitive in its reactions to different
influences, while neuroglial metabolism has a more non-specific,
supportive, trophic character.

The uptake of K^+ ions and amino acids is clearly an important
function of the neuroglia. Apart from the activation of glial metabolism described above, such uptake is also important in the buffer
role of glial cells. GABA, as well as a number of other amino
acids are considered to be neurotransmitters (for literature, see:
Hebb, 1970; Curtis and Johnston, 1974). The uptake of K^+ and amino
acids from the extracellular space is thus important in limiting
their duration of action on the synapse. A similar role may be
played by glial phosphodiesterase, which hydrolyzes and thus
inactivates cyclic nucleotides. Such a buffer role may also be
seen for a familiar transmitter like acetylcholine. Judging by the
choline acetyltransferase activity, ACh formation occurs mainly in
the neurones. Acetylcholinesterase is found in both neurones and
glia, while butyrylcholinesterase is mainly glial. The latter
enzyme is less active than acetylcholinesterase, but is not subject
to substrate inhibition. This leads to the suggestion that when
ACh release is low, it is hydrolyzed primarily by neuronal ACh, but
when its release exceeds normal levels, generalized excitation of
the nervous system is prevented by inactivation of ACh by the glial
butyrylcholinesterase.

Hydén has put forward a hypothetical mechanism for the regulation of the neuroglial influence on the synaptic apparatus (Hydén,
1974; Haglid et al., 1974; Hansson et al., 1975). This concerns the
protein S-100, which is of glial origin. Histochemical studies on
the localization of S-100 in rabbit brain, using anti-S-100 serum
conjugated either with fluorescin-isocyanate (for light microscopy)
or with horseradish peroxidase type VI (for electron microscopy)
showed that S-100 occurred on the post-synaptic membrane, both on
the surface, and directly under it. Much smaller amounts of S-100
were found on the presynaptic membrane, and none at all in the
synaptic cleft. Electron microscopic studies of microdissected
neurones showed that the inner surface of the neuronal membrane is
covered with a net of twisted, paired spiral filaments each filament
having a diameter of 2 nm. These filaments were shown to be actin-
like, and addition of 2 mM $CaCl_2$ resulted in a reversible untwisting
of the spirals. Since S-100 is known to have a high affinity for
calcium, Hydén and his co-workers propose that S-100 competes with
the actin-like protein of the post-synaptic membrane for Ca^{2+} ions.
When the S-100 level is high it binds the bulk of the Ca^{2+}; the
actin-like filaments assume a spiral form, and the post-synaptic
membrane contracts, resulting in a widening of the synaptic cleft.

Contrariwise, when there is little S-100 the Ca^{2+} ions bind to the actin-like protein, the filaments unwind, and the post-synaptic membrane relaxes (Fig. 58). In the first case activation of the synapse will be difficult, in the second case, easy (Hydén, 1974). Now data on the accumulation of S-100 during learning was presented in Chapter Three (Fig. 17), and it may be that the formation of a new behavioral pattern, the consolidation of memory, is dependent on the facilitation of conduction in wide synapses. Thus the excitation threshold of the synapse is high until it is incorporated into the spatial organization of nerve connections, by means of the accumulation of S-100 protein. These ideas are highly conjectural, and are in need of direct investigation. S-100 is synthesized in neuroglia, and the role of this protein in neuronal function is one of great interest. We should also mention here the data of Michetti et al. (1976). They showed that incubation of cell nuclei from brain of newborn rat or of 11-day chick embryo in a medium containing S-100 markedly activated RNA polymerase. Since S-100 is found mainly in the glia, while macromolecule synthesis occurs most rapidly in the neurones, this forms the basis for a possible glial regulation of neuronal metabolism.

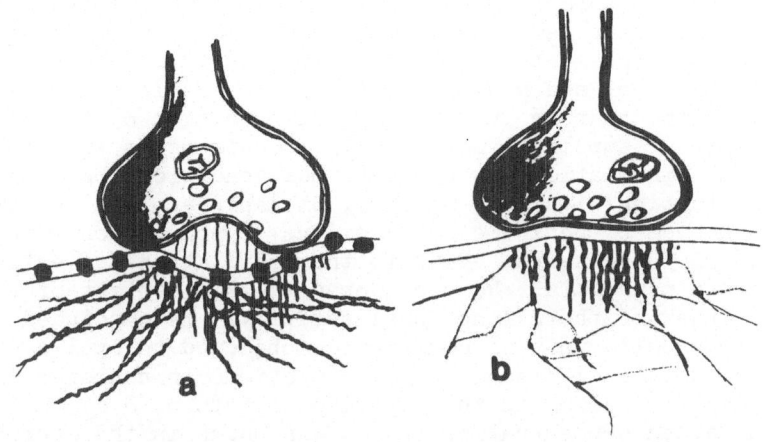

Fig. 58. Suggested effect of S-100 protein on synaptic state (Hydén, 1974). a- synapse in the presence of excess S-100, which binds Ca^{++}, leading to a twisting of the actin-like protein fibers, and a widening of the synaptic cleft; b- synapse when little S-100 is present, and Ca^{++} binds to the actin-like protein fibers which untwist, causing a narrowing of the synaptic cleft.

    Taking together all the data from electrophysiological,
histological, biochemical and cytochemical investigations (including
the modest part carried out in our laboratory) it can be seen that
neither the neurones nor the neuroglia may be considered as a
completely independent component. We consider that these two cell
types form a single metabolic system, the neurone-neuroglia unit.
This leads us to the conclusion that the neurone doctrine of the
structure and function of the nervous system is obsolete. The basic
functional entity, the structural unit of nervous tissue must be
thought of not as the neurone alone, but as the neurone and its
corresponding glial cells, that is the neurone-neuroglia unit.

    Thus the scheme set out in Fig. 56 reflects the participation
of the neuroglia in functional-biochemical transitions. As can be
seen, these transitions form a kind of cycle, in which the neurone-
neuroglia unit, disturbed from its resting condition, passes through
a series of steps to re-attain its initial state.

    The cyclic character of these changes leads to the suggestion
of the existence of a negative feedback regulatory system within
the neurone-neuroglia unit (Saudargene and Pevzner, 1969; Pevzner,
1970b; Shelepin, 1970). The mathematical analysis of the sequence
of events within this cycle carried out by Shelepin (1970) is very
interesting. In his original work the suggestion is made -
unexpected at first glance - of a similarity of such quite different
biological systems as the cell system of neurone-neuroglia and the
ecological system of predator-prey. In the same way as the food
supply of the predator is dependent on the population of its prey,
while the food of the latter, on the basis of natural resources is
considered as stable, and non-limiting, the energetic and biosynthetic demands of the neurone are tied to the condition of the glial
cells, while their supplies and energetic requirements from the
cerebral blood circulation may under normal conditions be considered
unlimited. Starting with this analogy, Shelepin applied Volterra's
well known theorem of periodic oscillations of population numbers of
predator and prey (Volterra, 1931) to the oscillations of activity
of neurones and neuroglia. He considered the RNA content of the
cells as an index of their functional activity. He concluded that
the neurone-neuroglia unit is constantly subjected to cyclic changes,
and that the functional-biochemical fluctuations in neurones and glia
differ in phase. Indeed, in the experiments both of Hydén and of
our group a raised concentration of RNA was shown in the neurones
and a lower level in the glia as a result of activation of function.
Verpintsev's group (D'yakonova, 1972; Bocharova et al., 1972) have
also presented interesting data showing that with excitation of the
giant identified neurones of invertebrates an opposite relationship
is observed: a lowering of neuronal RNA synthesis and its activation
in the perineuronal glia.

    The interrelationships described above of nucleic acid

biosynthesis in the neurone-neuroglia unit seem also to be valid for protein metabolism. To what degree these interdependences apply to other types of metabolism (carbohydrate, lipid, oxidative, etc.) remains to be demonstrated.

Work on the functional biochemistry of the neuroglia has so far been concerned primarily with the glial cell-satellites, which are primarily oligodendroglia. Other types of glia, such as astrocytes and ependymal cells have been rather neglected, with a few exceptions such as Hamberger's work (1963) on the oxidative processes in the pericapillary glia, and that of Norton's group on the comparison of the lipid metabolism in astrocytes and oligodendroglia (Norton and Poduslo, 1971, 1973; Poduslo and Norton, 1972; Abe and Norton, 1974). In addition a number of authors have presented data on astrocyte-oligodendroglia differences in the metabolism of other classes of compound (Austoker et al., 1972, 1973; Cohen and Bernsohn, 1973; Dittman et al., 1973; Hertz et al., 1973; Margolis and Margolis, 1974; Dadoune and Baudrimont, 1975). These studies however, involve only animals under resting conditions. No comparison of the metabolic responses of astrocytes and oligodendroglia to changes in the functional state of the nervous system have been carried out. As far as ependymal cells are concerned, we have compared functionally conditioned changes in protein and RNA content in these cells with those occurring in oligodendroglia (Pevzner, 1969e, 1972a, 1974, 1976). However, these are isolated cases and the detailed study of different kinds of glia is work for the future.

In conclusion, it is necessary to remark that the problem of the neurone-neuroglia unit - above all its specificity - is only a part of a general cytological problem of cell association. Two different cell types may be distinguished. The first is the cell bearing the basic functional load, that is to say carrying out specifically that activity which is characteristic of the given tissue. The second type is the cell-satellite, the auxilliary cell, whose function is to guarantee conditions necessary for the primary cell. By comparison with the neurone-neuroglia unit, other types of cell-association have been studied far less. However it seems that the characteristics of biochemical interrelationships between neurones and neuroglia described above are also true for other types of cell associations. Thus studies of the oocyte-trophocyte system in the ovary of insects has shown that during vitellogenesis the basic synthesis of RNA and protein (from autoradiographic data) occurs in the nucleus of the auxilliary cell - the trophocyte - while accumulation of RNA and protein are noted in the primary cell structure of the ovary - the oocyte (Sirlin and Jacob, 1960; Bier, 1963, 1964; Przelecka, 1966a, 1966b). It is suggested that during vitellogenesis RNA and protein migrate from the trophocyte to the oocyte. This is supported by the electron-microscopic picture of the membrane structure which separates these two cell types (Bier and Ramamurty, 1964). In myocardium, study of the compensatory processes showed

that most of the accumulation of nucleic acids occurred not in the
heart muscle cells themselves, but in the surrounding connective
tissue cells (Meerson, 1967; Chernukh et al., 1968). Studies on
functional morphology of receptors also indicate the high metabolic
activity of the surrounding supporting cells, presumably assisting
the specific function of the visual (for reviews, see: Vinnikov,
1966) and olfactory receptor cells (Bronshtein, 1966).

SUMMARY

The present work has demonstrated changes in the content of
RNA in neuronal and glial cells in various functional states of the
nervous system. All the changes in state used were major ones -
the transition from rest to marked excitation, from excitation to
general fatigue and exhaustion, and thence to the reparative
processes, restoring the nervous system to its original condition.
This was because the total RNA content could not be shown to change
significantly with brief or mild changes in the state of cells
being studied. Under the conditions used, then, the RNA metabolism
in the nervous system was occurring under deliberately unfavorable
conditions: hypoxia, adrenalectomy, administration of antimetabolites.

Analysis of the data obtained clearly confirmed that the glial
cell-satellites are not a supporting, metabolically inert cell
structure in the nervous system. Changes in the functional state
of the nervous system were marked by substantial changes in content
of RNA, occurring both in neuronal and glial cells.

At the same time, it is impossible to consider that the metabolic characteristics of neurones and glia are identical: moderate
excitation of the nervous system leads to changes in the RNA content
only of the neurones in the absence of significant changes in the
glia. Upon cessation of such an excitation the neuronal RNA content
gradually returns to normal in parallel with the animal's behavior.
This again confirms that it is the neurone which is the primary
functional unit in nerve tissue and that its metabolism is directly
involved in the molecular mechanisms underlying specific nervous
activity.

A number of extreme factors (hypoxia and several antimetabolites) led only to changes in content of neuronal RNA, while RNA
content in the glia changed little or not at all. This indicates
the high degree of metabolic stability of the glial cells by contrast
with the neurones.

In defined stages of excitation of the nervous system raised
RNA content was observed in the neuronal cytoplasm, but was lowered
in the glial cell-satellites. In this case, biochemical analysis
showed that RNA levels in homogenate of whole nervous tissue in
these regions of the nervous system showed no marked changes in
total RNA content. These results are in indirect confirmation of
the possibility of migration of RNA from the glia to the neuronal
cell body.

With the cessation of stimuli which lead to an overall biochemical shift in the neurone-neuroglia system, the restoration of
the initial quantity of RNA in the glia proceeds more rapidly than
in the neurones. It may be concluded that the reparative capacity
of glial cells is considerably higher than that of neurones. This
is in agreement with the fact that the majority of neurones are
characterized by a predominance of cytoplasm over nucleus, while in
the glia (particularly in the oligodendrogliocytes and fibrillary
astrocytes) the nucleus constitutes the main bulk of the cell. It
is the nucleus which is the site of the regualtion of biosynthesis
of nucleic acid and protein in the cell. A high rate of synthetic
processes in the glia is clearly linked to the high activity of the
pentose cycle for oxidation of glucose, which generates the coenzymes
and intermediate products necessary for biosynthetic reactions.
This also seems to lead to the greater resistance of the neuroglia
to the influence of inhibitors or of hypoxia.

This greater reparative biosynthetic capacity is probably the
most important feature of the glial cells and possibly of cell
satellites in general.

A series of humoral factors (including also those which may be
called neurotrophic) cause changes in RNA content specifically in
the glial cells. We may think that the cell satellite, as the
intermediate link between neurones and capillaries, plays a major
role in the reaction of the nervous system to neurotrophic influences.

We have repeatedly observed secondary changes in glial RNA content that occur 1-3 days after cessation of the experimental
influence. By this time the state of the experimental animals has
in other respects normalized, and cytochemistry shows that the RNA
content in neurones has returned to the normal levels. However,
the content of glial RNA returns to the norm shortly after the
cessation of the experimental influence and then changes again at
this later time. We propose that these secondary changes lie at

the basis of the mechanism of subsequent consolidation of adaptive
processes in the nervous system at the cellular level.

The data obtained are concerned mainly with the metabolism of
RNA in the neurone-neuroglia unit. In so far as the primary function
of nucleic acid in the cell is the control of protein synthesis,
we may think that similar changes would also be expected to occur
for protein metabolism. In fact the results of preliminary studies
confirm that the trend of the changes in protein content in the
neurone-neuroglia unit with changes in its functional state is on
the whole similar to the changes in content of nucleic acids.

Thus the results of the present study show that glial cells
actively participate in the functioning of the neurone. In this
context, the neuroglia play a particularly important role in intense
functional activity of the nerve cell, and also in the period of
reparation. The glial cell-satellites are, evidently, also the site
of action of various trophic influences on the nervous system. It
is also possible that the neuroglia may be involved in the initial
processes of reconstruction of nervous system metabolism which
permits subsequent cell adaptation.

Nervous tissue functions as a highly organized self-regulating
system, in which continual changes of functional state occur side by
side with homeostatic processes. We may affirm that the morphological substrate of the complex homeostatic processes in the nervous
system is above all the neuroglial cell.

REFERENCES

Abdel-Latif, A.A., Yau, S.-J., Smith, J.P. (1974). "Effect of
    neurotransmitters on phospholipid metabolism in rat cerebral
    cortex slices. Cellular and subcellular distribution,"
    J. Neurochem., 22, 383-393.

Abe, T., and Norton, W.T. (1974). "The characterization of sphingo-
    lipids from neurones and astroglia of immature rat brain,"
    J. Neurochem., 23, 1025-1036.

Abood, L.G., Gerard, R.W., Banks, J., and Tschirgi, R.D. (1952).
    "Substrate and enzyme distribution in cells and cell fractions
    of the nervous system," Amer. J. Physiol., 168, 728-738.

Ádám, G., (ed.) (1971). Biology of Memory, Akademiai Kaido,
    Budapest.

Adams, C.W.M. (1965). "Histochemistry of the cells in the nervous
    system," in: Neurohistochemistry, Amsterdam, pp. 253-331.

Adams, C.W. (1969). "Enzyme histochemistry," in: Handbook of
    Neurochemistry, A. Lajtha (ed.), Plenum Press, New York,
    Vol. II, pp. 525-537.

Adams, C.W.M., and Davison, A.N. (1965). "Morphology of the myelin
    sheath," in: Neurohistochemistry, Amsterdam, pp. 332-400.

Adzhimolaev, T.A., Bezruchko, S.M., and Vozhenina, N.I. (1972).
    "Effect of electrical stimulation on the incorporation of
    ^{3}H-lysine into isolated cells of Tritonia nervous system,"
    Biofizika, 17, 77-84.

Agroskin, L.S. (1958). "A modern apparatus for cytospectrophoto-
    metry," Biofizika, 3, 343-354.

Agroskin, L.S. (1964). "Cytospectrophotometry of cells in the
    visible region of the spectrum," Biofizika, 9, 456-462.

Agroskin, L.S. (1967) "Compensated two-wavelength cytophotometry,"
    Tsitologiya, 9, 746-751.

Agroskin, L.S., Brodskii, V.Ya., Gruzdev, A.D., and Korolev, N.V.
    (1960). "Some questions on quantitative spectrophotometric
    measurements on cells," Tsitologiya, 2, 337-352.

Agroskin, L.S., Papayan, G.B., and Rautian, L.P. (1970). "A
    recording shearing microspectrophotometer," Tsitologiya, 12,
    548-558.

Aladzhalova, N.A., and Kol'tsova, A.V. (1964). "Electrical
    activity in a region with accumulation of glial cells in
    medulla oblongata (the area postrema)," Byul. Éksperim. Biol.
    i Med., 58 (12), 9-12.

Albrecht, J., and Smialek, M. (1975). "Effect of hypoxia, ischemia
    and carbon monoxide intoxication on in vivo protein synthesis
    in neurone and glia cell enriched fractions from rat brain,"
    Acta neuropathol., 31, 257-262.

Aleksandrovskaya, M.M. (1950). Neuroglia in Various Psychoses
    (in Russian), Moscow.

Aleksandrovskaya, M.M. (1965). "Changes in neurone-glia inter-
    relationships as one of the structural bases of barrier
    mechanisms in the central nervous system," in: Problems of
    the blood-brain barrier, Moscow, pp. 288-293.

Aleksandrovskaya, M.M. (1969). "Morphological correlates of
    functional changes in the central nervous system shown by
    investigation of neuronal-glial inter-relationships," Zh.
    Vyssh. Nervn. Deyat., 19, 156-161.

Aleksandrovskaya, M.M., Brazovskaya, F.A., Geinisman, Yu.Ya.,
    Kazakova, P.B., Larina, V.N., and Mats, V.N. (1968).
    "Morphological reconstruction of neuroglia under conditions
    of increased function of nerve centers," Dokl. Akad. Nauk SSSR,
    180, 719-722.

Aleksandrovskaya, M.M., and Chizhenkova, R.A. (1973). "Analysis
    of changes in glial cell number in acoustic regions of cortex
    with sound stimulation of varying intensity," Fiziol. Zh.
    SSSR, 59, 870-874.

Aleksandrovskaya, M.M., Geinisman, Yu.Ya., Lapina, V.N., and Mats, V.N. (1967). "Changes in neuronal RNA following momentary functional load," Byul. Éksperim. Biol. i Med., 64 (7), 103-106.

Aleksandrovskaya, M.M., Geinisman, Yu.Ya, and Mats, V.N. (1965). "Morphological investigation of neuronal-glial interactions during increase in neuronal function," Zh. Nevropatol. i Psikhiatrii, 65, 161-167.

Aleksandrovskaya, M.M., Geinisman, Yu.Ya., and Samoilova, L.G. (1964). "Structural and metabolic changes in the brain in animals with altered higher nervous activity after repeated administration of aminazine," Byul. Éksperim. Biol. i Med., 58, 80-86.

Aleksandrovskaya, M.M., Kruglikov, R.I., and Shevtsov, V.V. (1972). "Structure and protein metabolism of the cerebral cortex after repeated electroshock," Dokl. Akad. Nauk SSSR, 203, 251-253.

Aleksanyan, A.M. (1964). "Influence of the sympathetic nervous system on brain function," in: Evolution of Functions, Moscow and Leningrad, pp. 128-134.

Aleksidze, N.G. (1970). "Comparative data on the fractionation of soluble proteins of neuronal and glial cells of various brain-stem nuclei of rabbit," Zh. Evol. Biokhim. i Fiziol., 6, 381-383.

Aleksidze, N.G. (1974). "Metabolic inter-relationships of neurones and neuroglia studied by microchemical methods on isolated cells," in: Achievements in Neurochemistry, Nauka, Leningrad, pp. 132-140.

Aleksidze, N.G., Akhalkatsi, R.G., Balavadze, M.V., and Dolidze, N.I. (1974). "A comparative study of AChE and BuChE in enriched fractions of neurones and glial cells from rabbit cerebral cortex," Soobshch. Akad. Nauk Gruz. SSSR, 75, 701-704.

Akelsidze, N.G., and Blomstrand, C. (1968). "The influence of hydroxylamine, thiosemicarbazide and gamma-aminobutyric acid upon succinate oxidation by Deiters' nerve cells and neuroglia," Brain Res., 11, 717-719.

Aleksidze, N.G., and Blomstrand, C. (1969). "Effect of potassium ions on neuronal and neuroglial respiration in lateral vestibular nucleus of the rabbit," Dokl. Akad. Nauk SSSR, 186, 1429-1430.

Aleksidze, N.G., and Khaglid, K. (1970). "Lactate dehydrogenase in neurones and neuroglia of lateral vestibular nucleus of the rabbit," Dokl. Akad. Nauk SSSR, 190, 972-974.

Alfert, M., and Geschwind, I.I. (1953). "A selective staining
method for the basic proteins of cell nuclei," Proc. Nat.
Acad. Sci. USA, 39, 991-999.

Alipova, N.G. (1952). "Effect of aurantin on the pathological
picture of surviving ependymoblasts of white mice receiving
carcinogenic substances," in: Aurantin - An Antitumor,
Antibiotic Preparation of the Actinomycin Group, Moscow,
pp. 171-179.

Altman, J. (1962). "Autoradiographic study of degenerative and
regenerative proliferation of neuroglia cells with tritiated
thymidine," Exp. Neurol., 5, 302-318.

Altman, J. (1963). "Autoradiographic investigation of cell proliferation in the brains of rats and cats," Anat. Rec., 145,
573-577.

Altman, J. (1972). "Autoradiographic examination of behaviorally
induced changes in the protein and nucleic acid metabolism of
the brain," in: Macromolecules and Behavior, 2nd Ed., New
York, pp. 305-334.

Altman, J., and Das, G.D. (1964). "Autoradiographic examination of
the effects of enriched environment on the rate of glial multi-
plication in the adult rat brain," Nature, 204, 1161-1163.

Andén, N.-E., Dahlstrom, A., Fuxe, K., and Larsson, K. (1965).
"Further evidence for the presence of nigroneostriatal dopamine
neurones in the rat," Amer. J. Anat., 116, 329-334.

Andreeva, L.F., Dondua, A.K., and Zavarzin, A.A. (1963). "Studies
of RNA synthesis during cell differentiation using differential
extraction and autoradiography," Collection of the Work of the
Institute of Cytology of the Academy of Sciences of the USSR,
5, 102-120.

Ansell, G.B., and Spanner, S. (1968). "Plasmalogenase activity in
normal and demyelinating tissue of the central nervous system,"
Biochem. J., 108, 207-209.

Antanitus, D.S., Choi, B.H., and Lapham, L.W. (1975). "Immuno-
fluorescence staining of astrocytes in vitro using antiserum to
glial fibrillary acidic protein," Brain Res., 89, 363-367.

Antanitus, D.S., Choi, B.H., and Lapham, L.W. (1976). "The demonstration of glial fibrillary acidic protein in the cerebrum of
the human fetus by indirect immunofluorescence," Brain Res.,

_103_, 613-616.

Antonov, L.M., and Pevzner, L.Z. (1968). "Nucleic acids of neurones
    and neuroglia in cerebellum and hippocampus of rat after
    adrenalectomy and administration of hydrocortisone," Tsitologiya,
    _10_, 599-603.

Appel, E. (1968). "Aspecte morfo-functional ale nevrogliei,"
    Stud. Cercet. Neurol., _13_, 421-440.

Appel, S.H. (1972). "Macromolecular synthesis in synapses," in:
    Macromolecules and Behavior, 2nd Ed., J. Gaito (ed.), Appleton-
    Century-Crofts, New York, pp. 185-203.

Appel, S.H., and Parrot, B.L. (1970). "Hexose monophosphate path-
    way in synapses," J. Neurochem., _17_, 1619-1626.

Arbogast, B.W., and Arsenis, C. (1974). "The enzymatic ontogeny
    of neurones and glial cells isolated from postnatal rat cerebral
    gray matter," Neurobiology, _4_, 21-37.

Arbuzov, S.Ya., and Nikiforov, M.I. (1967). Systemic Nervous
    Narcosis (in Russian), Moscow.

Ashmarin, I.P. (1975). Enigmas and Revelations in the Biochemistry
    of Memory (in Russian), State University Press, Leningrad.

Austoker, J., Cox, D., and Mathias, A.P. (1972). "Fractionation
    of nuclei from brain by zonal centrifugation and a study of the
    ribonucleic acid polymerase activity in the various classes of
    nuclei," Biochem. J., _129_, 1139-1155.

Austoker, J., Cox, D., and Mathias, A.P. (1973). "The synthesis
    of ribonucleic acid _in vivo_ in the nuclei of rat brain
    fractionated by zonal centrifugation," Biochem. J., _132_, 813-
    819.

Baethmann, A., and Van Harreveld, A. (1973). "Water and electro-
    lyte distribution in gray matter rendered edematous with a
    metabolic inhibitor," J. Neuropathol. and Exp. Neurol., _32_,
    408-423.

Bakharev, F.M., Davydova, M.I., Zarubina, I.L., Popov, A.I., Skvortso
    Skvortsov, G.E., and Smirnov, V.A. (1964). "A microspectro-
    photometer for UV and visible regions of the spectrum (MUF-5),"
    Tsitologiya, _6_, 114-120.

Balázs, R., Patel, A.J., and Richter, D. (1973). "Metabolic
    compartments in the brain: their properties and relation to
    morphological structures," in: Metabolic Compartmentation in

the Brain, R. Balazs and J.E. Cremer (Eds.), McMillan, London, pp. 167-184.

Banks-Schlegel, S.P., and Johnson, T.C. (1975). "RNA metabolism in isolated mouse brain nuclei during early post-natal development," J. Neurochem., 24, 947-952.

Baranov, M.N. (1962). "Nucleic acid content in various layers of cerebral cortex," Biokhimiya, 27, 702-707.

Baranov, M.N. (1963). "Concentration of nucleic acids of various layers of rat cerebral cortex as a function of various functional states,(in Russian)," 3rd All-Union Conference on the Biochemistry of the Nervous System, Erevan, pp. 319-325.

Baranov, M.N., and Pevzner, L.Z. (1963a). "Microchemical and microspectrophotometric studies on the intralaminar distribution of nucleic acids in the brain cortex under various experimental conditions," J. Neurochem. 10, 279-283.

Baranov, M.N., and Pevzner, L.Z. (1963b). "Nucleic acid content of superior cervical sympathetic ganglion under normal conditions and during excitation," Biokhimiya, 28, 958-963.

Barer, R. (1956). "Phase-contrast and interference microscopy in cytology," in: Physical Techniques in Biological Research, 3, pp. 29-30.

Barlow, C.F., Domer, N.S., and Goldberg, M.A. (1961). "Extracellular brain space measured by S^{35} sulphate," Arch. Neurol. 5, 102-110.

Barondes, S.H. (1973). "Neuronal proteins: synthesis, transport, and neuronal regulation," in: Proteins of the Nervous System, D.J. Schneider et al. (Eds.), Raven Press, New York, pp. 215-226.

Barrnett, R.J., and Seligman, A.M. (1954). "Histochemical demonstration of the sulfhydryl and disulfide groups of protein," J. Nat. Cancer Inst., 14, 769-803.

Benda, P., Lightbody, J., Sato, G., Levine, L., and Sweet, W. (1968). "Differentiated rat glial cell strain in tissue culture," Science, 161, 370-371.

Benda, P., Mori, T., and Sweet, W.H. (1970). "Demonstration of an astrocytospecific cerebroprotein by an immunofluorescence study of human brain tumors," J. Neurosurg. 33, 281-286.

Benjamins, J.A., Guarnieri, M., Miller, K., Sonneborn, M., and
    Mckhann, G.M. (1974). "Sulphatide synthesis in isolated
    oligodendroglial and neuronal cells," J. Neurochem, 23, 751-
    757.

Beritashvili, I.S. (1968). Memory in Vertebrates - Its Character-
    istics and Origin, (in Russian), Tbilisi.

Berl, S., Nicklas, W.J., and Clarke, D.D. (1968). "Compartmenta-
    tion of glutamic acid metabolism in brain slices," J. Neurochem.
    15, 131-140.

Berl, S., Takagaki, G., Clarke, D.D., and Waelsch, H. (1962).
    "Metabolic compartments in vivo. Ammonia and glutamic acid
    metabolism in brain and liver," J. Biol. Chem., 237, 2562-
    2569.

Bezruchko, S.M., Adzhimolaev, T.A., Timkin, V.N., Mezentsev, A.N.,
    and Vozhenina, N. (1970). "Effect of electrical stimulation
    on nucleic acid and protein turnover in the nervous system of
    Tritonia diamedia," in: Neuronal Mechanisms of Learning,
    Moscow Univ. Press, Moscow, pp. 76-77.

Bier, K. (1963). "Syntese, intersellulärer Transport, und Abbau
    von Ribonukleinsäure im Ovar der Stubenfliege Musca domestica,"
    J. Cell Biol. 16, 436-440.

Bier, K. (1964). "Gerichteter Ribonukleinsäuretransport durch das
    Cytoplasma," Naturwissensh., 51, 418.

Bier, K., and Ramamurty, P.S. (1964). "Elektronenoptische Unter-
    suchungen zur Einlagerung der Dotterprotein in die Oocyte,"
    Naturwissensch., 51, 223-224.

Bignami, A., and Dahl, D. (1973). "Differentiation of astrocytes
    in the cerebellar cortex and the pyramidal tracts of the newborn
    rat. An immunofluorescence study with antibodies to a protein
    specific to astrocytes," Brain Res., 49, 393-402.

Bignami, A., and Dahl, D. (1974). "Astrocyte-specific protein and
    neuroglial differentiation. An immunofluorescence study with
    antibodies to the glial fibrillary acidic protein," J. Comp.
    Neurol., 153, 27-37.

Bignami, A., Eng, L.F., Dahl, D., and Uyeda, C.T. (1972). "Locali-
    zation of the glial fibrillary acidic protein in astrocytes by
    immunofluorescence," Brain Res., 43, 429-435.

Binaglia, L., Goracci, G., Porcellati, G., Roberti, R., and Woelk,
    H. (1973). "The synthesis of choline and ethanolamine

phosphoglycerides in neuronal and glial cells of rabbit in vitro," J. Neurochem., 21, 1067-1082.

Birks, R., Katz, B., and Miledi, R. (1960). "Physiological and structural changes at the amphibian myoneural junction, in the course of nerve degeneration," J. Physiol., 150, 145-168.

Black, M., and Ansley, H. (1966). "Histone specificity revealed by ammoniacal silver staining," J. Histochem. Cytochem., 14, 177-188.

Bloch, D.P., and Godman, G.C. (1955). "A microphotometric study of the synthesis of desoxyribonucleic acid and nuclear histone," J. Biophys. Biochem. Cytol., 1, 17-28.

Blomstrand, C., (1970). "Effects of hypoxia on protein metabolism in neurone and neuroglia cell enriched fractions from rabbit brain," Exp. Neurol., 29, 175-188.

Blomstrand, C., (1971). Studies on Protein Metabolism in Neuronal and Glial Cell-Enriched Fractions from Brain Tissue, Goteborg.

Blomstrand, C., and Hamberger, A. (1969). "Protein turnover in cell-enriched fractions from rabbit brain," J. Neurochem., 16, 1401-1407.

Blomstrand, C., and Hamberger, A. (1970). "Amino acid incorporation in vitro into protein of neuronal and glial cell-enriched fractions," J. Neurochem., 17, 1187-1195.

Blomstrand, C., Hamberger, A., Sellström, Å., and Steinwall, O. (1975). "Protein bound radioactivity in neuronal and glial fractions following intra-carotid ^{3}H-leucine perfusion," Neurobiology, 5, 178-187.

Bocharova, L.S., Borovyagin, V.L., Dyakonova, T.L., Warton, S.S., and Veprintsev, B.N. (1971). "Ultrastructure and RNA synthesis in a molluscan giant neurone under electrical stimulation," Brain Res., 36, 371-384.

Bogoch, S. (1969). "Proteins," in: Handbook of Neurochemistry, New York, Vol. I, pp. 75-92.

Bondy, S.C. (1966). "The ribonucleic acid metabolism of the brain," J. Neurochem., 13, 955-959.

Bondy, S.C., and Roberts, S. (1969). "Developmental and regional variations in ribonucleic acid synthesis on cerebral chromatin," Biochem. J., 115, 341-349.

Borgström, L., Norberg, K., and Siesjö, B.K. (1976). "Glucose consumption in rat cerebral cortex in normoxia, hypoxia and hypercapnia," Acta physiol. scand., 96, 569-574.

Bowery, N.G., Brown, D.A., and Yamini, G. (1975). "Autoradiographic observations on the transport of amino acids by neurones and glia in isolated rat superior cervical ganglia," J. Physiol. (Gr. Brit.), 246, P23-P24.

Bradford, H.F., and Rose, S.P.R. (1967). "Ionic accumulation and membrane properties of enriched preparations of neurone and glia from mammalian cerebral cortex," J. Neurochem., 14, 373-375.

Brattgård, S.-O. (1956). "Microscopical determinations of the thickness of histological sections," Stain Technol., 31, 52.

Brazovskaya, F.A. (1968). "Quantitative changes in the neurone-neuroglia unit with changes in the level of afferentation," Dokl. Akad. Nauk SSSR, 178, 968-970.

Brightman, M.W. (1965a). "The distribution within the brain of ferritin injected into cerebrospinal fluid compartments. I. Ependymal distribution," J. Cell Biol., 26, 99-123.

Brightman, M.W. (1965b). "The distribution within the brain of ferritin injected into cerebrospinal fluid compartments. II. Parenchymal distribution," Amer. J. Anat., 117, 193-220.

Brodskii, V.Ya. (1955). "Quantitative determination of substances in various structures of animal cells (for example cytophotometry of DNA in nuclei of liver cells)," Dokl. Akad. Nauk SSSR, 102, 357-360.

Brodskii, V.Ya. (1956). "Cytophotometry (quantitative cytochemical methods)," Usp. Sovr. Biol., 42, 87-107.

Brodskii, V.Ya. (1957). "Nucleic acids in motoneurones of spinal cords under the influence of barbamyl and urethane (quantitative analysis)," Dokl. Akad. Nauk SSSR, 112, 753-755.

Brodskii, V.Ya. (1958). "On the qualitative heterogeneity of ribonucleic acid in animal cells," Dokl. Akad. Nauk SSSR, 123, 546-549.

Brodskii, V.Ya. (1960). "On methods of fixation and preparation of material for quantitative cytochemical analysis," Tsitologiya, 2, 605-613.

Brodskii, V.Ya. (1961). "Some cytological characteristics of the nuclear type of cell protein synthesis as studied in retinal ganglion cells," Tsitologiya, 3, 312-326.

Brodskii, V.Ya. (1964). "Somatic polyploidy as a type of physiological regeneration of structural units of organs during development in vertebrates," Zh. Obshch. Biol., 25, 39-50.

Brodskii, V.Ya. (1965). "Cytophotometry in visible and UV regions of the spectrum," in: Manual of Cytology, Vol. I (in Russian), Moscow and Leningrad, pp. 85-98.

Brodskii, V.Ya. (1966). Trophics of the Cells (in Russian), Moscow.

Brodskii, V.Ya. (1970). "Possible mechanisms of renewal of neuronal cytoplasm during ontogenesis (from cytochemical investigations of protein metabolism in retinal ganglion cells) (in Russian)," Fifth All-Union Conference on Neurochemistry, Tbilisi, pp. 114-128.

Brodskii, V.Ya., Aref'eva, A.M., and Kuznetsova, L.P. (1966). "On the destructive phase of physiological regeneration of the neurone (interferometric and electrophysiological studies)," Tsitologiya, 8, 662-664.

Brodskii, V.Ya., and Kuznetsova, A.F. (1961). "Interference microscopy of retinal ganglion cells under various functional states," Tsitologiya, 3, 89-91.

Brodskii, V.Ya., and Neverova, M.E. (1968). "Kinetic changes of nuclear dry weight of ganglion cells during light stimulation of chick at different ages," Dokl. Akad. Nauk SSSR, 181, 217-220.

Brodskii, V.Ya., and Peizulaev, Sh.I. (1955). "On the magnitude of the error in quantitative histochemical determination of substances by photographic and photoelectric cytophotometry," Izv. Akad. Nauk SSSR, Ser. Biol., 6, 100-108.

Brodskii, V.Ya., and Polyakov, N.I. (eds.) (1969). Introduction to Cytochemistry (in Russian), Moscow.

Brodskii, V.Ya., and Suetina, I.A. (1960). "Effect of X-irradiation on nucleic acid content in cell nuclei of the bone marrow," Dokl. Akad. Nauk SSSR, 134, 439-442.

Bronstein, A.A. (1966). "Some data on the fine structure and cytochemistry of olfactory receptors," in: Primary Processes in Receptor Elements of Sensory Organs, Moscow and Leningrad, pp. 65-83.

Broun, R.G., and Goncharova, V.P. (1962). "Quantitative determination of nucleic acid in brain tissue," Ukrainsk. Biokhim. Zh., 34, 734-740.

Brue, F., Joanny, P., Bertharion, G., Morcellet, J.L., and Corriol, J. (1972). "Métabolisme du glucose dans les tissues nerveux isolés sous oxygène hyperbare; stimulation du cycle des pentoses phosphates et de la glycolyse," J. de Physiol., 65, 366.

Brumberg, V.A. (1968a). "Effect of swimming for various lengths of time on RNA content in neurones and neuroglia of motor and sensory regions of spinal cord," Dokl. Akad. Nauk SSSR, 182, 228-230.

Brumberg, V.A. (1968b). "RNA content in neurones and glial cells of spinal cord during muscular effort and subsequent rest," Tsitologiya, 10, 1193-1197.

Brumberg, V.A. (1969). "On the changes in volume of cell bodies in motor and sensory neurones of spinal cord and their surrounding glial cells under various conditions of motor activity," Dokl. Akad. Nauk SSSR, 184, 1231-1234.

Brumberg, V.A., and Pevzner, L.Z. (1966). "Cytophotometric analysis of some methods for selective extraction of RNA from sections of nerve tissue," Tsitologiya, 8, 437-439.

Brumberg, V.A., and Pevzner, L.Z. (1968). "RNA content in motor and sensory neurones and their surrounding neuroglia of spinal cord under conditions of hypoactivity and subsequent normalization," Tsitologiya, 10, 1452-1459.

Brumberg, V.A., and Pevzner, L.Z. (1971). "Nucleic acids in the neurone-neuroglia unit under various functional conditions of the nervous system," Tsitologiya, 13, 129-147.

Brumberg, V.A., and Pevzner, L.Z. (1972). "On the use of amido black for cytophotometric studies of cell protein," Tsitologiya, 14, 674-676.

Brumberg, V.A., and Pevzner, L.Z. (1975a). Neurochemistry of Isoenzymes (in Russian), Nauka, Leningrad.

Brumberg, V.A., and Pevzner, L.Z. (1975b). "Cytospectrophotometric studies on the lactate dehydrogenase isoenzymes in functionally different neurone-neuroglia units. I. Optimal procedure conditions and topochemical comparisons in normal mice," Acta Histochem., 54, 218-229.

Burdman, J.A. (1970). "Incorporation in vivo of radioactive leucine into neuronal and glial nuclear proteins of rat brain," J. Neurochem., 17, 1555-1562.

Burdman, J.A. (1972). "The relationship between DNA synthesis and the synthesis of nuclear proteins in rat brain during development," J. Neurochem., 19, 1459-1469.

Burnet, F.R., Ter Haar, M.B., and Mackinnon, P.C.B. (1974). "The effects of exogenous oestradiol on protein synthetic activity in discrete areas of the rat brain," Proc. Roy. Soc. Med., 67, 39-40.

Burovina, I.V., and Skul'skii, I.A. (1965). "Alkaline metals and chemical specificity of brain tissues," Dokl. Akad. Nauk SSSR, 160, 236-239.

Burovina, I.V., Skul'skii, I.A., and Fleishman, D.G. (1967). "On the interdependence of the distribution of stable Cs and ^{137}Cs with the distribution of two alkaline elements in brain and muscle of reindeer," Zh. Evol. Biokhim. i Fiziol., 3, 281-286.

Busnyuk, M.M., and Pigareva, Z.D. (1974). "Activities of some enzymes in the visual region of brain with early deprivation of visual function," Zh. Nevropatol. i Psikhiatri, 74, 1345-1349.

Caley, D.W., and Butler, A.B. (1974). "Formation of central and peripheral myelin sheaths in the rat: an electron microscopic study," Amer. J. Anat., 140, 339-347.

Calissano, P. (1973). "Specific properties of the brain-specific protein S-100," in: Proteins of the Nervous System, D.J. Schneider et al. (Eds.), Raven Press, New York, pp. 13-26.

Cammermeyer, J. (1960a). "Reappraisal of the perivascular distribution of oligodendrocytes," Amer. J. Anat., 106, 197-232.

Cammermeyer, J. (1960b). "Distribution of oligodendrocytes in cerebral gray matter and white matter of several animals," Amer. J. Anat., 107, 107-127.

Carton, H.C., and Appel, S.H. (1973). "The contribution of axoplasmic nerve and protein synthesis within the optic tectum to synaptic membrane proteins of the chick optic tectum," J. Neurochem., 20, 1707-1717.

Caspersson, T. (1936). "Über den chemischen Aufbau der Structuren des Zellkernes," Skand. Arch. Physiol., 73, Suppl., 8.

Caspersson, T. (1940). "Methods for the determination of the
    absorption spectra of cell sturctures," J. Roy. Microscop. Soc.,
    60, 8-25.

Caspersson, T. (1950). Cell Growth and Function, New York.

Caspersson, T. (1955). "Quantitative cytochemical methods for the
    study of cell metabolism," Experientia, 11, 45-60.

Cavalieri, L.F., and Bendich, A. (1950). "The ultraviolet absorp-
    tion spectra of pyrimidines and purines," J. Amer. Chem. Soc.,
    72, 2587-2594.

Cavalieri, L.F., Bendich, A., Tinker, J.F., and Brown, G.B. (1948).
    "Ultraviolet absorption spectra of purines, pyrimidines, and
    triazolopyrimidines," J. Amer. Chem. Soc., 70, 3875-3880.

Chang, J.J., and Hild, W. (1959). "Contractile responses to
    electrical stimulation of glial cells from the mammalian
    central nervous system cultivated in vitro," J. Cell. and Comp.
    Physiol., 53, 139-144.

Chapman-Andresen, C. (1965). "The induction of pinocytosis in
    amoebae," Arch. Biol., 76, 189-207.

Chaubal, K.A., Lodin, Z., and Pilný, J. (1967). "A new approach
    to the determination of thickness of biological specimens by
    two-wavelength method using interference microscopy," Acta
    Histochem., 26, 131-143.

Chernigovskii, V.N. (1971). "Some developments in problems of
    physiology," Izv. Akad. Nauk SSSR, Ser. Biol., No. 5, 661-672.

Chernukh, A.M., Aleksandrov, P.N., Alekhina, G.M., Pshennikova, M.G.,
    and Meerson, F.Z. (1968). "DNA content in muscle cell nuclei
    of myocardium during compensatory hypertrophy of the heart,"
    Dokl. Akad. Nauk SSSR, 178, 255-256.

Chetverikov, D.A. (1966). "Phospholipid metabolism in rat brain
    during acute oxygen starvation," in: Problems in Neurochemistry,
    Moscow and Leningrad, pp. 141-147.

Chetverikov, D.A. (1967). "Modes of influence of some environmental
    factors on brain phospholipid metabolism," in: Biochemistry
    and Function of the Nervous System, Leningrad, pp. 160-167.

Chetverikov, D.A., Gasteva, S.V., Dvorkin, V.Ya., Shmelev, A.A., and
    Bobkov, V.A. (1970). "Phospholipid metabolism in various
    regions of the central nervous system following hypoxia,"(in
    Russian), Fifth All-Union Conference on Neurochemistry, Tbilisi,

pp.   274-284.

Chitre, V.S., Chopra, S.P., and Talwar, G.P.   (1964).   "Changes in
    the ribonucleic acid content of the brain during experimentally
    induced convulsions," J. Neurochem., 11, 439-448.

Cicero, T.J., Cowan, W.M., Moore, B.W., and Suntzeff, V.   (1970).
    "The cellular localization of the two brain-specific proteins,
    S-100 and 14-3-2," Brain Res., 18, 25-34.

Cicero, T.J., Ferrendelli, J.A., Suntzeff, V., and Moore, B.W.
    (1972).   "Regional changes in CNS levels of the S-100 and
    14-3-2 proteins during development and aging of the mouse,"
    J. Neurochem., 19, 2119-2125.

Cicero, T.J., and Moore, B.W.   (1970).   "Turnover of the brain
    specific protein, S-100," Science, 169, 1333-1334.

Cohen, S.R., and Bernsohn, J.   (1973).   "Incorporation of 1-^{14}C
    labelled fatty acids into isolated neuronal soma, astroglia
    and oligodendroglia from calf brain," Brain Res., 60, 521-525.

Collewijn, H., and Schadé, J.P.   (1965).   "Changes in the size of
    astrocytes and oligodendrogliocytes during anoxia, hypothermia
    and spreading depression," Progr. Brain Res., 15, 184-195.

Colmant, H.J.   (1968).   "Allgemeine Histopathologie der Glia," Acta
    neuropathol. suppl. 4, 61-76.

Commoner, B., and Lipkin, D.   (1949).   "The application of the
    Beer-Lambert law to optically anisotropic systems," Science,
    110, 41-43.

Coxon, R.V.   (1964).   "The blood-brain barrier system in various
    species with special reference to urea," in: Comparative
    Neurochemistry, Oxford, pp.   261-274.

Cremer, J.E., Johnston, P.V., Roots, B.I., and Trevor, A.J.   (1968).
    "Heterogeneity of brain fractions containing neuronal and glial
    cells," J. Neurochem., 15, 1361-1370.

Cummins, J., and Hydén, H.   (1962).   "Adenosine triphosphate levels
    and adenosine triphosphatases in neurones, glia and neuronal
    membranes of the vestibular nucleus," Biochim. biophys. acta,
    60, 271-283.

Curtis, D.R., and Johnston, G.A.   (1974).   "Amino acid transmitters
    in the mammalian central nervous system," Ergeb. Physiol., 69,
    97-188.

Cymborowski, B., and Dutkowski, A. (1969). "Circadian changes in RNA synthesis in the neurosecretory cells of the brain and suboesophageal ganglion of the house cricket," J. Insect Physiol., 15, 1187-1197.

Dadoune, J.P., and Baudrimont, M. (1975). "Etude radioautographique de l'incorporation de leucine tritiée dans les neurones et les cellules gliales de la souris," C.R. Soc. bio., 169, 851-855.

Dahl, D. (1976). "Glial fibrillary acidic protein from bovien and rat brain. Degradation in tissues and homogenates," Biochim. et biophys. acta., 420, 142-154.

Dahl, D., and Bignami, A. (1973). "Immunochemical and immuno-fluorescence studies of the glial fibrillary acidic protein in vertebrates," Brain Res., 61, 279-293.

Dahlström, A. (1971). "Axoplasmic transport (with particular respect to adrenergic neurones," Phil. Trans. Roy. Soc. London, B261, 325-358.

Dahlström, A., and Fuxe, K. (1964). "Evidence for the existence of monoamine neurones in the central nervous system. I. Demonstration of monoamines in the cell bodies of brain stem neurones," Acta Physiol. scand., 62 suppl. 232, 1-55.

Dahlström, A., Häggendal, J., Heilbronn, E., Heiwall, P.-O., and Saunders, N.R. (1974). "Proximodistal transport of acetylocholine in peripheral cholinergic neurones," in: Dynamics of Degeneration and Growth of Neurones, K. Fuxe et al. (Eds.), Pergamon Press, Oxford, pp. 275-289.

Dalton, M.M., Hommes, O.R., and Leblond, C.P. (1968). "Correlation of glial proliferation with age in the mouse brain," J. Comp. Neurol., 134, 397-400.

Daneholt, B., and Brattgård, S.-O. (1966). "A comparison between RNA metabolism of nerve cells and glia in the hypoglossal nucleus of the rabbit," J. Neurochem., 13, 913-924.

Danielli, J.F. (1950). "Studies on the cytochemistry of proteins," Cold Spring Harbor Symposia on Quantitative Biology, 14, 32-39.

Dannies, P.S., and Levine, L. (1969). "Demonstration of subunits in beef brain acidic protein," Biochem. Biophys. Res. Commun., 37, 587-592.

Dannilova, O.A. (1958). "UV-microscopy of several regions of rabbit cerebral cortex in various functional states," Izv. Akad. Nauk SSSR, Ser. Biol., 23, 161-169.

Davies, H.G. (1958). "The determination of mass and concentration by microscope interferometry," Gen. Cytochem. Methods, 1, 55-161.

Davison, A.N., Cuzner, M.L., Banik, N.L., and Oxberry, J. (1967). "Myelinogenesis in the rat brain," Nature, 212, 1373-1374.

Davison, H., and Bradbury, M. (1965). "The extracellular space of brain," Progr. Brain Res., 15, 124-134.

Dawson, H., and Spaziani, E. (1959). "The blood-brain barrier and the extracellular space of brain," J. Physiol., 149, 135-143.

Deanin, G.G., and Gordon, M.W. (1973). "Chloramphenicol- and cycloheximide-sensitive protein synthetic systems on brain mitochondrial and nerve-ending preparations," J. Neurochem., 20, 55-68.

Deitch, A.D. (1960). "On the susceptibility of whole cells to deoxyribonuclase treatment," J. Histochem. Cytochem., 8, 350-355.

Dement'eva, S.P. (1961). "Influence of barbamyl on the functional state of the nervous system and content of gamma-aminobutyric acid in brain," Vestn. Leningr. Univ., 21, 135-137.

Demin, N.N., Nechaeva, G.A., and Pevzner, L.Z. (1968). "Influence of adrenaline on some aspects of nucleic acid metabolism in nerve tissue," in: Hormones and the Brain (in Russian), Kiev, pp. 121-124.

Demin, N.N., and Rubinskaya, N.L. (1974). "Protein and RNA content in neurones of supraoptic nucleus of rat brain and their glial cell-satellites after deprivation of paradoxical sleep for 24h," Dokl. Akad. Nauk SSSR, 214, 940-942.

Dennis, M.J., and Gerschenfeld, H.M. (1969). "Some physiological properties of identified mammalian neuroglial cells," J. Physiol., 203, 211-222.

Dermietzel, R. (1974). "Junctions in the central nervous system of the cat. I. Membrane fusion in central myelin," Cell and Tissue Res., 148, 565-576.

De Robertis, E., Alberici, M., and De Lores Arnaiz, G.R. (1969). "Astroglial swelling and phosphohydrolases in cerebral cortex of Metrazol convulsant rats," Brain Res., 12, 461-466.

De Robertis, E., and Gerschenfeld, H.M. (1961). "Submicroscopic morphology and function of glia cells," Intern. Rev. Neurobiol., 3, 1-65.

De Robertis, E., Gerschenfeld, H.M., and Wald, F. (1958). "Cellular mechanism of myelination in the central nervous system," J. Biophys. Biochem. Cytol., 4, 651-658.

De Robertis, E., Gerschenfeld, H.M., and Wald, F. (1960). "Ultrastructure and function of glial cells," in: Structure and Function of the Cerebral Cortex, Elsevier, Amsterdam, pp. 69-80.

Deshmukh, D.S., Flynn, T.J., and Pieringer, R.A. (1974). "The biosynthesis and concentration of galactosyl diglyceride in glial and neuronal enriched fractions of actively myelinating rat brain," J. Neurochem., 22, 479-485.

De Vellis, J., and Inglish, D. (1973). "Age-dependent changes in the regulation of glycerolphosphate dehydrogenase in the rat brain and in a glial cell line," Progr. Brain Res., 40, 320-330.

De Vellis, J., and Kukes, G. (1973). "Regulation of glial cell functions by hormones and ions," Tex. Repts. Biol. Med., 31, 271-293.

Diamond, M.C., Krech, D., and Rosenzweig, M.R. (1964). "The effect of enriched environment on the histology of the rat cerebral cortex," J. Comp. Neurol., 123, 111-120.

Diamond, M.C., Law, F., Rhodes, H., Lindner, B., Rosenzweig, M.R., Krech, D., and Bennett, E. (1966). "Increases in cortical depth and glia numbers in rats subjected to enriched environment," J. Comp. Neurol., 128, 117-125.

Dimova, R. (1966). "Démonstration histochimique d'une orientation précoce du métabolisme de la névroglie vers la synthèse de protéines dans la cicatrisation du tissue nerveux central," Acta neurol. psychiat. belg., 66, 527-534.

Dimova, R. (1971). "Sur les particularités histoenzymologiques de la névroglie de la moelle épiniere," Bull. Assoc. Anat., 146, 463-466.

Dimova, R., Duchesne, P.Y., and Csillik, B. (1966). "Cholinestérase vasculaire, macroglie et barrière hematoencephalique: étude de quelques organes circumventriculaires," Compt. Rend. Soc. Biol., 160, 1325-1326.

Di Stefano, H.S. (1952). "Perchloric acid extraction of ribose nucleic acid from cytological preparations," Science, 115, 316.

Dittmann, L., Sensenbrenner, M., Hertz, L., and Mandel, P. (1973). "Respiration by cultivated astrocytes and neurones from the cerebral hemispheres," J. Neurochem., 21, 191-198.

Doemin, N.N., see Demin, N.N.

Domǎnska-Janik, K., and Wideman, J. (1974). "Regulation of thiols in the brain. 3. Glucose metabolism in the cytoplasmic fraction of rat brain during stimulation of the hexose monophosphate shunt in vitro," Resuscitation, 3, 43-50.

Droz, B. (1967). "Synthèse et transfert des protéines céllulaires dans les neurones ganglionnaires. Etude radioautographique quantitative en microscopie électronique," J. de Microscopie, 6, 201-228.

Droz, B. (1973). "Renewal of synaptic proteins," Brain Res., 62, 383-394.

Droz, B., Burner, J., and Gerschenfeld, H. (1974). "Transport axonal de protéines marquées dans le motoneurone géant de l'écrevisse," Biochimie, 56, 1613-1620.

D'yakonova, T.L. (1972). "Activation of RNA synthesis in glial cell-satellites during generation of neuronal action potentials. Tsitologiya 14, 1147-1155.

Edström, A. (1964a). "The ribonucleic acid in the Mauthner neurone of the goldfish," J. Neurochem., 11, 309-314.

Edström, A. (1964b). "Effect of spinal cord transection on the base composition and content of RNA in the Mauthner nerve fibre of the goldfish," J. Neurochem., 11, 557-559.

Edström, J.-E. (1953). "Ribonucleic acid mass and concentration in individual nerve cells; new methods for quantitative determinations," Biochim. biophys. acta, 12, 361-386.

Edström, J.-E. (1958). "Quantitative determination of ribonucleic acid in the micromicrogram range," J. Neurochem., 3, 100-106.

Edström, J.-E. (1960). Extraction, hydrolysis, and electrophoretic analysis of ribonucleic acid from microscopic tissue unit (microphoresis)," J. Biochys. Biochem. Cytol., 8, 39-51.

Edström, J.-E., Eichner, D., and Edström, A. (1962). "Ribonucleic acid of axons and myelin sheaths from Mauthner neurones," Biochim. biophys. acta, 61, 178-184.

Edström, J.-E., and Hydén, H. (1954). "Ribonucleotide analysis of individual nerve cells," Nature, 174, 128-129.

Egyhazi, E., and Hydén, H. (1961). "Experimentally induced changes in the base composition of the ribonucleic acids of isolated nerve cells and their oligodendroglial cells," J. Biophys. Biochem. Cytol., 10, 403-410.

Egyházi, E., and Hydén, H. (1966). "Biosynthesis of rapidly labeled RNA in brain cells," Life Sci., 5, 1215-1223.

Eik-nes, K.B., and Brizzee, K.R. (1967). "Concentration of tritium in brain tissue of dogs given [1,2-^{3}H$_2$] cortisol intravenously," Biochim. biophys. acta, 97, 320-333.

Einarson, L. (1935). "Histological analysis of the Nissl-pattern and substance of nerve cells," J. Comp. Neurol., 61, 101-133.

Einarson, L. (1951). "On the theory of gallocyanin-chromalum staining and its application for quantitative estimation of basophilia. A selective staining of exquisite progressivity," Acta pathol. microbiol. scand., 28, 82-102.

Einarson, L. (1957). "Cytological aspects of nucleic acid metabolism," in: Metabolism of the Nervous System, London, pp. 403-421.

Engberg, I., and Ryall, R.W. (1966). "The inhibitory action of noradrenaline and other monoamines on spinal neurones," J. Physiol., 185, 298-322.

Epstein, M.H. (1970). "The localization of respiration in the normal nervous system and during hypoglycemia," Johns Hopkins Med. J., 126, 267-275.

Erenpreis, Ya.G. (1965). "Histochemical studies of basic cell proteins," Arkh. Anat. Gistol. i Embriol., 49 (12), 3-8.

Essman, W.B. (1971). "Isolation-induced behavioural modification: some neurochemical correlates," in: Brain Development and Behaviour, M.B. Sterman et al. (Eds.), Academic Press, New York and London, pp. 265-276.

Faivre-Bauman, A., Rossier, J., and Benda, P. (1974). "Glutamate accumulation by a clone of glial cells," Brain Res., 76, 371-375.

Feulgen, R., and Rossenbeck, H. (1924). "Microscopisch-chemischer Nachweis einer Nucleinsaure vom Typus der Thymonucleinsäure und die darauf beruhende elektive Farbung von Zellkernen in mikroskopischen Praparaten," Z. physiol. Chem., 135, 203-248.

Fewster, M.E., Blackstone, S.C., and Ihrig, T.J. (1973). "The preparation and characterization of isolated oligodendroglia from bovine white matter," Brain Res., 63, 263-271.

Fewster, M.E., Einstein, E.R., Csejtey, J., and Blackstone, S.C. (1974). "Proteins in the bovine oligodendroglia cells at various stages of brain development," Neurobiology, 4, 388-401.

Filipchenko, R.Ye., Pevzner, L.Z., and Slonim, A.D. (1975). "RNA content of neurones and glia of hypothalamic nuclei after interrupted cooling," Dokl. Akad. Nauk SSSR, 223, 252-255.

Fischer, J. (1972). "Attempt at an ultrastructural analysis of blood-barrier damage in the epileptogenic zone round experimental cobalt-gelatine necrosis in the rat," Physiol. bohemosl., 21, 171-176.

Flangas, A.L. and Bowman, R.E. (1968). "Neuronal perikarya of rat brain isolated by zonal centrifugation," Science, 161, 1025-1027.

Flangas, A.L., and Bowman, R.E. (1970). "Differential metabolism of RNA in neuronal-enriched and glial-enriched fractions of rat cerebrum," J. Neurochem., 17, 1237-1245.

Fleischhauer, K. (1968). "Postnatale Entwicklung der Neuroglia," Acta neuropathol.,Suppl., 20-32.

Franck, G. (1970). "Échanges cationiques au niveau des neurones et des cellules gliales du cerveau," Arch. Intern. Physiol. Biochim., 78, 613-866.

Freysz, L., Bieth, R., Judes, C., Jacob, M., and Sensenbrenner, M. (1967). "Répartition des divers phospholipides dans les neurones et les cellules gliales isolés du cortex cérébral de rat," J. de physiol. (France), 59, 239.

Freysz, L., Beith, R., Judes, C., Sensenbrenner, M., Jacob, M., and Mandel, P. (1968). "Distribution quantitative des divers phospholipides dans les neurones et les cellules gliales isolés du cortex cérébral de rat adulte," J. Neurochem., 15, 307-313.

Freysz, L., Beith, R., and Mandel, P. (1969). "Kinetics of the biosynthesis of phospholipids in neurones and glial cells isolated from rat brain cortex," J. Neurochem., 16, 1417-1424.

Freysz, L., and Mandel, P.   (1974).   "Étude comparative de la
    biosynthèse des phosphatidylcholines dan les neurones et les
    cellules gliales du cerveau de poulet," FEBS Lett., 40, 110-
    113.

Friede, R.L.   (1965).   "Enzyme histochemistry of neuroglia," Progr.
    Brain Res., 15, 35-47.

Friede, R.L.   (1966).   Topographic Brain Chemistry, New York.

Friede, R.L.   (1967).   "A comparative histochemical mapping of the
    distribution of butyryl cholinesterase in the brains of four
    species of mammals, including man," Acta anat., 66, 161-177.

Friede, R.L.   (1970).   "Die Bedeutung der Glia-Saugfüsschen für das
    Elektrolytgleichgewicht im Gehirn," Triangel, 9, 165-173.

Gaevskaya, M.S., Portugalov, V.V., Nosova, E.A., and Gershtein, L.M.
    (1966).   "Characteristics of nitrogen metabolism in brain after
    restoration of its function following clinical death," in:
    Problems of Neurochemistry (in Russian), Moscow and Leningrad,
    pp. 88-96.

Gaito, J. (1969).   Molecular Psychobiology (in Russian), Moscow.

Gaito, J. (1972).   DNA Complex and Adaptive Behaviour, Prentice-
    Hall, Englewood Cliffs, New Jersey.

Galambos, R.   (1961).   "A glia-neural theory of brain function," Proc.
    Nat. Acad. Sci. USA, 47, 129-136.

Galambos, R.   (1965).   "Introductory discussion on glial function,"
    Progr. Brain Res., 15, 267-277.

Garfinkel, D.   (1973).   "Possible correlations between morphological
    structures in the brain and the compartmentations indicated by
    stimulation," in: Metabolic Compartmentation in the Brain,
    R. Balázs and J.E. Cremer (Eds.), MacMillan Press, London,
    pp. 129-136.

Gasteva, S.V., Dvorkin, V.Ya., and Chetverikov, D.A.   (1966).   "On
    the causes of the reduction of metabolic rate of various
    phospholipid fractions of rat brain during oxygen starvation of
    the organism," Dokl. Akad. Nauk SSSR, 169, 978-981.

Gazenko, O.G., Demin, N.N., Malkin, V.B., and Pevzner, L.Z.   (1968).
    "On the adaptive changes in the cerebellum and motor structures
    of spinal cord with various regimes of hypoxia," Dokl. Akad.
    Nauk SSSR, 179, 997-1000.

Geiger, R.S. (1963). "The behavior of adult mammalian brain cells in culture," Intern. Rev. Neurobiol., 5, 1-52.

Geinisman, Yu.Ya. (1966). "Data for the analysis of functionally-conditioned changes in size of neuronal cell body and nucleus," Tsitologiya, 8, 348-358.

Geinisman, Yu.Ya. (1974). Structural and Metabolic Manifestations of Neuronal Function (in Russian), Nauk, Moscow.

Geinisman, Yu.Ya., Larina, V.N., and Mats, V.N. (1970a). "Synaptic influence as a cause of quantitative changes in neuronal RNA," Dokl. Akad. Nauk SSSR, 192, 232-234.

Geinisman, Yu.Ya., Larina, V.N., and Mats, V.N. (1970b). "Changes in RNA content in neuronal and glial cells during normal and artifical stimulation of the neurone," Tsitologiya, 12, 1028-1 1038.

Geinisman, Yu.Ya., Larina, V.N., Mats, V.N., Mikheyeva, T.S., and Persina, I.S. (1969). "Methodical prerequisites for a study of changes in volume of nerve cells," in: Modern Methods for Morphological Investigation of the Brain (in Russian), Moscow, pp. 30-35.

Georgiev, G.P., and Mant'eva, V.L. (1962). "Messenger and ribosomal RNA of the nuclear-chromosomal apparatus - methods of separation and nucleotide composition," Biokhimiya, 27, 949-957.

Gerebtzoff, M.A. (1959). Cholinesterases. A Histochemical Contribution to the Solution of Some Functional Problems, London.

Gerebtzoff, M.A. (1966). "Détection histochimique d'isoenzymes de la lactate déshydrogénase dans le nerf et le ganglion spinal," Compt. Rend. Socl. Biol., 160, 1323-1325.

Gerebtzoff, M.A. (1968). "Contribution histochimique à l'étude de la lactate déshydrogénase et de ses isoenzymes," Path. Biol., 16, 601-608.

Gerebtzoff, M.A. (1970). "Recherches histochimiques et histo-enzymologiques sur la synergie métabolique entre neurones et névroglie dans la chaine nerveuse ventrale de la sangsue Hirudo medicinalis," Bull. Acad. roy. med. Belg., 10, 337-361.

Geren, B.B. (1954). "The formation from the Schwann cell surface of myelin in the peripheral nerves of chick embryos," Exp. Cell Res., 7, 558-562.

Gerhardt, H.W. (1968). "Lactate dehydrogenase isoenzymes in the central nervous system," Danish Med. Bull., 15, suppl. 1.

Gerschenfeld, H.M., Wald, F., Zadunaisky, J.A., and De Rombertis, E.D.P. (1959). "Function of astroglia in the water-ion metabolism of the central nervous system; an electron microscope study," Neurobiology, 9, 412-425.

Gershtein, L.M., and Vavilov, A.N. (1969). "Methods of cytophotometric measurements of total protein in histochemical preparations of the brain," in: Modern Methods of Morphological Investigation of the Brain (in Russian), Moscow, pp. 100-102.

Geyer, G. (1960). "Zur Eiweissfarbung mit Amidoschwärtz 10B," Acta histochem., 10, 286-292.

Giacobini, E. (1956). "Histochemical demonstration of AChE activity in isolated nerve cells," Acta physiol. scand., 36, 276-290.

Giacobini, E. (1959). "Distribution and localization of cholinesterases in nerve cells," Acta physiol. scand., 45, suppl. 156.

Giacobini, E. (1961). "Localization of carbonic anhydrase in the nervous system," Science, 134, 1524-1525.

Giacobini, E. (1962). "A cytochemical study of the localization of carbonic anhydrase in the nervous system," J. Neurochem., 9, 169-177.

Giacobini, E. (1964). "Metabolic relations between glia and neurones studied in single cells," in: Morphological and Biochemical Correlates of Neural Activity, New York, pp. 15-38.

Gielen, W., and Hinzen, D. (1974). "Acetylneuraminat-Cytidyltransferase und Sialytransferase in isolierten Neuronal- und Gliazellen des Rattengehirns," Z. physiol. Chem., 355, 895-901.

Giorgi, P.P. (1971). "Preparation of neurones and glial cells from rat brain by zonal centrifugation," Exp. Cell Res., 68, 273-282.

Giuditta, A. (1974). "Functional aspects of nucleic acid metabolism in brain," in: Central Nervous System Studies on Metabolic Regulation and Function, E. Genazzani and H. Herkeni (Eds.), Springer, Berlin, pp. 70-77.

Giuditta, A. Rutigliano, B., Casola, L., and Romano, M. (1972). "Biosynthesis of RNA in two nuclear classes separated from rat cerebral cortex," Brain Res., 46, 313-328.

G"l"bov, G. (Ed.): "Spinal cord in norm and in experiment I" (Bulg.) Sofia.

G"l"bov, G. (Ed.): "Spinal cord in norm and experiment II" (Bulg.) Sofia.

Glebov, R.N. (1970). "Biochemistry of synapses," Usp. Sovr. Biol., 70, 26-40.

Glebov, R.N. (1974). "Synthesis of protein and RNA localised in nerve terminals and dependent on axoplasmic flow," in: Achievements in Neurochemistry (in Russian), Nauka, Leningrad, pp. 39-49.

Glees, P. (1955). Neuroglia Morphology and Function, Springfield.

Glees, P. (1963). "Neuere Ergebnisse auf dem Gebiet der Neuro-histologie: Nissl-Substanz, corticale Synapsen, Neuroglia und intercellularer Raum," Deutsch. Z. Nervenheilk., 164, 607-631.

Glees, P. (1973). "The neuroglial compartments at light microscopic and electron microscopic levels," in: Metabolic Compartmentation in the Brain, R. Balázs and J.E. Cremer (Eds.), MacMillan Press, London, pp. 167-184.

Glick, D. (1953). "A critical survey of current approaches in quantitative histo- and cytochemistry," Intern. Rev. Cytol., 2, 447-474.

Glushchenko, T.S., and Demin, N.N. (1971). "Activity of proteolytic enzymes of various brain regions during natural sleep, and deprivation of its paradoxical phase," Dokl. Akad. Nauk SSSR, 197, 1222-1224.

Gombos, G., Fillipowicz, W., and Vincendon, G. (1971). "Fast and slow components of S-100 protein fraction: regional distribution in bovine central nervous system," Brain Res., 26, 475-479.

Gomirato, G., and Hydén, H. (1963). "A biochemical glia error in the Parkinson Disease," Brain, 86, 773-780.

Gomori, G. (1952). Microscopic Histochemistry, Chicago.

Goodman, F.R., Weiss, G.B., and Alderdice, M.T. (1973). "On the measurement of extracellular space in slices prepared from different rat brain areas," Neuropharmacology, 12, 867-873.

Goracci, G., Blomstrand, C., Arienti, G., Hamberger, A., and
    Porcellati, G. (1973). "Base-exchange enzymic system for the
    synthesis of phospholipids in neuronal and glial cells and their
    subfractions: a possible marker for neuronal membranes," J.
    Neurochem., 20, 1167-1180.

Goracci, G., Francescangeli, E., Piccinin, G.L., Binaglia, L.,
    Woelk, H., and Porcellati, G. (1975). "The metabolism of
    labelled ethanolamine in neuronal and glial cells of the
    rabbit in vivo," J. Neurochem., 24, 1181-1186.

Goridis, C., Massarelli, R., Sensenbrenner, M., and Mandel, P.
    (1974). "Guanylcyclase in chick embryo brain cell cultures:
    evidence of neuronal localization," J. Neurochem., 23, 135-138.

Gorizontov, P.D. (1940). Importance of the Brain in Cholesterol
    Metabolism (The Role of the Neuroglia in the Pathogenesis of
    Cholesterinaemia) (in Russian), Moscow.

Govardovskii, V.I., Pevzner, L.Z., and Agroskin, L.S. (1969).
    "On the testing of the suitability of staining histochemical
    reactions for cytophotometry," Tsitologiya, 11, 655-656.

Gracheva, N.D. (1963). "Autoradiographic demonstration of DNA
    synthesis in cellular elements of the nervous system using
    ^{3}H-thymidine after general Roentgen radiation," Radiobiologiya,
    3, 81-87.

Gracheva, N.D. (1968). Autoradiography of Nucleic Acid and the
    Protein Synthesis in the Nervous System (in Russian), Leningrad.

Gracheva, N.D. (1969). "An autoradiographic study of the prolifer-
    ation of subependymal cells of rat brain," Tsitologiya, 11, 1
    1521-1527.

Grenell, R.G., Hazama, H., Kakazawa, M., and Einberg, E. (1968).
    "Effects of gravitational changes on RNA of cerebral neurons
    and glia. I. RNA changes of Deiters' cells and glia," Brain
    Res., 9, 115-125.

Grinyavichus, K., Saudargene, D., and Shalnene, A. (1966). "Two-
    wavelength microspectrophotometric method for quantitative
    determination of nucleic acids in cells," Mater. XVI Nauchn.
    Konf. Prepodavatelei Kaunasskogo Med. Inst. (in Russian),
    Kaunas, pp. 72-73.

Grossman, R.G., Whiteside, L., and Hampton, T.L. (1969). "The time
    course of evoked depolarization of cortical glial cells," Brain
    Res., 14, 401-415.

Guerra Martinieri, R.M. (1967). "Exploración de las vias meta-
    bolicas de la glucosa en cerebro de rata," Rev. Fac. Farmac.
    y Bioquim., 29 (101), 5-21.

Gurr, E. (1959). Methods of Analytical Histology and Histochemistry,
    Baltimore.

Guth, L., and Watson, P.K. (1968). "A correlated histochemical
    and quantitative sutdy on cerebral glycogen after brain injury
    in the rat," Exp. Neurol., 22, 590-602.

Gutiérrez, M. (1960). "Investigación nuclear con extracción del
    RNA en citohematología," Laboratorio, 30, 401-405.

Haas, R.J., Werner, J., and Fliedner, T.M. (1970). "Cytokinetics
    of neonatal brain cell development in rats as studied by the
    complete ^{3}H-thymidine labelling method," J. Anat., 107, 421-
    437.

Haglid, K.G. (1973). Acidic Marker Proteins for Normal Brain and
    Tumours of the Nervous System of Man, Rat, and Dog, University
    of Göteborg, Göteborg.

Haglid, K.G., and Hamberger, A. (1973). "Cellular and subcellular
    distribution of protein-bound radioactivity in the rat brain
    during maturation after incorporation of ^{3}H-leucine in vitro,"
    Brain Res., 52, 277-287.

Haglid, K., Hamberger, A., Hansson, H.-A., Hydén, H., Persson, L.,
    and Ronnback, L. (1974). "S-100 protein in synapses of the
    central nervous system," Nature, 251, 532-534.

Hakin, A.M., Moss, G., and Gollomp, S.M. (1976). "The effect of
    hypoxia on the pentose phosphate pathway in brain," J.
    Neurochem., 26, 683-688.

Hale, A.J. (1958). The Interference Microscope in Biological
    Research, Livingstone, Edinburgh and London.

Haljamäe, H., and Hamberger, A. (1971). "Potassium accumulation
    by bulk prepared neuronal and glial cells," J. Neurochem., 18,
    1903-1912.

Hallén, O. (1955). "On the cutting and thickness determination of
    microtome sections. A study in quantitative cytology," Acta
    anat., 25, suppl. 25.

Hallen, O. (1962). "Quantitative analysis of sectioned biological
    material," J. Histochem. Cytochem., 10, 98-101.

Hamberger, A. (1961). "Oxidation of tricarboxylic acid cycle intermediates by nerve cell bodies and glial cells," J. Neurochem., $\underline{8}$, 31-35.

Hamberger, A. (1963). "Difference between isolated neuronal and vascular glia with respect to respiratory activity," Acta physiol. scand., $\underline{58}$, 203.

Hamberger, A. (1971). "Amino acid uptake in neuronal and glial cell fractions from rabbit cerebral cortex," Brain Res., $\underline{31}$, 169-178.

Hamberger, A., Babich, J.A., Blomstrand, C., Hansson, H.-A., and Sellström, Å. (1975). "Evidence for differential function of neuronal and glial cells in protein metabolism and amino acid transport," J. Neurosci. Res., $\underline{1}$, 37-56.

Hamberger, A., Blomstrand, C., and Lehninger, A. (1970). "Comparative studies on mitochondria isolated from neurone-enriched and glia-enriched fractions of rabbit and beef brain," J. Cell Biol., $\underline{45}$, 221-234.

Hamberger, A., Blomstrand, C., and Yanagihara, T. (1971). "Subcellular distribution of radioactivity in neuronal and glial enriched fractions after incorporation of [^{3}H] leucine in vivo and in vitro," J. Neurochem., $\underline{18}$, 1469-1478.

Hamberger, A., and Henn, F.A. (1973). "Some aspects of the differential biochemistry and functional relationships between neurones and glia," in: Metabolic Compartmentation in the Brain, R. Balázs and J.E. Cremer (Eds.), MacMillan Press, London, pp. 305-318.

Hamberger, A., and Hydén, H. (1963). "Inverse enzymatic changes in neurones and glia during increased function and hypoxia," J. Cell Biol., $\underline{16}$, 521-525.

Hamberger, A., Hydén, H., and Lange, P.W. (1966). "Enzyme changes in neurones and glia during barbiturate sleep," Science, $\underline{151}$, 1394-1395.

Hamberger, A., Løvtrup, S. (1964). "The effect of brain dehydration on the activity of respiratory enzymes in isolated neurones, neuroglial cells and in brain mitochondria," J. Neurochem., $\underline{11}$, 687-694.

Hamberger, A., and Rinder, L. (1966). "Experimental brain concussion: the early effect of sudden increase in intracranial pressure on the succinoxidase activity of isolated neurones and glial cells from the lateral vestibular nucleus of the rabbit," J. Neuropathol. Exp. Neurol., 25, 68-75.

Hamberger, A., and Röckert, H. (1964). "Intracellular potassium in
    isolated nerve cells and glial cells," J. Neurochem., 11, 757-
    760.

Hamberger, A., and Sjostrand, J. (1966). "Respiratory enzyme
    activities in neurones and glial cells of the hypoglossal
    nucleus during nerve regeneration," Acta physiol. scand., 67,
    76-88.

Hamberger, A., and Svennerholm, L. (1971). "Composition of Ganglio-
    sides and phospholipids of neuronal and glial cell enriched
    fractions," J. Neurochem., 18, 1821-1829.

Hansson, H.-A., Hydén, H., and Rönnbäck, L. (1975). "Localization
    of S-100 protein in isolated nerve cells by immunoelectron
    microscopy," Brain Res., 93, 349-352.

Hazama, H., Ito, M., Hirano, M., and Uchimura, H. (1976). "Mono-
    amine oxidase activities in neuronal and glial fractions from
    regional areas of rat brain," J. Neurochem., 26, 417-420.

Hazama, H., and Uchimura, H. (1973). "Bulk separation of neurones
    and glia from small amounts of brain tissue," Exp. Cell Res.,
    82, 452-454.

Hebb, C. (1970). "CNS at the cellular level: identity of trans-
    mitter agents," Ann.Rev. Physiol., 32, 165-192.

Hemminki, K. (1970). "An improved method for preparing brain cell
    suspensions," FEBS Lett., 9, 290-292.

Hemminki, K. (1972). "Preparation of viable and morphologically
    intact cells from a newborn rat brain," Exp. Cell Res., 75,
    379-384.

Hemminki, K. (1973). "Incorporation of ^{3}H-leucine into insoluble
    proteins of neuronal and glial cell fractions in vitro," Acta
    Chem. scand., 27, 336-344.

Hemminki, K., Hemminki, E., and Giacobini, E. (1973). "Activity of
    enzymes related to neurotransmission in neuronal and glial
    fractions," Int. J. Neurosci., 5, 87-90.

Hemminki, K., and Holmila, E. (1971). "Characterization of
    neuronal and glial fractions separated in sucrose and ficoll
    media," Acta physiol. scand., 82, 135-142.

Hemminki, K., Huttunen, M.O., and Järnefelt, J. (1970). "Some properties of brain cell suspensions prepared by a mechanical-enzymic method," Brain Res., 23, 23-34.

Henn, F.A., Goldstein, M.N., and Hamberger, A. (1974). "Uptake of the neurotransmitter candidate glutamate by glia," Nature, 249, 663-664.

Henn, F.A., Haljamäe, H., and Hamberger, A. (1972). "Glial cell function: active control of extracellular K^+ concentration," Brain Res., 43, 437-443.

Henn, F.A., and Hamberger, A. (1971). "Glial cell function: uptake of transmitter substances," Proc. Nat. Acad. Sci. USA, 68, 2686-2690.

Hensel, H. (1973). "Neural processes in thermoregulation," Physiol. Rev., 53, 948-1017.

Herman, C.J., and Lapham, L.W. (1968). "DNA content of neurones in the cat hippocampus," Science, 160, 537.

Hertz, L. (1965). "Possible role of neuroglia: a potassium-mediated neuronal-neuroglial-neuronal impulse transmission system," Nature, 206, 1091-1094.

Hertz, L. (1966). "Neuroglial localization of potassium and sodium effects on respiration in brain," J. Neurochem., 139, 1373-1387.

Hertz, L. (1968). "Potassium effects on ion transport in brain slices," J. Neurochem., 15, 1-16.

Hertz, L., Dittmann, L., and Mandel, P. (1973). "K^+-induced stimulation of oxygen uptake in cultured cerebral glial cells," Brain Res., 60, 517-520.

Heslop, J.P. (1975). "Axonal flow and fast transport in nerves," Adv. Comp. Physiol. Biochem., 6, 75-163.

Hess, A. (1958). "The fine structure of nerve cells and fibers, neuroglia, and sheaths of the ganglion chain in the cockroach (Periplaneta americana)," J. Biophys. Biochem. Cytol., 4, 731-742.

Hess, H.H., and Thalheimer, C. (1965). "Microassay of biochemical structural components in nervous tissues. I. Extraction and partition of lipids and assay of nucleic acids," J. Neurochem., 12, 193-204.

Higashida, H., Miyake, A., Tarao, M., and Watanabe, S. (1971).

"Membrane potential changes of neuroglial cells during spreading depression in the rabbit," Brain Res., 32, 207-211.

Hild, W., and Tasaki, I. (1962). "Morphological and physiological properties of neurones and glia cells in tissue cultures," J. Neurophysiol., 25, 277-304.

Hild, W., Tasaki, I., and Chang, J.J. (1958). "Electrical responses of astrocytic glia from the mammalian central nervous system cultivated in vitro," Experientia, 14, 220-221.

Hinzen, D.H., and Gielen, W. (1973). "Distribution and composition of gangliosides and galactocerebrosides in neuronal and glial cells isolated from rabbit cerebral cortex," Fourth Intern. Meet. of the Intern. Soc. for Neurochem., Abstracts, Y. Tsukada (Ed.), Tokyo, pp. 200.

Hinzen, D.H., Müller, U., Sobotka, P., and Lang, R. (1972). "Über den O2-Verbrauch isolierter Neuron und Glia der Grosshirnrinde des Hundes nach prolongierter Ischämie," Europ. J. Physiol., 332, suppl. 90.

Hirsch, H.E. (1972). "Differential determination of hexosaminidases A and B of two forms of β-galactosidase, in the layers of the human cerebellum," J. Neurochem., 19, 1513-1517.

Hirsch, H.E., and Robins, E. (1962). "Distribution of γ-amino-butyric acid in the layers of the cerebral and cerebellar cortex. Implications for its physiological role," J. Neurochem., 9, 63-70.

Hochberg, I. "On the ultraviolet absorption and the structure of motor anterior horn cells in the spinal cord from exhausted rabbit," Acta pathol. microbiol. scand., 42, 289-301.

Hogenhuis, L.A.H., Spaulding, S.W. (1967). "Autoradiography of longterm RNA metabolism in rabbit neurones," Nature, 215, 281-283.

Hogenhuis, L.A.H., Spaulding, S.W., and Engel, W.K. (1967). "Neuronal RNA metabolism in infantile spinal muscular atrophy (Werdnig-Hoffmann's disease) studied by radioautography: a new technique in the investigation of neurological disease," J. Neuropathol. Exp. Neurol., 26, 335-341.

Holter, H. (1959). "Pinocytosis," Intern. Rev. Cytol., 8, 481-504.

Holter, H. (1961). "The Cartesian diver," Gen. Cytochem. Methods, 2, 93-129.

Holter, H., and Linderström-Lang, K. (1951). "Micromethods and their application in the study of enzyme distribution in tissues and cells," Physiol. Rev., 31, 432-448.

Horrocks, L.A. (1968). "Composition of mouse brain myelin during development," J. Neurochem., 15, 483-488.

Horstmann, E., and Meves, H. (1959). "Die Feinstruktur des molekularen Rindengraues und ihre physiologische Bedeutung," Z. Zellforsch., 49, 569-604.

Hultborn, R., and Hydén, H. (1974). "Microspectrophotometric determination of nerve cell respiration at high potassium concentration," Exp. Cell Res., 87, 346-350.

Hydén, H. (1943). "Protein metabolism in the nerve cell during growth and function," Acta physiol. scand., 6 suppl. 17.

Hydén, H. (1947). "Protein and nucleotide metabolism in the nerve cell under different functional conditions," Symp. Soc. Exp. Biol., 1, 152-162.

Hydén, H. (1955). "The chemistry of single neurones: a study with new methods," in: Biochemistry of Developing Nervous System, New York, pp. 358-370.

Hydén, H. (1959a). "Quantitative assay of compounds in isolated, fresh nerve cells and glial cells from control and stimulated animals," Nature, 184, 433-435.

Hydén, H. (1959b). "Biochemical changes in glial cells and nerve cells at varying activity," Fourth Intern. Biochem. Congr., London, 3, 64-89.

Hydén, H. (1960). "The neuron," in: The Cell, Vol. 4, New York, pp. 215-323.

Hydén, H. (1961). "Satellite cells in the nervous system," Sci. Amer., 205, 62-70.

Hydén, H. (1962). "Cytophysiological aspects of the nucleic acids and proteins of nervous tissue," in: Neurochemistry, Springfield, pp. 331-375.

Hydén, H. (1963). "The metabolic and functional interaction between the neuron and its glia," in: The Effect of Use and Disuse on Neuromuscular Functions, Prague, pp. 184-196.

Hydén, H. (1964). "Biochemical and functional interplay between neurons and glia," Rec. Adv. in Biol. Psychiatry, 6, 31-54.

Hydén, H. (1967a). "Dynamic aspects of the neuron-glia relationship," in: Neuron, Amsterdam, pp. 179-217.

Hydén, H. (1967b). "Behavior, neural function, and RNA," Progr.
Nucleic Acid. Res. Molec. Biol., 6, 187-218.

Hydén, H. (1970). "The question of a molecular basis for the
memory trace," in: Biology of Memory, New York and London,
pp. 101-119.

Hydén, H. (1972). "Macromolecules and Behaviour", Arthur Thomson
Lectures, G.B. Ansell and P.B. Bradley (Eds.), MacMillan Press,
Birmingham.

Hydén, H. (1974). "A calcium-dependent mechanism for synapse and
nerve cell membrane modulation," Proc. Nat. Acad. Sci. USA,
71, 2965-2969.

Hydén, H., and Egyhazi, E. (1962). "Nuclear RNA changes of nerve
cells during a learning experiment in rats," Proc. Nat. Acad.
Sci. USA, 48, 1366-1373.

Hydén, H., and Egyhazi, E. (1963). "Glial RNA changes during a
learning experiment in rats," Proc. Nat. Acad. Sci. USA, 49,
618-623.

Hydén, H., and Egyhazi, E. (1968). "The effect of tranylcypromine
on synthesis of macromolecules and enzyme activities in neurons
and glia," Neurology, 18, 732-736.

Hydén, H., and Lange, P. (1961). "Differences in the metabolism of
oligodendroglia and nerve cells in the vestibular area," in:
Regional Neurochemistry, Oxford, pp. 190-199.

Hydén, H., and Lange, P.W. (1962). "A kinetic study of the neuron-
glia relationship," J. Cell Biol., 13, 233-237.

Hydén, H., and Lange, P.W. (1965). "Rhythmic enzyme changes in
neurons and glia during sleep," Science, 149, 654-658.

Hydén, H., and Lange, P.W. (1966). "A genetic stimulation with
production of adenic-uracyl rich RNA in neurons and glia in
learning. The question of transfer of RNA from glia to
neurons," Naturwissensch., 53, 64-70.

Hydén, H., and Lange, P.W. (1969). "Synthesis of acidic proteins
in nerve cells during establishment of new behavior," in:
Cellular Dynamics of the Neuron, New York and London, pp. 335-
350.

Hydén, H., and Lange, P.W. (1970). "Correlation of the S-100 brain protein with behavior," Exp. Cell Res., 62, 125-132.

Hydén, H., and Lange, P.W. (1971). "Do specific biochemical correlates to learning processes exist in brain cells?," in: Biology of Memory, Budapest, pp. 69-86.

Hydén, H., Løvtrup, S., and Pigon, A. (1958). "Cytochrome oxidase and succin-oxidase activities in spinal ganglion cells and in glial capsule cells," J. Neurochem., 2, 304-311.

Hydén, H., and McEwen, B. (1966). "A glial protein specific for the nervous system," Proc. Nat. Acad. Sci. USA, 55, 354-358.

Hydén, H., and Pigon, A. (1960). "A cytophysiological study of the functional relationship between oligodendroglial cells and nerve cells of Deiters' nucleus," J. Neurochem., 6, 57-72.

Hydén, H., and Ronnback, L. (1975). "S-100 on isolated neurons and glial cells from rat rabbit and guinea pig during early postnatal development," Neurobiology, 5, 291-302.

Idoyaga-Vargas, V., Santiago, J.C., Petiet, P.D., and Sellinger, O.Z. (1972). "The early postnatal development of the neuronal lysome," J. Neurochem., 19, 2533-2544.

Ingoglia, N.A., Grafstein, B., and McEwen, B.S. (1974). "Effect of actinomycin-D on labelled material in the retina and optic tectum of goldfish after intraocular injection of tritiated RNA precursors," J. Neurochem., 23, 681-687.

Iqbal, K., and Tellez-Nagel, I. (1972). "Isolation of neurons and glial cells from normal and pathological human brains," Brain Res., 45, 296-301.

Iurato, S., Luciano, L., Franke, K., Pannese, E., and Reale, E. (1974). "Histochemical localization of acetylcholinesterase activity in the cochlear and vestibular ganglion cells," Acta oto-laryngol., 78, 28-35.

Ivanitskii, G.P., Litinskaya, L.L., and Shikhmatova, V.L. (1967). Automatic Analysis of Micro-Objects (in Russian), Moscow and Leningrad.

Ivanov, V.B. (1961). "Use of procion stains in histochemistry," Dokl. Akad. Nauk SSSR, 137, 419-421.

Iversen, L.L., and Kelly, J.S. (1975). "Uptake and metabolism of γ-amino-butyric acid by neurones and glial cells," Biochem. Pharmacol., 24, 933-938.

Jacque, C.M., Jørgensen, O.S., and Bock, E. (1974). "Quantitative studies of the brain specific antigens S-100, GFA, 14-3-2, D1 D2, D3, and C1 in quanking mouse," FEBS Lett., 49, 264-266.

Jakoubek, B. (1974). Brain Function and Macromolecular Synthesis, Pion Ltd., London.

Jakoubek, B., and Semiginovský, B. (1970). "The effect of increased functional activity on the protein metabolism of the nervous system," Intern. Rev. Neurobiol., 13, 255-288.

Jarlstedt, J. (1962). "The distribution of RNA in the cerebellum of the rabbit," Exp. Cell Res., 28, 501-506.

Jarlstedt, J., and Hamberger, A. (1971). "Patterns and labelling characteristics in neuronal and glial RNA," J. Neurochem., 18, 921-930.

Jarlstedt, J., and Hamberger, A. (1972). "Experimental alcoholism in rats. Effect of acute ethanol intoxication on the in vitro incorporation of [^{3}H]leucine into neuronal and glial proteins," J. Neurochem., 19, 2299-2306.

Jeffrey, P.L., and Austin, L. (1973). "Axoplasmic transport," Progr. Neurobiol., 2, 205-255.

Jobst, K., and Sandritter, W. (1965). "Versuche zur quantitativen Erfassung von Nucleoproteiden an Thymuslymphocyten: Cyto-photometrische Messungen im ultravioletten Licht," Acta histochem., 21, 165-171.

Johnson, D.E., and Sellinger, O.Z. (1971). "Protein synthesis in neurons and glial cells of the developing rat brain: an in vivo study," J. Neurochem., 18, 1445-1460.

Johnston, P.V., and Roots, B.I. (1970). "Neuronal and glial perikarya preparations: an appraisal of present methods," Intern. Rev. Cytol., 29, 265-280.

Johnston, P.V., and Roots, B.I. (1972). Nerve Membranes. A Study of the Biological and Chemical Aspects of Neuron-Glia Relation-ships, Pergamon Press, Oxford.

Jones, J.P., Nicholas, H.J., and Ramsey, R.B. (1975). "Rate of sterol formation by rat brain glia and neurons in vitro and in vivo," J. Neurochem., 24, 123-126.

Jones, J.P., Ramsey, R.B., Aexel, R.T., and Nicholas, H.J. (1972). "Lipid biosynthesis in neuron-enriched glial-enriched fractions of rat brain: ganglioside biosynthesis," Life Sci., Pt. 1,

$\underline{11}$, 309-315.

Jones, J.P., Ramsey, R.B., and Nicholas, H.J. (1971). "Lipid biosynthesis in neuron-enriched and glial-enriched fractions of rat brain: sterol formation," Life Sci., Pt. 2, $\underline{10}$, 997-1003.

Jongking, J.F. (1967). "The quantitative histochemistry of hypothalamus. I. Pentose shunt enzymes in the activated supraoptic nucleus of the rat," J. Histochem. Cytochem., $\underline{15}$, 394-398.

Jouvet, D., Vimont, P., Delorme, F., and Jouvet, M. (1964). "Étude de la privation de las phase paradoxale de sommeil chez le chat," Compt. Rend. Soc. Biol., $\underline{158}$, 756-759.

Jouvet, M. (1967). "Neurophysiology of the states of sleep," Physiol. Rev., $\underline{47}$, 117-177.

Kaplan, L.L. (1965). "Interactions between components of the cerebral cortex," in: Morphology, Paths and Connections in the Central Nervous System (in Russian), Moscow and Leningrad, pp. 11-18.

Karlsson, J.-O., Hamberger, A., and Henn, F. (1973). "Polypeptide composition of membranes derived from neuronal and glial cells," Biochim. biophys. acta, $\underline{298}$, 219-229.

Kato, T., and Kurokawa, M. (1967). "Isolation of cell nuclei from the mammalian cerebral cortex and their assortment on a morphological basis," J. Cell Biol., $\underline{32}$, 649-662.

Kato, T., and Kurokawa, M. (1970). "Studies on ribonucleic acid and homopolyribonucleotide formation in neuronal, glial and liver nuclei," Biochem. J., $\underline{116}$, 599-609.

Katzman, R. (1961). "Electrolyte distribution in mammalian central nervous system. Are glia high sodium cells?" Neurology, $\underline{11}$ 27-36.

Katzman, R. (1970). "Ion movement," in: Handbook of Neurochemistry, $\underline{4}$, A. Lajtha (Ed.), Plenum Press, New York, pp. 313-328.

Kaye, G.I., Donahue, S., and Pappas, G.D. (1963). "Electron microscopical evidence for the uptake of colloidal particles by Schwann cells in situ," J. Microscopie, $\underline{2}$, 605-612.

Kazakhashvili, M.P. (1974). "Effect of learning on RNA content in neurones and neuroglia of rat hippocampus," (in Russian), $\underline{16}$, 988-992.

Kerkut, G.A. (1975). "Axoplasmic transport," Comp. Biochem.
    Physiol., 51A, 701-704.

Kessler, D., Levine, L., and Fasman, G. (1968). "Some conforma-
    tion and immunological properties of a bovine brain acidic
    protein (S-100)," Biochemistry, 7, 758-764.

Khachatryan, G.S. (1963). "New data on carbohydrate and glyco-
    lipid metabolism of the brain in different functional states,"
    Third All-Union Conference on the Biochemistry of the Nervous
    System (in Russian), Yerevan, pp. 431-445.

Khachatryan, G.S. (1967). Brain Biochemistry Under Normal
    Conditions. The Hexose-monophosphate Shunt in the Brain (in
    Russian), Yerevan.

Khachatryan, G.S. (1968). "The role of alternative pathways of
    carbohydrate metabolism in biosynthetic reactions in the brain,"
    Proceedings of the Vth All-Union Conference of Neurochemistry
    (in Russian), Tbilisi, pp. 77-79.

Khaidarliu, S.Kh. (1967a). "Cell volume changes in sensory and
    motor neurones of spinal cord in various functional states of
    the nervous system," Tsitologiya, 9, 644-651.

Khaidarliu, S.Kh. (1967b). "Nucleic acid content content in small
    samples and separated cells of spinal cord in various functional
    states," Biokhimiya, 32, 677-682.

Khesin, Ya.E. (1967). Nuclear Size and Functional State of the
    Cell (in Russian), Moscow.

Khrust, Yu.R., Litinskaya, L.L., Yakovashvili, M.M., Agroskin, L.S.,
    and Kaminir, L.B. (1971). "A scanning and integrating micro-
    photometer for cytochemical investigations," Tsitologiya, 13,
    1525-1530.

Kimura, H., Naito, K., Nakagawa, K., and Kuriyama, K. (1974).
    "Activation of hexose monophosphate pathway in brain by elec-
    trical stimulation in vitro," J. Neurochem., 23, 79-84.

King, J.S., and Schwyn, R.C. (1970). "The fine structure of neuro-
    glial cells and pericytes in the primate red nucleus and
    substantia nigra," Z. Zellforsch., 106, 309-321.

Kirby, K.S. (1961). "Separation of deoxyribonucleic acid from
    ribonucleic acid without the use of enzymes," Biochim.
    biophys. acta, 47, 18-26.

Kiseli, D. (1962). Practical Microtecniques and Histochemistry, Budapest.

Klatzo, J., and Miquel, J. (1960). "Observations on pinocytosis in nervous tissue," J. Neuropathol. Exp. Neurol., 19, 475-487.

Klement'ev, B.I., Grinkevich, L.N., Glushchenko, T.S., Repin, V.S., and Pevzner, L.Z. (1975). "Production of a conditional reflex of passive avoidance and turnover of 14-3-2 and S-100 proteins in rat hippocampus," Dokl. Akad. Nauk SSSR, 221, 243-246.

Knobler, R.L., and Stempak, J.G. (1973). "Serial section analysis of myelin development in the central nervous system of albino rat: an electron microscopical study of early axonal ensheathment," Progr. Brain Res., 40, 407-423.

Koch, A., Ranck, J.B., and Newman, B.L. (1962). "Ionic content of the neuroglia," Exp. Neurol., 6, 186-200.

Koenig, E. (1965). "Synthetic mechanisms in the axons. II. RNA in myelin-free axons of the cat," J. Neurochem., 12, 357-361.

Koenig, E. (1967). "Synthetic mechanisms in the axon. IV. In vitro incorporation of H^3-precursors into axonal protein and RNA," J. Neurochem., 14, 437-446.

Koenig, E. (1972). "A molecular basis for regional differentiation of the excitable membrane," in: Macromolecules and Behavior, J. Gaito (Ed.), 2nd Ed., Appleton-Century-Crofts, New York, pp. 167-183.

Koenig, H. (1958). "Histochemical study of nucleoprotein metabolism in the central nervous system with radioisotopes," J. Histochem. Cytochem., 6, 93-94.

Koenig, H., and Barron, K.D. (1962). "Morphological and enzymic alterations in reacting glia," Acta neurol. scand., 38, suppl. 1, 72-73.

Koenig, H., and Stahlecker, H. (1951). "Further studies on the differential extraction of nucleic acids from mammalian nerve cells with perchloric acid," J. Nat. Cancer Inst., 12, 237-238.

Koenig, H., and Stahlecker, H. (1952). "Use of perchloric acid for nucleic acid histochemistry in mammalian nerve and liver cells," Proc. Soc. Exp. Biol. Med., 79, 159-163.

Kogan, A.B. (1970). "Neurochemical organisation of the processes of excitation and inhibition," in: Fifth All-Union Conference on Neurochemistry (in Russian), Tbilisi, pp. 39-51.

Kometiani, P.A., Klein, E.E., Gotsiridze, E.G., and Aleksidze, N.G. (1970). "On the proteins of the brain, sensitive to memory inhibitors and on the proteins containing metabolically active amide nitrogen," in: Fifth All-Union Conference for Neurochemistry (in Russian), Tbilisi, pp. 87-101.

Konecki, N.B., and Kozubska, M. (1961). "Zmiany dobowe w zawartości tigroidu w komórkach ruchowych rogów przednych rdzenia myszy," Fol. Morphol., 4, 279-283.

Korey, S.R. (1957). "Some characteristics of a neuroglial fraction," in: Metabolism of the Nervous System, London, pp. 87-90.

Korey, S.R. (1958). "Concentration of neuroglia cells," in: Biology of Neuroglia, Springfield, pp. 203-210.

Korey, S.R. (1960). "Glia, lipogenesis, and formation of myelin," Arch. Neurol., 2, 140-145.

Korey, S.R., and Orchen, M. (1959). "Relative respiration of neuronal and glial cells," J. Neurochem., 3, 277-285.

Korey, S.R., Orchen, M., and Brotz, M. (1958). "Studies of white matter. I. Chemical constitution and respiration of neuroglia and myelin-enriched fractions of white matter," J. Neuropathol. Exp. Neurol., 17, 430-438.

Korochkin, L.I. (1972). "A simple ultramicromethod for the separation of proteins and isoenzymes from isolated cells using polyacrylamide gel electrophesis," Tsitologiya, 14, 670-674.

Korr, H., Schultze, B., and Maurer, W. (1973). "Autoradiographic investigations of glial proliferation in the brain of adult mice. I. The DNA synthesis phase of neuroglia and endothelial cells," J. Comp. Neurol., 150, 169-175.

Kostenko, M.A., Geletyuk, V.I., and Veprintsev, B.N. (1974). "Completely isolated neurons in the mollusc, Lymnaea stagnalis. A new objective for nerve cell biology investigation," Comp. Biochem. Physiol., A49, 89-100.

Krasnov, I.B. (1967). "Quantitative cytochemical analysis of separate isolated tissue cells," Usp. Sovr. Biol., 64, 399-424.

Krech, D., Rosenzweig, M.R., and Bennett, E.L. (1960). "Effects of environmental complexity and training on brain chemistry," J. Comp. Physiol. Psychol., 53, 509-519.

Kreps, E.M., and Chenykaeva, E.Yu. (1955). "An investigation on
    the influence of hypoxia on nucleoprotein content in cells of
    rat cerebral cortex using the UV-microscope," Dokl. Akad. Nauk
    SSR, 104, 276-279.

Krichevskaya, A.A., Mogil'nitskaya, L.V., and Pevzner, L.Z. (1976).
    "Nuclear histone content of neurones and neuroglia of several
    hypothalamic regions during exposure to cold of varying dura-
    tion," Dokl. Akad. Nauk SSR, 222, 982-984.

Kristensson, K., and Olsson, Y. (1973). "Diffusion pathways and
    retrograde axonal transport of protein tracers in peripheral
    nerves," Progr. Neurogiol., 1, 85-109.

Kruglikov, R.I., Glebov, R.N., and Bazyan, A.S. (1975). "Incorpor-
    ation of radioactive label in vitro into synaptosomal proteins
    of rat hippocampus during the formation of conditional
    avoidance reflexes," Ukrainsk. Biokhim. Zh., 47, 280-283.

Kudryavina, N.A. (1954). "The course of artificial sleep, induced
    by barbamyl," Farmakol. i Toxicol., 17, (2), 22-28.

Kuffler, S.W. (1967). "Neuroglial cells: physiological properties
    and a potassium mediated effect of neuronal activity on the
    glial membrane potential," Proc. Roy. Soc. B., 168, 1-21.

Kuffler, S.W., and Nicholls, J.G. (1964). "Glia cells in the cen-
    tral nervous system of the leech; their membrane potential and
    potassium content," Arch. Exp. Pathol. a. Pharmacol., 248, 216-
    222.

Kuffler, S.W., and Nicholls, J.G. (1965). "How do materials
    exchange between blood and nerve cells in the brain?" Perspect.
    Biol. Med., 9, 69-76.

Kuffler, S.W., and Nicholls, J.G. (1966). "The physiology of
    neuroglial cells," Ergebn. Physiol., 57, 1-90.

Kuffler, S.W., Nicholls, J.G., and Orkand, R.K. (1966). "Physio-
    logical properties of glial cells in the central nervous
    system of amphibia," J. Neurophysiol., 29, 768-787.

Kuffler, S.W., and Potter, D.D. (1964). "Glia in the leech central
    nervous system: physiological properties and neuron-glia
    relationship," J. Neurophysiol., 27, 290-320.

Kuhlenbeck, H. (1970). The Central Nervous System of Vertebrates.
    Vol. 3, Part I. Structural Elements: Biology of Nervous
    Tissue, S. Karger, Basel.

Kuhlmann, D. (1969). "Bestimmung des DNS-Gehaltes in Sellkernen des Nervengewebes von Helix pomatta L. und Planorbarius corneus L. (Stylommatophora und Basommatophora, Gastropoda)," Experientia, 25, 848-849.

Kulenkampff, H., and Wüstenfeld, E. (1954). "Funktionsbedingte Veränderungen der Kerngrösse von Gliazellen in Grau des Rückenmarkes der weissen Maus," Z. Anat. u. Entwickl., 118, 97-101.

Kuntz, A., and Sulkin, N.M. (1947). "The neuroglia in the autonomic ganglia: cytologic structure and reactions to stimulation," J. Comp. Neurol., 86, 467-477.

Kurokawa, M., Kato, T., and Inamura, H. (1966). "Unequal distribution of ATP: NMN adenyltransferase activity among neuronal and glial cell nuclie," Proc. Japan Acad., 42, 1217-1222.

Kushch, A.A., and Yarygin, V.N. (1965). "Polyploid uninucleate and binucleate neurones of rabbit superior cervical ganglion," Tsitologiya, 7, 228-232.

Kuz'min, S.M., Timkin, V.N., Bezruchko, S.M., and Adzhimolaev, T.A. (1975). "Dynamic aspects of RNA metabolism in mollusc ganglia during electrical stimulation," Zh. Evol. Biokhim. i Fiziol., 11, 274-281.

Kuznetsova, A.F., and Brodskii, V.Ya. (1968). "Interferometric determination of the thickness of histological sections," Tsitologiya, 10, 392-402.

Laborit, H. (1964). "Les rapports neurono-névrogliques. Interpretation sur des bases experimentales de leurs roles en neurophysiologie et dans le mécanisme d'action des drogues psychotropes," Presse méd., 72, 1-5.

Laborit, H. (1965). Les régulations métaboliques. Aspects théorique expérimental pharmacologique et therapeutique. Paris.

Laborit, H., and Laborit, G. (1965). "Le sommeil physiologique," Agressologie, 6, 639-651.

Labourdette, G., and Marks, A. (1975). "Synthesis of S-100 protein in monolayer cultures of rat glial cells," Eur. J. Biochem., 58, 73-79.

Lajtha, A. (1957). "The development of the blood-brain barrier," J. Neurochem., 1, 216-227.

Lajtha, A., Latzkovits, L., and Toth, J. (1976). "Comparison of turnover rates of proteins of the brain, liver and kidney in mouse in vivo following long term labeling," Biochim. et biophys. acta, 425, 511-520.

Lapham, L.W. (1962). "Cytological and cytochemical studies of neuroglia. I. A study of the problem of amitosis in reactive protoplasmic astrocytes," Amer. J. Pathol., 41, 1-22.

Lapham, L.W. (1968). "Tetraploid content of Purkinje neurons of human cerebellar cortex," Science, 159, 310-312.

Lapham, L.W., and Johnstone, M.A. (1963). "Cytologic and cyto-chemical studies of neuroglia. II. The occurence of two DNA classes among glial nuclei in the Purkinje cell layer of normal adult human cerebellar cortex," Arch. Neurol., 9, 194-202.

Lapham, L.W., and Johnstone, M.A. (1964). "Cytologic and cyto-chemical studies of neuroglia. III. The DNA content of giant fibrous astrocytes with implications concerning the nature of these cells," J. Neuropathol. a. Exp. Neurol., 23, 419-430.

Lasansky, A. (1965). "Functional implication of structural findings in retinal glial cells," Progr. Brain Res., 15, 48-72.

Lasek, R.J., Gainer, H., and Przybylski, R.J. (1974). "Transfer of newly synthesized proteins from Schwann cells to the squid giant axon," Proc. Nat. Acad. Sci. USA, 71, 1188-1192.

Lazarewicz, J.W., Haljamäe, H., and Hamberber, A. (1974). "Calcium metabolism in isolated brain cells in subcellular fractions," J. Neurochem., 22, 33-45.

Lebovitz, R.M. (1970). "A theoretical examination of ionic inter-actions between neural and non-neural membranes," Biophys. J., 10, 423-444.

Lees, M.B., and Shein, H.M. (1970). "Sodium and potassium content of normal and neoplastic rodent astrocytes in cell culture," Brain Res., 23, 280-283.

Lehrer, G.M. (1973). "Distribution of sodium and potassium in the nucleus and cytoplasm of neurons," in: Metabolic Compartmen-tation in the Brain, R. Balázs and J.E. Cremer (Eds.) MacMillan Press, London, pp. 259-264.

Lentz, R.D., and Lapham, L.W. (1969). "A quantitative cytochemical study of the DNA content of neurons of rat cerebellar cortex," J. Neurochem., 16, 379-384.

Lentz, R.D., and Lapham, L.W. (1970). "Postnatal development of tetraploid DNA content in rat Purkinje cells: a quantitative cytochemical study," J. Neuropathol. a. Exp. Neurol., 29, 43-56.

Lesansky, A. (1965). "Functional implication of structural findings in retinal glial cells," Progr. Brain Res., 15, 48-72.

Lewin, E., and Hess, H.H. (1965). "Intralaminar distribution of cerebrosides in human frontal cortex," J. Neurochem., 12, 213-220.

Lierse, W. (1968). "Die Hirncapillaren und ihre Glia," Acta neuropathol., suppl. 4, 40-52.

Lightbody, J., Pfeiffer, S.E., Kornblith, P.L., and Herschman, H. (1970). "Biochemically differentiated clonal human glial cells in tissue culture," J. Neurobiol., 1, 411-417.

Linderström-Lang, K. (1937). "Principle of the Cartesian diver applied to gasometric technique," Nature, 140, 108.

Lipp, W., and Gubisch, W. (1961). "Ein einfaches Verfahren zur Bestimmung von Schnittund Schichtdicken in histologischen präparaten (Mikroskipie des senkrecht stehenden Schnitten)," Acta histochem., 11, 268-275.

Lisý, V., and Lodin, Z. (1973). "Incorporation of radioactive leucine into neuronal and glial proteins during postnatal development," Neurobiology, 3, 320-326.

Litinskaya, L.L., Pevzner, L.Z., and Khrust, Yu.R. (1976). "Scanning, integrating cytospectrophotometry of closely adjoining microscopic objects," Tsitologiya, 18, 470-475.

Lodin, Z. (1964). "Problémy kvantitativní cytochemie," Českosl. fysiol., 13, 126-145.

Lodin, Z., Booher, J., and Kasten, F.H. (1970). "Phase-contrast cinematographic study of dissociated neurons from embryonic chick dorsal root ganglia cultured in the Rose chamber," Exp. Cell Res., 60, 27-39.

Lodin, Z., Faltin, J., Kazakashvili, M., Hartman, J., and Müller, J. (1968). "Metabolism of nucleic acids and proteins at various functional states," in: Macromolecules and the Function of the Neuron, Amsterdam, pp. 373-383.

Lord, K.A., Gregory, G.E., and Burt, P.E. (1967). "The penetration of acetylcholine into the central nervous system of the cockroach Periplaneta americana L.," J. Exp. Biol., 46, 153-159.

Løvtrup-Rein, H.   (1970a).   "Protein synthesis in isolated nuclei
     of nerve and glial cells from rat brain," Brain Res., 19, 433-
     444.

Løvtrup-Rein, H.   (1970b).   "Synthesis of nuclear RNA in nerve and
     glial cells," J. Neurochem., 17, 853-863.

Løvtrup-Rein, H., and Grahn, B.   (1970).   "Newly synthesized RNA in
     nuclei isolated from nerve and glial cells," J. Neurochem., 17,
     845-852.

Løvtrup-Rein, H., and Grahn, B.   (1974).   "Polysomes and polysomal
     RNA from nerve and glial cell fractions," Brain Res., 72, 123-
     136.

Løvtrup-Rein, H., and McEwen, B.S.   (1966).   "Isolation and fraction-
     ation of rat brain nuclei," J. Cell Biol., 30, 405-416.

Lowry, O.H.   (1941).   "A quartz fiber balance," J. Biol. Chem., 140,
     183-190.

Lowry, O.H.   (1953).   "Quantitative histochemistry of brain; histolo-
     gical sampling," J. Histochem.   Cytochem., 1, 420-428.

Lowry, O.H.   (1955).   "A study of the nervous system with quantita-
     tive cytochemical methods," in: Biochemistry of the Developing
     Nervous System, New York, pp.   350-357.

Lowry, O.H.   (1957).   "Enzyme concentrations in individual cell
     bodies," in: Metabolism of the Nervous System, London, pp.   323
     -328.

Lowry, O.H.   (1962).   "The chemistry of the individual neuron," Bull.
     N.Y. Acad. Med., 38, 789-798.

Lowry, O.H.   (1967).   "Chemical investigation of individual nerve
     cells," in: Biochemistry and Function in the Central Nervous
     System (in Russian), Nauka, Leningrad, pp.   16-20.

Lowry, O.H., and Bessey, O.A.   (1946).   "The adaptation of the
     Beckman spectrophotometer to measurements on minute quantities
     of biological materials," J. Biol. Chem., 163, 633-639.

Lowry, O.H., Roberts, N.R., and Chang, M.-L.W.   (1956).   "The
     analysis of single cells," J. Biol. Chem., 222, 97-107.

Lowry, O.H., Roberts, N.R., Leiner, K.Y., Wu, M.-L., Farr, A.L., and Albers, R.W. (1954). "The quantitative histochemistry of brain. II. Ammon's horn," J. Biol. Chem., 207, 39-49.

Lowry, O.H., Rosebrough, N.J., Farr, A.L., and Randall, R.J. (1951). "Protein measurement with Folin phenol reagent," J. Biol. Chem., 193, 265-275.

Lubińska, L. (1964). "Axoplasmic streaming in regenerating and in normal nerve fibres," Progr. Brain Res., 13, 1-74.

Lubińska, L. (1971). "Acetylocholinesterase in mammalian peripheral nerves and characteristics of its migration," Acta neuropathol., 18, Suppl., 136-143.

Lumsden, C.E. (1955). "The cytology and cell physiology of the neuroglia and of the connective tissue in the brain with reference to the blood-brain barrier," Excerpta med., 8, 832-834.

Luse, S.A. (1956a). "Formation of myelin in the central nervous system of mice and rats, as studied with the electron microscope," J. Biophys. Biochem. Cytol., 2, 777-784.

Luse, S.A. (1956b). "Electron microscopic observations of the central nervous system," J. Biophys. Biochem. Cytol., 2, 531-542.

Luse, S.A. (1960). "The ultrastructure of normal and abnormal oligodendroglia," Anat. Rec., 138, 461-492.

Manina, A.A. (1966). "Neuronal ultrastructure," in: Functional Mechanisms of Central Neurones (in Russian), Moscow/Leningrad, pp. 34-37.

Mann, D.M.A., and Yates, P.O. (1973). "Polyploidy in the human nervous system. Part I. The DNA content of neurones and glia of the cerebellum," J. Neurol. Sci., 18, 183-196.

Manocha, S.L., and Bourne, G.H. (1968). "Histochemical mapping of lactate dehydrogenase and monoamine oxidase in the medulla oblongata and cerebellum of squirrel monkey (Saimiri sciureus)," J. Neurochem., 15, 1033-1040.

Manolov, S., and Davidov, M. (1976). Histochemistry of Cholinesterase in Nerve Tissue (in Bulgarian), Academy of Sciences Press, Sofia.

Margolis, F.L. (1971). "Search for S-100 protein variants in inbred strains and neurological mutants of the mouse," in:

Third Intern. Meet. of the Intern. Soc. for Neurochemistry, Budapest, pp. 25.

Margolis, R.K., Heller, A., and Moore, R.Y. (1968). "Effects of changes in cellular composition following neuronal degeneration on amino acids in brain," Brain Res., 11, 19-31.

Margolis, R.U., and Margolis, R.K. (1974). "Distribution and metabolism of mucopolysaccharides and glycoproteins in neuronal perikarya, astrocytes, and oligodendroglia," Biochemistry, 13, 2849-2852.

Mashanskii, V.F., Zaguskin, S.L., and Fedorenko, G.M. (1974). "Histochemical and electron microscopic study of the neurone-glia relationships in the stretch receptor of the fresh water crayfish," (in Russian), Tsitologiya 16, 770-772.

Mats, V.N., Larina, V.N., and Geinisman, Yu.Ya. (1970). "Comparative assessment of size changes of neuronal and glial cells under conditions of increase functioning of the nervous system," Tsitologiya, 12, 737-744.

Mazza, J.P., Hanker, J.S., and Dixon, A.D. (1973). "Ultrastructural localization of cholinesterase activity in the trigeminal ganglion of the rat," J. Anat., 115, 65-78.

Mchedlishvili, G.I. (1968). Function of Vascular Mechanisms in the Brain. Their Function in Regulation and Pathology of Brain Circulation (in Russian), Leningrad.

Mchedlishvili, G.I. (Ed.) (1969). Correlation of Blood Supply with Metabolism and Function (in Russian), Tbilisi.

Medzihradsky, F., Nandhasri, P.S., Idoyaga-Vargas, V., and Sellinger, O.Z. (1971). "A comparison of the ATPase activity of the glial cell fraction and the neuronal perikaryal fraction isolated in bulk from rat cerebral cortex," J. Neurochem., 18, 1599-1603.

Medzihradsky, F., Sellinger, O.Z., Nandhasri, P.S., and Santiago, J.C. (1972). "ATPase activity in glial cells and in neuronal perikarya of rat cerebral cortex during early postnatal development," J. Neurochem., 19, 543-545.

Meerson, F.Z. (1967). Plastic Maintenance of Function in the Organism (in Russian), Moscow.

Mellors, R.C. (Ed.) (1955). Analytical Cytology. Methods for Studying Cellular Form and Function. New York-Toronto-London.

Mendelsohn, M.L. (1958). "The two-wavelength method of microspectrophotometry. II. A set of tables to facilitate calculations," J. Biophys. Biochem. Cytol., 4, 415-424.

Mendelsohn, M.L., and Richards, B.M. (1958). "A comparison of scanning and two-wavelength microspectrophotometry," J. Biophys. Biochem. Cytol., 4, 707-709.

Menzel, E. (1959). "Optische Verfahren der Dickenmessung. II. Abbildende Interferometer mit homogener Vergleichwelle," Arch. techn. Messen., 281, 109-112.

Menzies, D.W. (1963). "Red-blue staining of hydrolised nucleic acids in paraffin sections," Stain Technol., 38, 157-160.

Merriam, R.W. (1957). "Determination of section thickness in quantitative microspectrophotometry," Lab. Invest., 6, 28-43.

Merritt, J.H., and Sulkowski, T.S. (1970). "Rhythmicity of RNA polymerase activity and RNA levels in nuclei of rat cerebral cortex," J. Neurochem., 17, 1327-1328.

Messier, B., and Leblond, C.P. (1960). "Cell proliferation and migration as revealed by radioautography after injection of thymidine-H^3 into male rats and mice," Amer. J. Anat., 106, 247-285.

Mezentsev, A.N., and Messinova, O.V. (1971). "Axonal transport and biosynthetic processes in the axon," Usp. Sovr. Biol., 72, 62-76.

Michetti, F., DeRenzis, G., Donato, R., and Miani, N. (1976). "Brain-specific effect of the S-100 protein on the RNA polymerase I activity in isolated nuclei," Brain Res., 105, 372-375.

Minchin, M.C.W. (1975). "Factors influencing the efflux of [^{3}H] gamma-aminobutyric acid from satellite glial cells in rat sensory ganglia," J. Neurochem., 24, 571-577.

Minchin, M.C.W., and Beart, P.M. (1975). "Compartmentation of amino acid metabolism in the rat dorsal root ganglion; a metabolic and autoradiographic study," Brain Res., 83, 437-449.

Minchin, M.C.W., and Iversen, L.L. (1974). "Release of [^{3}H] gamma-aminobutyric acid from glial cells in rat dorsal root ganglia," J. Neurochem., 23, 533-540.

Mizuno, N., and Okamoto, M. (1964). "Some observations on the astrocytes cultures in vitro: 'Flattened astrocyte' and pinocytosis," Arch. histol. Japon., 24, 347-367.

Monard, D., Solomon, F., Rentsch, M., and Gysin, R. (1973). "Glia-induced morphological differentiation in neuroblastoma cells (glial-neuronal interactions)," Proc. Nat. Acad. Sci. USA, 70, 1894-1897.

Monard, D., Stockel, K., Goodman, R., and Thoenen, H. (1975). "Distinction between nerve growth factor and glial factor," Nature, 258, 444-445.

Moore, B.W. (1969). "Acidic proteins," in: Handbook of Neuro-chemistry, Vol. I., New York, pp. 93-99.

Moore, B. (1972). "Chemistry and biology of two proteins, S-100 and 14-3-2, specific to the nervous system," Int. Rev. Neuro-biol., 15, 215-225.

Moore, B.W. (1973). "Brain-specific proteins," in: Proteins of the Nervous System, D.J. Schneider et al (Eds.), Raven Press, New York, pp. 1-12.

Moore, B.W. (1975). "Brain-specific proteins: S-100 protein, 14-3-2 protein, and glial fibrillary protein," Adv. Neurochem., 1, New York/London, 137-155.

Moore, B.W., Perez, V.J., and Gehring, M. (1968). "Assay and regional distribution of a soluble protein characteristic of the nervous system," J. Neurochem., 15, 265-272.

Mrosovsky, N. (1971). Hibernation and Hypothalamus, Appleton-Century-Crofts, New York.

Munro, H.N., and Fleck, A. (1966). "Recent developments in the measurement of nucleic acids in biological materials," Analyst, 91, 78-88.

Murai, S. (1973). "Experimental studies on the cerebral metabolism in head injury. With special reference to respiratory function of mitochondria and carbonic anhydrase activity of rabbits cerebral cortex fractions containing isolated, metabolically active neuronal and glial cells," Nihon Univ. Med. J., 32, 583-605.

Murone, I., and Ogata, K. (1973). "Studies on creatine kinase of skeletal muscle and brain with special reference to subcellular distribution and isozymes," J. Biochem., 74, 41-48.

Museridze, D.P., and Svanidze, I.K. (1975). "Quantitative analysis of DNA in neuronal and glial cells of cat cerebral cortex," Izd. Akad. Nauk Gruz. SSSR ser. Biol., 1, 494-498.

McEwen, B.S., and Hydén, H. (1966). "A study of specific brain proteins on the semi-micro scale," J. Neurochem., 13, 823-833.

McIlwain, H. (1956). "Anaerobic glycolysis of cerebral tissues and a second, electrically-induced, metabolic defect," Biochem. J., 63, 257-263.

McIlwain, H. (1962). "New factors connecting metabolic and electrical events in cerebral tissue," Res. Publ. Assoc. Nerv. Ment. Dis., 15, 43-56.

McIlwain, H. (1966). Biochemistry and the Central Nervous System, Third Edition, London.

McIlwain, H., and Bachelard, H.S. (1971). Biochemistry and the Central Nervous System, Fourth Edition, Edinburgh and London.

Nagata, Y., Mikoshiba, K., and Tsukada, Y. (1974). "Neuronal cell body enriched and glial cell enriched fractions from young and adult rat brains: preparation and morphological and biochemical properties," J. Neurochem., 22, 493-503.

Nagata, Y., Mikoshiba, K., and Tsukada, Y. (1976). "Neurochemical studies of bulk-separated neuronal cell body enriched and glial cell enriched fractions from young and adult rat brains," Asian Med. J., 19, 13-43.

Nakai, J. (Ed.) (1963). Morphology of Neuroglia, Tokyo.

Naora, H. (1961). "Microspectrophotometry and cytochemical analysis of nucleic acids," Science, 114, 279-280.

Nasonov, D.N. (1956). Local Reaction of Protoplasm and Spreading Excitation (in Russian), Moscow/Leningrad.

Nasonov, D.N. (1963). Some Questions of Cell Morphology and Physiology (in Russian), Izbr. Brudy, Moscow/Leningrad.

Nasonov, D.N., and Aleksandrov, V.Ya. (1940). Reaction of Organic Substance to External Influences (in Russian), Moscow/Leningrad.

Nayeemunnisa, and Rao, K.P. (1975). "Physiology of low temperature acclimation: triggering of biochemical changes by hormone-like action of injected nerve tissue extracts," Indian J. Biochem. Biophys., 12, 194-196.

Nicholls, J.C., and Kuffler, S.W. (1964). "Extracellular space as a pathway for exchange between blood and neurons in the central nervous system of the leech: ionic composition of glial cells and neurons," J. Neurophysiol., 27, 645-671.

Nicholls, J.G., and Kuffler, S.W. (1965). "Na and K content of glial cells and neurons determined by flame photometry in the central nervous system of the leech," J. Neurophysiol., 28, 519-525.

Nicholls, J.G., and Wolfe, D.E. (1967). "Distribution of C^{14}-labeled sucrose, inulin, and dextran in extracellular spaces and in cells of the leech central nervous system," J. Neurophysiol., 30, 1574-1592.

Niemierko, S., and Oderfeld-Nowak, B. (1968). "Injury induced synthesis of nucleic acids in peripheral nerve," in: Macromolecules and the Function of the Neuron, Amsterdam, pp. 148-155.

Nikulesku, I.T. (1963). (Ed.) Pathomorphology of the Nervous System, Bucharest.

Norton, W.T., and Poduslo, S.E. (1970). "Neuronal soma and whole neuroglia of rat brain: a new isolation technique," Science, 167, 1144-1145.

Norton, W.T., and Poduslo, S.E. (1971). "Neuronal perikarya and astroglia of rat brain: chemical composition during myelination," J. Lipid Res., 12, 84-90.

Norton, W.T., and Poduslo, S.E. (1973). "Myelination in rat brain: changes in myelin composition during brain maturation," J. Neurochem., 21, 759-773.

Obata, K., Otsuka, M., and Tanaka, Y. (1970). "Determination of gamma-aminobutyric acid in single nerve cells of cat central nervous system," J. Neurochem., 17, 697-698.

Obchinnikova, L.P., and Selivanova, G.B. (1964). "Validation of the use of gallocyanin chrome alum as a stain for photometric determination of cytoplasmic RNA," Tsitologiya, 6, 387-388.

O'Brien, J.S., and Sampson, E.L. (1965). "Lipid composition of the normal human brain: gray matter, white matter, and myelin," J. Lipid Res., 6, 537-544.

Ochs, S. (1966). "Axoplasmic flow in neurones," in: Macromolecules and Behaviour, New York, pp. 20-39.

Ochs, S. (1972). "Axoplasmic flow - the fast transport system in mammalia nerve fibers," in: Macromolecules and Behavior, New York, pp. 147-166.

Ochs, S. (1974). "Systems of material transport in nerve fibers (axoplasmic transport) related to nerve function and trophic control," Ann. New York Acad. Sci,, <u>228</u>, 202-223.

Ochs, S. (1975). "Retention and redistribution of proteins in mammalian nerve fibers by axoplasmic transport," J. Physiol., <u>253</u>, 459-475.

Oderfeld-Nowak, B., and Niemierko, S. (1969). "Synthesis of nucleic acids in the Schwann cells as the early cellular response to nerve injury," J. Neurochem., <u>16</u>, 235-248.

Ogur, M., and Rosen, G.U. (1949). "Extraction and estimation of desoxypentose nucleic acid (DNA) and pentose nucleic (PNA) from plant tissues," Feder. Proc., <u>8</u>, 234.

Ogur, M., and Rosen, G. (1950). "The nucleic acids of plant tissues. I. The extraction and estimation of desoxypentose nucleic acid and pentose nucleic acid," Arch. Biochem., <u>25</u>, 262-276.

Oksche, A. (1968). "Die pränatale und vergleichende Entwicklungs-geschichte der Neuroglia," Acta neuropathol. suppl., <u>4</u>, 4-19.

Oram, V. "The cytoplasmic basophilic substance of the exocrine pancreatic cells," Acta anat., suppl., <u>23</u>, 34-41.

Orbeli, L.A. (1938). Lectures on the Physiology of the Nervous System (in Russian), Leningrad.

Orkand, P.M., Bracho, R., and Orkand, R.K. (1973). "Glial metabo-lism: alteration by potassium levels comparable to those during neural activity," Brain Res., <u>55</u>, 467-471.

Orkand, R.K., Nicholls, J.G., and Kuffler, S.W. (1966). "Effect of nerve impulses on the membrane potential of glial cells in the central nervous system of amphibia," J. Neurophysiol., <u>29</u>, 788-806.

Ornstein, L. (1952). "The distributional error in microspectro-photometry," Lab. Invest., <u>1</u>, 250-262.

Otsuka, M., Obata, K., Miyata, Y., and Tanaka, Y. (1971). "Measure-ment of γ-aminobutyric acid in isolated nerve cells of cat central nervous system," J. Neurochem., <u>18</u>, 287-295.

Ovchinnikova, L.P., and Selivanova, G.V. (1964). "Testing the use of the chrome-gallocyanin stain for photometric measurement of RNA in cytoplasm," Tsitologiya, <u>6</u>, 387-388.

Packman, P.M., Blomstrand, C., and Hamberger, A. (1971). "Disc electrophoretic separation of proteins in neuronal, glial and subcellular fractions from cerebral cortex," J. Neurochem., <u>18</u>, 479-487.

Pakkenberg, H. (1958). "On cytoplasmic basophilia in the nerve cells of the cerebral cortex. I. Methodology," Acta anat., <u>35</u>, 85-106.

Pakkenberg, H. (1962). "Gallocyanin-chrome alum staining: a quantitative evaluation," J. Histochem. Cytochem., <u>10</u>, 367.

Palay, S.L., and Chan-Palay, V. (1973). "The structural heterogeneity of central nervous tissue," in: Metabolic Compartmentation in the Brain, R. Balázs and J.E. Cremer (Eds.), MacMillan Press, London, pp. 187-207.

Palladin, A.V. (1959). "Biochemical characterization of functionally distinct regions of the nervous system," Ukr. Biokhim. Zhurn., <u>31</u>, 765-779.

Palladin, A.V. (1963). "Protein metabolism in the nervous system," in: Third All-Union Conference on the Biochemistry of the Nervous System (in Russian), Yerevan, pp. 9-23.

Palladin, A.V. (1965). Questions on the Biochemistry of the Nervous System (in Russian), Kiev.

Palladin, A.V., Belik, Ya.V., and Polyakova, N.M. (1976). Proteins of the Brain and Their Metabolism, Plenum Press, New York.

Pannese, E. (1964). "Number and structure of perisomatic satellite cells of spinal ganglia under normal conditions or during axon regeneration and neuronal hypertrophy," Z. Zellforsch., <u>63</u>, 568-592.

Pappius, H.M. (1965). "The distribution of water in brain tissues swollen in vitro and in vivo," Progr. Brain Res., <u>15</u>, 135-154.

Patau, K. (1952). "Absorption microphotometry of irregular-shaped objects," Chromosoma, <u>5</u>, 341-362.

Patel, K.K., Hartmann, J.F., and Cohen, M.M. (1971). "Ultrastructural estimation of relative volume of extracellular space in brain slices," J. Neurol. Sci., <u>12</u>, 275-288.

Pavlin, R. (1969). "The hydrolysis of acetylcholine in single nerve cells isolated from the guinea pig - 240 days old," Second Intern. Meet. of Intern. Soc. for Neurochemistry, Milano, pp. 314.

Pavlin, R. (1971). "Holinesteraze v putamenu in retikularni formaciji človeka," Zdravstv. vestu., 40, 381-383.

Penfield, W. (1932). "Neuroglia: normal and pathological," in: Cytology and the Cellular Pathology of the Nervous System, Vol. II, New York, pp. 421-479.

Peters, A., Palay, S.L., and Webster, H. de F. (1970). The Fine Structure of the Nervous System, Harper and Row, New York.

Peterson, N.A., and Chaikoff, J.L. (1963). "Uptake of intra-venously-injected $4\text{-}C^{14}$-cortisol by adult rat brain," J. Neurochem., 10, 17-23.

Pevzner, L.Z. (1959). "Quantitative cytochemical studies on the influence of circulatory hypoxia on nucleic acid content of cells of cerebral cortex," Nauchn. Soobshch. Inst. Fiziol. im I.P. Pavlova A.N. SSSR, 2, 198-201.

Pevzner, L.Z. (1960). "UV cytophotometry of various functional regions of cerebral cortex," Tsitologiya, 2, 179-186.

Pevzner, L.Z. (1962a). "Content of cytoplasmic RNA in neurones of various cell layers of different functional regions of the cerebral cortex," Biokhimiya, 27, 663-669.

Pevzner, L.Z. (1962b). "Effect of hypoxia on the cytoplasmic RNA content of different cell layers in various functional regions of cerebral cortex," Dokl. Akad. Nauk SSSR, 145, 447-449.

Pevzner, L.Z. (1963a). "Content of cytoplasmic RNA in neurones of various cell layers of cerebral cortex under normal conditions and during hypoxia," Third All-Union Conference on the Biochemistry of the Nervous System (in Russian), Yerevan, pp. 327-338.

Pevzner, L.Z. (1963b). "The content of DNA in the cells of human brain tumours studied by means of ultraviolet microspectro-photometry," Folia Histochem. Cytochem., 1 (suppl. 1), 196-197.

Pevzner, L.Z. (1963c). "Nucleic acid content in nerve cells under various functional conditions (data from quantitative cyto-chemical studies)," Ukr. Biokhim. Zhurn., 35, 448-477.

Pevzner, L.Z. (1964). "Effect of injection of adrenaline on nucleic acid content in neuronal and glial cells of the sympathetic ganglion," Dokl. Akad. Nauk SSSR, 156, 1213-1214.

Pevzner, L.Z. (1965a). "Histochemical methods in electron micro-scopy," Arkhiv anat. Gistol. i Embriol., 48, 91-106.

Pevzner, L.Z.   (1965b)   "On the constancy of nuclear DNA content in
    neurones and macroglia," in: Morphology of Paths and Connec-
    tions in the Central Nervous System (in Russian), Moscow/
    Leningrad, pp. 135-140.

Pevzner, L.Z.   (1965c). "On the relative volumes of nucleus and
    cytoplasm in neurones of the autonomic ganglia of the rat,"
    Byull. Eksp. Biol. i Med., 59, 102-105.

Pevzner, L.Z.   (1965d). "Dry mass of nucleus and cytoplasm of
    neurones of the cat vegetative ganglia during rest and with
    excitation," Tsitologiya, 7, 253-257.

Pevzner, L.Z.   (1965e). "Topochemical aspects of nucleic acid and
    protein metabolism within the neuron-neuroglia unit of the
    superior cervical ganglion," J. Neurochem., 12, 993-1002.

Pevzner, L.Z.   (1966a). "Nucleic acid changes during behavioural
    events," in: Macromolecules and Behavior, New York, pp. 43-
    70.

Pevzner, L.Z.   (1966b). "Metabolism in the neurone," in: Functional
    mechanisms of Central Neurones (in Russian), Moscow/Leningrad,
    pp. 7-33.

Pevzner, L.Z.   (1966c). "Nucleic acid and protein content of glial
    cells in superior cervical ganglion during excitation," Ukr.
    Biokhim. Zhurn., 38, 123-127.

Pevzner, L.Z.   (1967). "The neuroglia as the site of action of
    trophic influences on the nervous system," in: Biochemistry
    and Function in the Nervous System (in Russian), Leningrad,
    pp. 49-59.

Pevzner, L.Z.   (1968). "Nucleic acids in the neuron-neuroglia unit
    in various functional states of the nervous system," in:
    Macromolecules and the Function of the Neurone, Amsterdam,
    pp. 353-364.

Pevzner, L.Z.   (1969a). "Biochemical studies on the neuroglia,"
    Vopr. Med. Khimii, 15, 211-223.

Pevzner, L.Z.   (1969b). "Functional-biochemical characteristics of
    neuroglia," Usp. Sovr. Biol., 68, 363-383.

Pevzner, L.Z.   (1969c). "Metabolism of nucleic acid and protein in
    neurones and neuroglia under various functional states of the
    nervous system," Proceedings of the IVth All-Union Conference
    on the Biochemistry of the Nervice System (in Russian), Tartu,
    pp. 109-121.

Pevzner, L.Z. (1969d). "Nucleic acids in neurones and neuroglia of various regions of the nervous system under the influence of antimetabolites in vivo," Tsitologiya, 11, 856-864.

Pevzner, L.Z. (1969e). "In vivo effects of some hormones and anti-metabolites on nucleic acid content in the neurones neuroglia and ependymal cells," Second International Meeting for the International Society for Neurochemistry, Milan, pp. 316-317.

Pevzner, L.Z. (1970a). "Quantitative changes in nucleic acids and protein of neurones and neuroglia with shifts in functional state," Fifth All-Union Conference on Neurochemistry (in Russian), Tbilisi, pp. 129-146.

Pevzner, L.Z. (1970b). "Neurone-neuroglia as a self-regulating system with negative feedback," Proceedings of the IVth All-Union Conference on Neurocybernetics (in Russian), Rostov, p. 108.

Pevzner, L.Z. (1971). "Topochemical aspects of nucleic acid and protein metabolism within the neurone-neuroglia unit of the spinal cord anterior horn," J. Neurochem., 18, 895-907.

Pevzner, L.Z. (1972a). "Changes in content of cytoplasmic RNA in ependymal cells with hormonal and hypoxic influences," Fiziol. Zhurn. SSSR, 58, 1275-1280.

Pevzner, L.Z. (1972b). "Macromolecular changes within neuron-neuroglia unit during behavioral events," in: Macromolecules and Behavior," 2nd Edition, New York, pp. 335-358.

Pevzner, L.Z. (1972c). "Nucleic acid and protein metabolism in the nervous system," in: Aims in Science and Technology. Physiology of Man and Animals (in Russian), Vol. 10, Moscow, pp. 124-199.

Pevzner, L.Z. (1974). "Analysis of the neurone-neuroglia unit using quantitative histochemistry," in: Achievements in Neurochemistry (in Russian), Nauka, Leningrad, pp. 140-153.

Pevzner, L.Z. (1976). "Cytophotometric comparison of changes in RNA and protein content in perineuronal oligodendroglia and ependymal cells of spinal cord under various experimental conditions," Tsitologiya, 18, 499-502.

Pevzner, L.Z., and Khaidarliu, S.Kh. (1967). "Nucleic acid content in sensory and motor neurones of the spinal cord and their glial cell satellites during various functional states of the nervous system," Tsitologiya, 9, 840-847.

Pevzner, L.Z., Koval', V.A., and Kuchin, A.A. (1964). "Cytospectro-
    photometric and interferometric studies on cells of the sympa-
    thetic ganglion during rest and excitation," Tsitologiya, 6,
    216-219.

Pevzner, L.Z., and Litinskaya, L.L. (1968). "Quantitative analysis
    of some morphological changes in neuroglia under the influence
    of continued injection of adrenaline," Tsitologiya, 10, 812-816.

Pevzner, L.Z., Litinskaya, L.L., and Khrust, Yu.R. (1973). "The
    circadian rhythm of RNA content in neurones and neuroglia of
    rat cerebellum shown by scanning integrating microspectro-
    photometry," Fiziol. Zh. 59, 883-889.

Pevzner, L.Z., Linitskaya, L.L., and Khrust, Yu.R. (1974).
    "Correlation of the circadian rhythms of RNA and protein in
    spinal motoneurones and their glial cell-satellites in the
    rat," Dokl. Akad. Nauk SSSR, 214, 1466-1468.

Pevzner, L.Z., Nozdrachev, A.D., Glushchenko, T.S., and Fedorova,
    L.D. (1973). "RNA content in the neurone-neuroglia unit of
    the sympathetic ganglion and the nature of synaptic trans-
    mission," Dokl. Akad. Nauk SSSR, 216, 1458-1460.

Pevzner, L.Z., and Saudargene, E.-D. (1971). "Two-wavelength visible
    cytospectrophotometry of nucleic acids and proteins in the
    motor and sensory neurons and their glial cell-satellites of
    rat spinal cord during corazol seizures," Acta histochem., 39,
    101-117.

Pevzner, L.Z., and Semeshina, T.M. (1976). "Topochemical and cyto-
    chemical studies on RNA content in the central nervous system
    of the active and hibernating ground squirrel, Citellus ery-
    throgenys," Brain Res., 108, 205-211.

Pevzner, L.Z., and Tomina, E.D. (1965). "Biochemical and cyto-
    chemical characteristics of brain tumours," Vopr. Med. Khim.,
    11, 3-17.

Pevzner, L.Z., Tomina, E.D., and Chaika, T.V. (1964). "Cytospectro-
    photometric investigation of DNA content in tumour cells from
    human brain," Vopr. Med. Khim., 10, 379-386.

Pfeiffer, S.E., Horschman, H.R., Lightbody, J., and Sato, G.
    "Synthesis by a clonal line of rat glial cells of a protein
    unique to the nervous system," J. Cell. Physiol., 75, 329-
    339.

Pigareva, Z.D. (1966). "Characteristics of energy metabolism in
    neurones and glia (Review)", Zh. Nevropatol. i Psikhiatrii, 66

1716-1721.

Piven, N.V. (1973). "Changes in cytoplasmic volume of neurones of various regions of the CNS during hypothermia and recovery," Tsitologiya, 15, 828-832.

Poduslo, S.E., and Norton, W.T. (1972). "Isolation and some chemical properties of oligodendroglia from calf brain," J. Neurochem., 19, 727-736.

Pogodayev, K.I. (1966). "Biochemical basis of fatigue and inhibition in the central nervous system," in: Problems in Neurochemistry (in Russian), Moscow/Leningrad, pp. 34-39.

Pogodayev, K.I. (1970). "Energy metabolism of the brain during overload of the nervous system," Fifth All-Union Conference on Neurochemistry (in Russian), Tbilisi, pp. 507-514.

Pogodayev, K.I., Turova, N.F., and Baryshnikov, V.A. (1968). "Oxidative phosphorylation in various brain regions of the rabbit during swimming," Ukr. Biokhim. Zh., 40, 496-499.

Pohle, W., and Matthies, H. (1974). "Incorporation of RNA-precursors into neuronal and glial cells of rat brain during a learning experiment," Brain Res., 65, 231-237.

Polak, M. (1965). "Morphological and functional characteristics of the central and peripheral neuroglia (light microscopical observations)," Progr. Brain Res., 15, 12-34.

Polenov, A.L. (1968). Hypothalamic Neurosecretion (in Russian), Leningrad.

Pollister, A.W., Himes, M., and Ornstein, L. (1951). "Localisation of substances in cells," Feder. Proc., 10, 629-639.

Pollister, A.W., and Ris, H. (1947). "Nucleoprotein determination in cytological preparations," Cold Spring Harbor Symp. on Quant. Biol., 12, 147-157.

Pomazanskaya, L.F. (1970). "Methods for obtaining fractions of neuronal and glial cells from brain tissue," Zh. Evolyuts. Biokhim. i Fiziol., 6, 477-484.

Pomazanskaya, L.F. (1974). "A method of separating cell fractions enriched in neuronal and glial cells, and biochemical data obtained from their use," in: Achievements in Neurochemistry (in Russian), Nauka, Leningrad.

Pomazanskaya, L.F., Fraiz, L., and Mandel, P. (1969). "Phospholipid fatty acids, isolated separately from neuronal and glial cells of rat brain gray matter," Zh. Evolyuts. Biokhim. i Fiziol. 5, 523-528.

Pomerat, C.M. (1952). "Dynamic neurogliology," Texas Rep. on Biol. Med., 10, 885-913.

Pope, A., and Hess, A. (1957). "Cytochemistry of neurones and neuroglia," in: Metabolism of the Nervous System, London, pp. 72-86.

Pope, A., Hess, H.H., and Allen, J.N. (1957). "Quantitative histochemistry of proteolytic and oxydative enzymes in human cerebral cortex and brain tumors," Progr. Neurobiol., 2, 182-194.

Porcellati, G. (1974). "Some aspects of phospholipid turnover in nervous membranes," in: Comparative Biochemistry and Physiology of Transport, L. Bolis et al (Eds.) North Holland Publishing Company, Amsterdam, pp. 281-287.

Portugalov, V.V., Il'ina-Kakueva, E.I., Artyukhina, T.V., Gotlib, V.Ya., and Starostin, V.I. (1968). "Changes in endocrine glands and in neurosecretory nuclei of the hypothalamus resulting from hypokinesis," in: Experimental Investigations of Hypokinesis, Changes in Gaseous Environment, Acceleration, Overwook and other Factors (in Russian), Moscow, pp. 29-33.

Privat, A. (1975). "Postnatal gliogenesis in the mammalian brain," Int. Rev. Cytol., 40, 281-323.

Promyslov, M.Sh. (1963). "Mucolipids and brain tumours," Third All-Union Conference on the Biochemistry of the Nervous System (in Russian), Yerevan, pp. 381-384.

Promyslov, M.Sh. (1966). "Neuroglial lipids," in: Problems in Neurochemistry (in Russian), Moscow/Leningrad.

Promyslov, M.Sh. (1967). "Glutamic acid and glutamine in glial tumours," Ukr. Biokhim. Zh., 39, 590-592.

Promyslov, M.Sh., Levchenko, L.I., and Lazareva, N.G. (1966). "Effect of aurantin on nucleic acid synthesis in mouse brain tumours," Vopr. Neirokhir., 30, 53-56.

Promyslov, M.Sh., Solov'eva, T.V., and Aniskina, R.I. (1968). "Features of γ-aminobutyric acid metabolism in brain tumours," Vopr. Med. Khim., 14, 619-622.

Promyslov, M.Sh., Solov'eva, T.V., and Sokovina, Ya.M. (1970).

"Properties of nitrogen metabolism in the neuroglia," Fifth
All-Union Conference on the Biochemistry of the Nervous System
(in Russian), Tbilisi, pp. 225-231.

Przelecka, A. (1966a). "Nucleic acid metabolism and cell interaction in the ovariole of Galeria mellonella," Folia histochem.
cytochem., 4, 223-235.

Przelecka, A. (1966b). "Incorporation of ^{14}C-sodium palmitate into
lipids and cell interaction in ovarioles of Galleria mollonella
(Lepidoptera)," Ann. Histochem., 11, 403-411.

Pyl'dvere, K.I. (1971). "Some key questions of histogenesis of
neuroglia in tissue culture," Arkh. Anat. Gistol. i Embriol.,
60, 27-38.

Pysh, J.J., and Khan, T. (1972). "Variations in mitochondrial
structure and content of neurons and neuroglia in rat brain:
an electron microscopic study," Brain Res., 36, 1-18.

Quastel, J.H., and Quastel, D.M.J. (1961). The Chemistry of Brain
Metabolism in Health and Disease, Springfield.

Radin, N.S., Brenkert, A., Arora, R.C., Sellinger, O.Z., and Flangas,
A.L. (1972). "Glial and neuronal localization of cerebroside-
metabolizing enzymes," Brain Res., 39, 163-169.

Raghavan, S.S., Rhoads, D.B., and Kanfer, J.N. (1972). "Acid
hydrolases in neuronal and glial enriched fractions of rat
brain," Biochim. Biophys. Acta, 268, 755-760.

Raigorodskaya, T.G., Agroskin, L.S., and Pevzner, L.Z. (1973).
"Cytospectrophotometric analysis of the correlation of arginine-
and lysine-rich proteins in neurone-neuroglia units with
different types of neuromediators," Tsitologiya, 15, 1390-1394.

Raine, C.S., Poduslo, S.E., and Norton, W.T. (1971). "The ultrastructure of purified preparations of neurons and glial cells,"
Brain Res., 27, 11-24.

Rapava, E.A., Kuz'mina, S.N., and Zbarskii, I.B. (1973). "Oxidative phosphorylation and ATPase activity of isolated nuclei
and nuclear envelopes of neurones and glia of rat brain,"
Biokhimiya, 38, 298-303.

Rechardt, L. (1969). "Electron microscopic and histochemical
observations on the supraoptic nucleus of normal and dehydrated
rats," Acta physiol. scand., suppl., 329.

Reinis, S. (1975). "Incorporation of [^{3}H] thymidine into brain

DNA after cerebellar damage," Pediat. Res., 9, 807-811.

Reznikov, K.Yu. (1968). "Incorporation of 3H-thymidine into brain
    cells of adult mice under normal conditions, with brain damage,
    and administration of RNA," Dokl. Akad. Nauk SSSR, 181, 467-
    469.

Reznikov, K.Yu. (1974). "On the ability of several types of nerve
    cells of adult mouse brain to incorporate 3H-thymidine after
    trauma," Arkh. Anat. Gistol. i Embriol., 67, 92-97.

Richardson, K., and Rose, S.P.R. (1971). "A diurnal rhythmicity
    in incorporation of lysine into rat brain regions," Nature
    New Biol., 233, 182-184.

Robert, J., Freysz, L., Sensenbrenner, M., Mandel, P., and Rebel, G.
    (1975). "Gangliosides of glial cells: a comparative study of
    normal astroblasts in tissue culture and glial cells isolated
    on sucrose-ficoll gradients," FEBS Lett., 50, 144-146.

Roberti, R., Binaglia, L., Francescangeli, E., Goracci, G., and
    Porcellati, G. (1975). "Enzymic synthesis of 1-alkyl-2-acyl-
    sn-glycero-3-phosphorylethanolamine through ethanolaminephos-
    photransferase activity in the neuronal and glial cells of
    rabbit in vitro," Lipids, 10, 121-127.

Roberts, E. (1962). "γ-Aminobutyric acid and neuronal function,"
    Res. Publ. Assoc. Nerv. Ment. Dis., 15, 288-299.

Roberts, P.J., and Keen, P. (1974). "^{14}C glutamate uptake and
    compartmentation in glia of rat dorsal sensory ganglion,"
    J. Neurochem., 23, 201-209.

Robin, E.D. (1965). "Some aspects of fluid and electrolyte meta-
    bolism in the brain," Anesthesiology, 26, p. 1: 791-804.

Robins, E., Smith, D.E., and Jen, M.K. (1957). "The quantitative
    distribution of eight enzymes of glucose metabolism and two
    citric acid cycle enzymes in the cerebellar cortex and its
    subjacent white matter," Progr. Neurobiol., 2, 205-214.

Roitbak, A.I. (1963). "On the nature of cortical inhibition,"
    Zh. Vyssh. Nerv. Deyat., 13, 859-869.

Roitbak, A.I. (1964). "Physiological effects of gamma-aminobutyric
    acid," in: The Role of Gamma-aminobutyric Acid in Nervous
    System Function (in Russian), Leningrad, pp. 66-79.

Roitbak, A.I. (1965). "Slow negative cortical potentials and the
    neuroglia," in: Current Problems in Physiology and Pathology

of the Nervous System (in Russian), Moscow, pp. 68-93.

Roitbak, A.I. (1968). "Current data and suggestions on nervous system function," in: Normal and Pathological Function of the Nervous System (in Russian), Tbilisi, pp. 79-96.

Roitbak, A.I. (1969). "A new hypothesis on the mode of formation of temporal connections," Neirofiziologiya, 1, 130-136.

Roitbak, A.I. (1970). "The negative component of the cortical evoked potential," Neirofiziologiya, 2, 339-348.

Roitbak, A.I. (1973). "Neuroglia and the formation of new nerve connections in cerebral cortex," in: Methods of Formation and Inhibition of Conditional Reflexes (in Russian), Nauka, Moscow, pp. 82-94.

Roitbak, A.I., and Fanardzhyan, V.V. (1973). "Intracellular potentials of cortical glial cells during electrical stimulation of the cortex," Dokl. Akad. Nauk SSSR, 211, 748-751.

Roots, B.I., and Johnston, P.V. (1964). "Neurons of ox brain nuclei: their isolation and appearance by light and electron microscopy," J. Ultrastruct. Res., 10, 350-361.

Roots, B.I., and Johnston, P.V. (1965). "Lipids of isolated neurons," Biochem. J., 94, 61-63.

Rose, S.P.R. (1965). "Preparation of enriched fractions from cerebral cortex containing isolated, metabolically active neuronal cells," Nature, 206, 621-622.

Rose, S.P.R. (1967). "Preparation of enriched fractions from cerebral cortex containing isolated, metabolically active neuronal and glial cells," Biochem. J., 102, 33-43.

Rose, S.P.R. (1968a). "The biochemistry of neurones and glia," in: Applied Neurochemistry, Oxford/Edinburgh, pp. 332-355.

Rose, S.P.R. (1968b). "Glucose and amino acid metabolism in isolated neuronal and glial cell fractions in vitro," J. Neurochem., 15, 1415-1429.

Rose, S.P.R. (1969). "Neurones and glia: separation techniques and biochemical interrelationships," in: Handbook of Neurochemistry, Vol. II, New York, pp. 183-193.

Rose, S.P.R. (1970). "The compartmentation of glutamate and its metabolites in fractions of neurone cell bodies and neuropil; studied by intraventricular injection of [U-^{14}C] glutamate,"

J. Neurochem., <u>17</u>, 809–816.

Rose, S.P.R. (1973). "Cellular compartmentation of metabolism in the brain," in: Metabolic Compartmentation in the Brain, R. Balázs and J.E. Cremer (Eds.) MacMillan Press, London, pp. 287–304.

Rose, S.P.R., and Cory, H.T. (1970). "Glutamate metabolism in octopus brain <u>in vivo</u>; absence of a Waelsch effect," J. Neurochem., <u>17</u>, 817–820.

Rose, S.P.R., and Sinha, A.K. (1969). "Some properties of isolated neuronal cell fractions," J. Neurochem., <u>16</u>, 1319–1328.

Rose, S.P.R., and Sinha, A.K. (1974). "Incorporation of amino acids into proteins in neuronal and neuropil fractions of rat cerebral cortex: presence of a rapidly labelling neuronal fraction," J. Neurochem., <u>23</u>, 1065–1076.

Rose, S.P.R., Sinha, A.K., and Broomhead, S. (1973). "Precursor incorporation into cortical protein during first exposure of rats to light; cellular localization of effects," J. Neurochem., <u>21</u>, 539–546.

Rosenberg, P. (1970). "Function of phospholipids in axons: depletion of membrane phosphorus by treatment with phospholipase C," C. Toxicon., <u>8</u>, 235–243.

Rosenbluth, J., and Wissing, S.L. (1964). "The distribution of exogeneous ferritin in toad spinal ganglia and the mechanism of its uptake by neurons," J. Cell Biol., <u>23</u>, 307–326.

Rosenzweig, M.R. (1970). "Evidence for anatomical and chemical changes in the brain during primary learning," in: Biology of Memory, New York/London, pp. 69–85.

Rosenzweig, M.R., Bennett, E.L., and Diamond, M.C. (1972). "Chemical and anatomical plasticity of brain: replication and extensions," in: Macromolecules and Behavior, 2nd Edition, New York, pp. 205–277.

Rubinskaya, N.L. (1971). "RNA content in spinal motoneurones, cerebellar Purkyně cells and their glial cell-satellites during picrotoxin convulsions," Vopr. Med. Khim., <u>17</u>, 306–311.

Rubinskaya, N.L. (1973). "RNA in neurones and their glial cell-satellites of red nucleus of rat brain during natural sleep, deprivation of paradoxical sleep, and phenamine insomnia," Tsitologiya, <u>15</u>, 1471–1475.

Ruščák, M., Hager, H., and Orlicky, J. (1976). "Alanine formation and alanine aminotransferase activity in the nerve tissue with proliferating macroglia," Acta neuropathol., 34, 149-155.

Ruščák, M., and Ruščáková, D. (1971). Metabolism of the Nerve Tissue in Relation to Ion Movements in vitro and in situ, Publ. House of the Slovak. Acad. of Sciences and University Park Press, Baltimore and Bratislava.

Ruščák, M., Ruščáková, D., and Hager, H. (1968). "The role of the neuronal cell in the metabolism of the rat cerebral cortex," Physiol. bohemosl., 17, 113-121.

Ruščáková, D. (1969). "Über den Einfluss hypertonischer Kalium-lösungen auf die Astroglia der Grosshirnrinde von Ratten," Verh. Anat. Ges., 63, 423-428.

Rutter, W.J., Blostein, R., Woodfin, B.M., and Weber, G.S. (1963). "Enzyme variants and metabolic diversification," Adv. Enzyme Regul., 1, 39-56.

Rutter, W.J., Rajkumar, T., Penhoet, E., and Kochman, M. (1968). "Aldolase variants: structure and physiological significance," Ann. New York Acad. Sci., 151, 102-117.

Salem, R.D., Hammerschlag, R., Bracho, H., and Orkand, R.K. (1975). "Influence of potassium ions on accumulation and metabolism of [^{14}C] glucose by glial cells," Brain Res., 86, 499-503.

Salway, J.G., Kai, M., and Hawthorne, J.N. (1967). "Triphospho-inositide phosphomonoesterase activity in nerve cell bodies, neuroglia and subcellular fractions from whole rat brain," J. Neurochem., 14, 1013-1024.

Sandritter, W. (1966). "Methods and results of quantitative cyto-chemistry," in: Introduction to Quantitative Cytochemistry, G.L. Weid (Ed.), Academic Press, New York/London, pp. 159-182.

Sandritter, W., Diefenbach, H., and Krantz, F. (1954). "Über die quantitative Bindung von Ribonukleinsäure mit Gallocyaninchrom-alaun," Experientia, 10, 210-215.

Sandritter, W., Kiefer, G., and Rick, W. (1963). "Über die Stochio-metrie von Gallocyaninchromalaun mit Desoxyribonukleisäure," Histochemie, 3, 315-340.

Sandritter, W., Kiefer, G., and Rick, W. (1966). "Gallocyanin chrome alum," in: Introduction to Quantitative Cytochemistry, G.L. Weid (Ed.), Academic Press, New York/London, pp. 295-326.

Sandritter, W., Kifer, G., and Rik, V. (1969). "Staining with chrome-gallocyanin," in: Introduction to Quantitative Cyto-chemistry (in Russian), Moscow, pp. 240-262.

Sandritter, W., and Krygier, A. (1959). "Cytophotometrische Bestimmung von proteingebundenen Thiolen in der Mitose und Interphase von HeLaZellen," Z. Krebsforsch., 62, 596-610.

Sarkisov, S.A., and Bogolepov, N.N. (1967). Electron Microscopy of the Brain (in Russian), Moscow.

Satake, M., and Abe, S. (1966). "Preparation and characterization of nerve cell perikaryon from rat cerebral cortex," J. Biochem., 59, 72-75.

Saudargene, D.S. (1969a). "Dynamic changes in levels of RNA in neurones and neuroglia of spinal cord during corazole convul-sions and subsequent recovery," Tsitologiya, 11, 642-646.

Saudargene, D.S. (1969b). "Changes in the content of several types of protein in neurones and neuroglia of spinal cord during corazole convulsions and subsequent recovery," Tsitologiya, 11, 1034-1038.

Saudargene, D.S., and Pevzner, L.S. "Cytophotometric analysis of nucleic acid and protein metabolism in motor and sensory neurones of spinal cord and their neuroglia during corazole convulsions," Tsitologiya, 11 , 1275-1285.

Schachner, M. (1974). "NS-1 (nervous system antigen-1), a glial cell-specific antigenic component of the surface membrane," Proc. Nat. Acad. Sci. USA, 5, 1795-1799.

Schachner, M., and Carnow, T.B. (1975). "Nervous system antigen-2 (NS-2), an antigenic cell surface somponent expressed on a murine glioblastoma," Brain Res., 88, 394-402.

Schmidt, G., and Thannhauser, S.J. (1945). "A method for the deter-mination of desoxyribonucleic acid, ribonucleic acid, and phosphoproteins in animal tissues," J. Biol. Chem., 161, 83-89.

Schneider, W.C. (1945). "Phosphorus compounds in animal tissues. I. Extraction and estimation of DNA and PNA," J. Biol. Chem., 161, 293-303.

Schon, F., and Kelly, J.S. (1974a). "Autoradiographic localisation of [3H] GABA and [3H] glutamate over satellite glial cells," Brain Res., 66, 275-288.

Schon, F., and Kelly, J.S. (1974b). "The characterisation of [3H] GABA uptake into the satellite glial cells of rat sensory ganglia," Brain Res., <u>66</u>, 289-300.

Schon, F., and Kelly, J.S. (1975). "Selective uptake of [^{3}H] β-alanine by glia: association with the glial uptake system for GABA," Brain Res., <u>86</u>, 243-257.

Schousboe, A., Booher, J., and Hertz, L. (1970). "Content of ATP in cultivated neurons and astrocytes exposed to balanced and potassium-rich media," J. Neurochem., <u>17</u>, 1501-1504.

Schultz, R.L., Maynard, E.A., and Pease, D.C. (1957). "Electron microscopy of neurons and neuroglia of cerebral cortex and corpus callosum," Amer. J. Anat., <u>100</u>, 369-407.

Schubert, D. (1973). "Protein secretion by clonal glial and neuronal cell lines," Brain Res., <u>56</u>, 387-391.

Schwartz, A.M., Lapham, L.W., and van den Noort, S. (1966). "Cytologic and cytochemical studies on neuroglia. IV. Experimentally induced protoplasmic astrocytosis in the Bergmann glia of cerebellum," Neurology, <u>16</u>, 1118-1126.

Sellinger, O.Z., and Azcurra, J.M. (1974). "Bulk separation of neuronal cell bodies and glial cells in the absence of added digestive enzymes," in: Research Methods in Neurochemistry, Vol. 2, N. Marks and M. Rodnight (Eds.), Plenum Press, New York, pp. 3-38.

Sellinger, O.Z., Azcurra, J.M., Johnson, D.E., Ohlsson, W.G., and Lodin, Z. (1971). "Independence of protein synthesis and drug uptake in nerve cell bodies and glial cells isolated by a new technique," Nature, <u>230</u>, 253-256.

Sellinger, O.Z., Johnson, D.E., Santiago, J.C., and Idoyaga-Vargas, V. (1973). "A study of the biochemical differentiation of neurons and glia in the rat cerebral cortex," Progr. Brain Res., <u>40</u>, 331-347.

Sellström, A., and Hamberger, A. (1975). "Neuronal and glial systems for γ-aminobutyric acid transport," J. Neurochem., <u>24</u>, 847-852.

Sellström, A., Sjöberg, L.-B., and Hamberger, A. (1975). "Neuronal and glial systems for γ-aminobutyric acid metabolism," J. Neurochem., <u>25</u>, 393-398.

Sellström, A., Sjöberg, L.-B., and Hamberger, A. (1975). "Neuronal
    and glial systems for γ-aminobutyric acid metabolism," J.
    Neurochem., 25, 393-398.

Selye, H. (1952). The Story of the Adaptation Syndrome (told in
    the form of informal, illustrated lectures), Med. Publ.,
    Montreal.

Selye, H. (1957). The Stress of Life, Longmans Green, London/New
    York/Toronto.

Selye, G. (1960). Essays on the Adaptive Syndrome (in Russian),
    Moscow.

Shabadash, A.L. (1958). "Structural-chemical basis of the biolo-
    gical organisation of the nervous system," Arkh. Anat. Gistol.
    i Embriol., 41, 4-12.

Shabadash, A.L. (1963). "Cytochemical basis and demonstration of
    the biological reactivity of cells in the nervous system," Third
    All-Union Conference on the Biochemistry of the Nervous System
    (in Russian), Yerevan, pp. 283-295.

Shabadash, A.L. (1966). "Cytochemical demonstration of functional
    states in the nervous system," in: Problems in Neurochemistry
    (in Russian), Moscow/Leningrad, pp. 53-61.

Shalina, N.M., Nikol'skaya, I.I., and Tikhonenko, T.I. (1967).
    "Dependence of the magnitude of hyperchromism on the method
    of degradation of phage DNA," Biokhimiya, 32, 711-715.

Shanta, T.R., Woods, W.D., Waitzman, M.B., and Bourne, G.H. (1966).
    "Histochemical method for localization of cyclic 3', 5'-nucleo-
    tide phosphodiesterase," Histochemie, 7, 177-190.

Sharobaiko, V.I. (1958). "Effect of ionizing radiation on RNA
    metabolism in nerve tissue," Tr. Voyenno-med. Akad., 98, 103-1
    118.

Shelepin, Yu.E. (1970). "Mathematical interpretation of neurone-
    glia interactions," Dokl. Akad. Nauk SSSR, 192, 698-701.

Shelikhov, V.N., Dergachev, V.V., Poletaev, A.B., and Naumova, T.S.
    (1975). "On the possible role of the neuroglia in nervous sys-
    tem function," Uspekhi Fiziol. Nauk, 6, 90-109.

Sherudilo, A.I. (1968). "Quantitative determination of optically
    dense substances in significantly heterogenous objects,"
    Biofizika, 13, 741-744.

Shivers, R.R. (1976). "Trans-glial channel-facilitated transloca-
tion of tracer protein across ventral nerve root sheaths of
crayfish, Brain Res., 108, 47-58.

Shtark, M.B. (1970). The Brain of the Hibernating Animal (in
Russian), Nauka, Siberian Section, Novosibirsk.

Shtark, M.B. (1972). The Brain of Hibernating Animals, NASA,
Springfield/Washington.

Silver, A. (1967). "Cholinesterases of the central nervous system
with special reference to the cerebellum," Intern. Rev. Neurobiol., 10, 57-109.

Sinha, A.K., and Rose, S.P.R. (1971). "Bulk separation of neurones
and glia: a comparison of techniques," Brain Res., 33, 205-217.

Sinha, A.K., and Rose, S.P.R. (1972a). "Monoamineoxidase and
cholinesterase activity in neurons and neuropil from the rat
cerebral cortex," J. Neurochem., 19, 1607-1610.

Sinha, A.K., and Rose, S.P.R. (1972b). "Compartmentation of
lysosomes in neurones and neuropil and a new neuronal marker,"
Brain Res., 39, 181-196.

Sinha, A.K., and Rose, S.P.R. (1973). "β-N-acetyl D-galactos-
aminidase in bulk separated neurons and neuropil from rat
cerebral cortex," J. Neurochem., 20, 39-44.

Sirlin, J.L., and Jacob, J. (1960). "Cell function in the ovary of
Drosophila. II. Behaviour of RNA," Exp. Cell Res., 20, 283-
293.

Sjöstrand, J. (1965). "Proliferative changes in glial cells during
nerve regeneration," Z. Zellforsch., 68, 481-493.

Skul'skii, I.A., Leont'ev, V.G., and Burovina, I.V. (1968). "The
content, distribution, and chemical composition of sodium,
potassium, rubidium, and caesium in various tissues of white
rat," Izv. Akad. Nauk SSSR, Ser. Biol., 33, 830-836.

Slagel, D.E., Hartmann, H.A., and Edström, J.E. (1966). "The
effect of iminodipropionitril on the ribonucleic acid content
and composition of mesencephalic V cells, anterior horn cells,
glial cells and axonal balloons," J. Neuropathol. Exp. Neurol.,
25, 244-253.

Slonim, A.D., and Shvetsova, E.I. (1973). "Chemical thermoregulation after 'accelerated' cold-adaptation," Fiziol. Zh. SSSR,
59, 1262-1267.

Smart, J. (1961). "The subependymal layer of the mouse brain and its cell production as shown by radioautography after thymidine-H³ injection," J. Comp. Neurol., 116, 325-347.

Smart, J., and LeBlond, C.P. (1961). "Evidence for division and transformation of neuroglia cells in the mouse brain, as derived from radioautography after injection of thymidine-H³," J. Comp. Neurol., 116, 349-367.

Snesarev, P.E. (1959). "Neuroglia," in: Multivolume Handbook on Neurology (in Russian), Vol. 1, Moscow, pp. 222-267.

Soetens, A. (1968). "Le metabolisme de l'eau et des electrolytes en neurologie et en neurochirurgie," Canad. Anaesth. Soc. J., 15, 1-14.

Soga, K., and Takahashi, Y. (1976). "Transcription of repeated and unique DNA sequences in brain nuclei," J. Neurochem., 26, 89-94.

Sokolov, E.N. (1962). "Nature of the background rhythms of cerebral cortex," in: Basic Questions in the Electrophysiology of the Central Nervous System (in Russian), Kiev, pp. 157-188.

Somjen, G.G. (1975). "Electrophysiology of neuroglia," Ann. Rev. Physiol., 37, 163-190.

Spirin, A.S., Gavrilova, L.P., and Belozerskii, A.N. (1959). "Methods of quantitative assessment of the 'hyperchromic effect' of nucleic acids," Biokhimiya, 24, 600-611.

Squire, L.R., and Barondes, S.H. (1972). "Inhibitors of cerebral protein or RNA synthesis and memory," in: Macromolecules and Behaviour, 2nd Edition, New York, pp. 61-82.

Stambolova, M.A., Cox, D., and Mathias, A.P. (1973). "The activity of deoxyribonucleic acid polymerase and deoxyribonucleic acid synthesis in nuclei from brain fractionated by zonal centrifugation," Biochem. J., 136, 685-695.

Steinman, R.M., Brodie, S.E., and Cohn, Z.A. (1976). "Membrane flow during pinocytosis. A stereologic analysis," J. Cell Biol., 68, 665-687.

Streicher, R., Ferris, P.J., Prokop, J.D., and Klatzo, I. (1964). "Brain volume and thiocyanate space in local cold injury," Arch. Neurol., 11, 444-448.

Sulkin, N.M., and Kuntz, A. (1950). "Histochemical determination of ribose nucleic acid in vertebrate tissues following extraction with perchloric acid," Proc. Soc. Exp. Biol. Med., 73,

413-415.

Suzuki, O., and Kato, T. (1973). "DNA polymerase activities in
    isolated neuronal, glial and liver nuclei," in: Fourth Intern.
    Meet. of the Intern. Soc. for Neurochem. Abstracts, Tokyo,
    p. 196.

Svaetichin, G., Negishi, K., Fatehchand, R., Drujan, B., and Selvin
    de Testa, A. (1965). "Nervous function based on interaction
    between neuronal and non-neuronal elements," Progr. Brain Res.,
    15, 243-266.

Svanidze, I.K. (1967). "Quantitative cytophotometry of DNA in
    nuclei of neurones and glial cells of rat cerebral cortex in
    various stages of development," Zh. Obshch. Biol., 28, 697-707.

Svanidze, I.K., and Berishvili, V.G. (1970). "Quantitative changes
    in nuclear DNA content in neurones and glia in onto-philogenesis'',
    Fifth All-Union Conference on Neurochemistry (in Russian), Tbi
    Tbilisi, pp. 147-158.

Svanidze, I.K., Didimova, Ye.V., Museridze, D.P., Bregvadze, I.A.,
    and Sheresheva, N.B. (1972). "Effect of functional activity
    on the distribution of cytoplasmic RNA in pyramidal neurones
    and neuroglia of motor cortex of brain," Vopr. Biokhim. Nervn.
    i Mysh. Sistem., Tbilisi, 2, 483-488.

Svanidze, I.K., and Museridze, D.P. (1974). Quantitative Analysis
    of Nucleic Acids in Pyramidal Cells of Cerebral Cortex,(in
    Russian), Metsniereva, Tbilisi.

Svanidze, I.K., Roitbak, A.I., and Didimova, E.V. (1973). "Effect
    of potassium ions on motility of cortical glial cells in
    tissue culture," Dokl. Akad. Nauk SSSR, 211, 1450-1452.

Sviridov, S.M., Korochkin, L.I., Ivanov, V.N., Maletskaya, E.I., and
    Bakhtina, T.K. (1972). "Immunohistochemical studies of S-100
    protein during postnatal ontogenesis of the brain of two strains
    of rats," J. Neurochem., 19, 713-718.

Svoboda, P., and Lodin, Z. (1972). "Postnatal development of some
    mitochondrial enzyme activities of cortical neurons and glial
    cells," Physiol. bohemosl., 21, 457-465.

Svorad, D., and Novakova, V. (1960). "Effect of experimentally-
    induced insomnia on the neurotic state in rats," Fiziol. Zh.
    SSSR, 46, 57-63.

Sweat, M.L., (1954). "Silica gel microcolumn for chromatographic
    resolution of cortical steroids," Analyt. Chem., 26, 1964-1967.

Sytinsky, I.A. (1969). "The gamma-aminobutyric acid (GABA) system of the cerebral white matter and of brain tumors," Neuropatol. polska., 7, 371-375.

Sytinskii, I.A., Chaika, T.V., and Bernshtam, V.A. (1968). "Υ-Amino-butyric acid and glutamatedecarboxylase in human brain tumors," Vopr. Med. Khim., 14, 434-436.

Szijan, I., and Burdman, J.A. (1974). "DNA polymerase activity in fractions from brain cell nuclei," Brain Res., 73, 563-567.

Szydlowska, H., and Kaluza, J. (1972). "A comparative histochemical study on the functional groups in proteins and some oxidizing-reducing enzymes in reactive glia and glial tumours. II. Reactive glia," Neuopatol. polska., 10, 285-293.

Takahashi, Y., Hsü, C.S., and Honma, S. (1970). "Potassium and glutamate effect on protein synthesis in isolated neuroglial cells," Brain Res., 23, 284-287.

Talwar, G.P., Sadasivudu, B., and Chitre, V.S. (1961). "Changes in the pentose-nucleic acid content of sub-cellular fractions of the brain of the rat during 'Metrasol' convulsions," Nature, 191, 1007-1008.

Tasaki, I. (1965). "Excitability of neurons and glial cells," Progr. Brain Res., 15, 234-242.

Tencheva, Ts.S., and Pevzner, L.Z. (1973). "RNA and protein content in cortical neurones and their glial cell-satellites under the influence of mescalin," Tsitologiya, 15, 783-788.

Tencheva, Ts.S., and Pevzner, L.Z. (1974). "Effect of electro-convulsive shock on RNA and protein content of neurones and neuroglia of hippocampus in rats of various ages," Fiziol. Zh. SSSR, 60, 37-41.

ter Haar, M.B., and MacKinnon, P.C.B. (1972). "An investigation of cerebral protein synthesis in various states on neuroendo-crine activity," Progr. Brain Res., 38, 211-223.

Terner, J.Y., and Clark, G. (1960). "Gallocyanin-chrome alum. II. Histochemistry and specificity," Stain Technol., 35, 305-311.

Thompson, R.J. (1973). "Studies on RNA synthesis in two populations of nuclei from the mammalian cerebral cortex," J. Neurochem., 21, 19-40.

Tiplady, B., and Rose, S.P.R. (1971). "Amino acid incorporation into protein in neuronal cell body and neuropil fractions in vitro," J. Neurochem., 18, 549-558.

Tiplady, B., Glushchenko, T.S., and Pevzner, L.Z. (1974). "Effect of forced motor activity on RNA content of neurones and glial cells of brain and spinal cord," Dokl. Akad. Nauk SSSR, 214, 973-976.

Tobias, J.M. (1960). "Further studies on the nature of the excitable system in nerve," J. Gen. Physiol., 43, 57-71.

Tomina, E.D. (1970). "The content of DNA in chick embryo skeletal muscle fibres differentiating in vitro," Arkh. Anat. Gistol. i Embriol., 59, 41-45.

Tomina, E.D., and Pevzner, L.Z. (1965). "Protein content in nuclei of human brain tumour cells," Byull. Eksp. Biol. i Med., 59, 83-87.

Tonkikh, A.V. (1964). "The mechanism of trophic-adaptive influences of the sympathetic nervous system," in: Functional Evolution (in Russian), Moscow/Leningrad, pp. 135-141.

Torack, R., Gordon, J., and Prokop, J. (1970). "Pathobiology of acute triethyltin intoxication," Intern. Rev. Neurobiol., 12, 45-48.

Torack, R.M., Terry, R.D., and Zimmerman, H.M. (1959). "The fine structure of cerebral fluid accumulation. I. Swelling secondary to cold injury," Amer. J. Pathol., 35, 1135-1147.

Trachtenberg, M.C., and Pollen, D.A. (1970). "Neuroglia: biophysical properties and physiologic function," Science, 167, 1248-1252.

Treff, W.M. (1963). "Interferometrische Dickenbestimmung von Hirnschnitten," J. Hirnforsch., 6, 71-78.

Troshin, A.D. (1956). Problems of Cell Permeability (in Russian), Moscow/Leningrad.

Tsanev, R.G., and Markov, G.G. (1960). "Quantitative determination of nucleic acids," Biokhimiya, 25, 151-159.

Ule, G. (1968). "Ultrastruktur der Astroglia und des Status spongiosus," Acta Neuropathol., suppl., 4, 98-105.

Ungar, G. (Ed.) (1970). Molecular Mechanisms in Memory and
    Learning, Plenum Press, New York/London.

Ungar, G., and Romano, D.V. (1958). "Sulfhydryl groups in resting
    and stimulated rat brain; their relationship with protein
    structure," Proc. Soc. Exp. Bio. Med., 97, 324-326.

Utley, J.D. (1963). "Gamma animobutyric acid and 5-hydroxy-
    tryptamine concentrations in neurons and glial cells in the
    medial geniculate body of the cat," Biochem. Pharmacol., 12,
    1228-1230.

Utley, J.D. (1964). "Glutamine synthetase, glutamotransferase, and
    glutaminase in neurons and non-neural tissue in the medial
    geniculate body of the cat," Biochem. Pharmacol., 13, 1383-
    1392.

Uyemura, K., Vincendon, G., Gombos, G., and Mandel, P. (1971).
    "Purification and some properties of S-100 protein fractions
    from sheep and pig brains," J. Neurochem., 18, 429-438.

Uzorin, E.K. (1972). "Some features of neuronal RNA metabolism.
    Nuclear-cytoplasmic and neuronal-glial relationships in the
    nervous system of Limnaea stagnalis," Dokl. Akad. Nauk SSSR,
    202, 1212-1215.

Valenzuela y Chacón, J. (1969). "Étude de l'activité acétylcholin-
    estérasique dans les cellules gliales du cervelet. Cytoarchi-
    tectonie gliale," Acta histochem., 32, 376-399.

Van den Berg, C.J., Kržalić, Lj., Mela, P., and Waelsch, H. (1969).
    "Compartmentation of glutamate metabolism in brain. Evidence
    for the existence of two different tricarboxylic acid cycles in
    brain," Biochem.J., 113, 281-290.

Van der Vies, I., Bakker, R.F.M., and De Weid, D. (1960).
    "Correlated studies on plasma free corticosterone and on adrenal
    steroid formation rate in vitro," Acta endocrinol., 34, 513-
    523.

Van Liere, E.J., and Stickney, J.C. (1963). Hypoxia, Chicago.

Van Nieuw Amerongen, A. Roukema, P.A., and van Rossum, A.L. (1974).
    "Immunofluorescence study on the cellular localization of
    GP-350, a sialoglycoprotein from brain," Brain Res., 81, 1-19.

Varon, S. (1975). "Neurons and glia in neural cultures," Exper.
    Neurol., 48, 93-134.

Varon, S., Rainborn, C., and Burnham, P.A. (1974). "Implication of a nerve growth factor-like antigen in the support derived by ganglionic neurons from their homologous glia in dissociated cultures," Neurobiology, 4, 317-327.

Varon, S., and Saier, M. (1975). "Culture techniques and glial-neuronal interrelationships in vitro," Exper. Neurol., 48, Part 2, 135-162.

Vejlsted, H., and Pakkenberg, H. (1972). "The specificity of gallocyaninechromalum staining in nerve cells," Acta anatom., 81, 139-147.

Venkov, L., and Pevzner, L.Z. (1975). "Biochemical and cytospectro-photometric analysis of extraction of RNA from cell structures of spinal cord," Tsitologiya, 17, 858-861.

Venkov, L., Rosental, L., and Manolova, M. (1976). "Subcellular distribution of LDH isoenzymes in neuronal- and glial-enriched fractions," Brain Res., 109, 323-333.

Venkov, L., and Rusanov, E. (1976). "Differences in enzyme activities in mitochondria from neuronal and glial fractions," Ukr. Biokhim. Zh., 48, 215-222.

Veprintsev, B.N. (Ed.) (1976). Handbook of Tissue Culture of Nerve Tissue. Methods. Techniques. Problems (in Russian), Nauka, Moscow.

Vernadakis, A. (1975). "Neuronal-glial interactions during development and aging," Fed. Proc., 34, 89-95.

Vernadakis, A., and Gibson, D.A. (1973). "Chemical properties of neuronal and glial fractions isolated from chicks early post-hatching," Fourth Intern. Meet. of the Intern. Soc. for Neurochem. Abstracts, Tokyo, p. 205.

Vernadakis, A., and Woodbury, D.M. (1965). "Cellular and extra-cellular spaces in developing rat brain," Arch. Neurol., 12, 284-293.

Villegas, J., Villegas, L., and Villegas, R. (1966). "Sodium, potassium, and chloride concentrations in the Schwann cell and axon of the squid nerve fiber," J. Gen. Physiol., 49, Part 1, 1-8.

Vinnikov, Ya.A. (1966). "Structural and cytochemical bases for the functional mechanism of sensory organ receptors," in: The Nerve Cell (in Russian), Leningrad, pp. 7-100.

Vladimirov, G.E. (1953). "Functional biochemistry of the brain (some conclusions and perspectives)," Fiziol. Zh. SSSR, 39, 3-16.

Vladimirov, G.E., Baranov, M.N., Pevzner, L.Z., and Wang, T.-Y. (1961). "On differences in metabolism existing in some areas and layers of brain cortex," in: S.S. Kety and J. Elkes (Eds.), Regional Neurochemistry, Pergamon, Oxford, pp. 126-134.

Vladimirova, E.A. (1956). "A chamber for biochemical study of rat brain during conditional reflex food-drive and defensive reactions," Vopr. Med. Khim., 2, 229-233.

Volpe, P., and Guiditta, A. (1967). "Biosynthesis of RNA in neuron- and glia-enriched fractions," Brain Res., 6, 228-240.

Volterra, V. (1931). In: RN.N Chapman (Ed.), Animal Ecology, Academic Press, New York/London, pp. 409-448.

Voronka, G.Sh. (1971). "Influence of duration of phenamine insomnia and subsequent sleep on protein content in neurones and their glial cell-satellites," Fiziol. Zh. SSSR, 57, 962-968.

Varonka, G.Sh., Demin, N.N., Rubinskaya, N.L., and Solov'yeva, I.A. (1972). "RNA content in neurones and their glial cell-satel-lites of supraoptic nucleus of rat brain during natural sleep, deprivation of paradoxical sleep, and phenamine insomnia," Ukr. Biokhim. Zh., 44, 712-717.

Voronka, G.Sh., Demin, N.N., and Pevzner, L.Z. (1971). "Content of total and basic protein in neurones and neuroglia of supra-optic and red nuclei of rat brain during natural sleep and paradoxical sleep deprivation," Dokl. Akad. Nauk SSSR, 198, 974-977.

Voronka, G.Sh., and Pevzner, L.Z. (1972). "Influence of barbamyl narcosis on protein content of neurones and neuroglia of supra-optic and red nuclei of the brain," Vopr. Med. Khim., 18, 418-424.

Wachsmuth, E.D., Thoner, M., and Pfleiderer, G. (1975). "The cellular distribution of aldolase isoenzymes in rat kidney and brain determined in tissue sections by the immuno-histochemical method," Histochemistry, 45, 143-161.

Walsh, R.N., Budtz-Olsen, O.E., Penny, J.E., and Cummins, R.A. (1969). "The effects of environmental complexity on the histology of the rat hippocampus," J. Comp. Neurol., 137, 361-366.

Waelsch, H. (1959). "Some problems of metabolism in relation to the structure of the nervous system," Fourth Intern. Biochem. Congr., Pergamon, New York, Vol. III, pp. 36-45.

Waelsch, H., Berl, S., Rossi, C.A., Clarke, D.D., and Purpura, D.P. (1964). "Quantitative aspects of CO_2 fixation in mammalian brain in vivo," J. Neurochem., 11, 717-728.

Walker, B.E., and LeBlond, C.P. (1958). "Sites of nucleic acid synthesis in the mouse visualized by radioautography after administration of C^{14}-labelled adenine and thymidine," Exp. Cell Res., 14, 510-531.

Warecka, K., Moller, H.J., Voge, H.-M., and Tripatzis, I. (1972). "Human brain-specific alpha$_2$-glycoprotein: purification by affinity chromatography and detection of a new component; localization in nervous cells," J. Neurochem., 19, 719-725.

Watson, W.E. (1965a). "An autoradiographic study of the incorporation of nucleic acid precursors by neurones and glia during nerve stimulation," J. Physiol., 180, 754-765.

Watson, W.E. (1965b). "An autoradiographic study of the incorporation of nucleic acid precursors by neurones and glia during nerve regeneration," J. Physiol., 180, 741-753.

Watson, W.E. (1972). "A quantitative study of some neuroglial responses to neuronal stimulation," J. Physiol., 225, 54P-56P.

Watson, W.E. (1974). "Physiology of neuroglia," Physiol. Revs., 54, 245-271.

Wechsler, W., and Kleihues, P. (1968). "Protein metabolism and cytodifferentiation in the nervous system. An autoradiographic and electron microscopic study," in: Macromolecules and the Function of the Neuron, Amsterdam, pp. 73-90.

Weight, F.F., and Salmoiraghi, G.C. (1967). "Motoneurone depression by norephinephrine," Nature, 213, 1229-1230.

Weiss, P.A. (1969). "Neuronal dynamics and neuroplasmic (axonal) flow," in: Cellular Dynamics of the Neuron, New York/London, pp. 3-34.

Weiss, P.A. (1974). "Dynamics and mechanics of neuroplasmic flow," in: Dynamics of Degeneration and Growth in Neurons, K. Fuxe et al (Eds.), Pergamon Press, Oxford, pp. 203-213.

Wender, M., Kozik, M., Mularek, O., and Ożarzewska, E. (1974). "Incorporation of ^{3}H-thymidine into neuroglial cells in the course of myelinogenesis," Folia histochem. cytochem., 12, 115-123.

Weid, G.L. (Ed.) (1966). Introduction to Quantitative Cytochemistry, Academic Press, New York/London.

Woelk, H., Goracci, G., Gaiti, A., and Porcellati, G. (1973). "Phospholipase A1 and A2 activities of neuronal and glial cells of the rabbit brain," Z. Physiol. Chem., 354, 729-736.

Woelk, H., Kanig,K., and Peiler-Ichikawa, K. (1974). "Incorporation of ^{32}P into the phospholipids of neuronal and glial cell enriched fractions isolated from rabbit cerebral cortex: effect of norepinephrine," J. Neurochem., 23, 1057-1063.

Wollemann, M. (1974). Biochemistry of Brain Tumours, Académiai kiadó, Budapest.

Wollemann, M., and Dévényi, T. (1963). "The γ-aminobutyric acid content and glutamate decarboxylase activity of brain tumors," J. Neurochem., 10, 83-88.

Woodward, D.L., Reed, D.J., and Woodbury, D.M. (1967). "Extracellular space of rat cerebral cortex," Amer. J. Physiol., 212, 367-370.

Wroński, A., and Von den Decken, A. (1975). "Synthesis of brain-specific acidic proteins in rat and mouse cerebral slices," Acta physiol. scand., 95, 482-493.

Yakovlev, N.N. (1955). Essays in the Biochemistry of Sport, Leningrad.

Yakovlev, N.N. (1956). "Dynamics of labile phosphorus compounds in brain during muscular function of varying duration," Vopr. Med. Khim., 2, 140-149.

Yaklovev, N.N. (1963). "γ-Aminobutyric acid metabolism in cerebrum during muscular function of varying duration," Ukr. Biokhim. Zh., 35, 175-187.

Yanagihara, T. (1974a). "Cerebral anoxia: effect on transcription and translation," J. Neurochem., 22, 113-117.

Yanagihara, T. (1974b). "RNA metabolism in rabbit brain: study with neuron-glia and subcellular fractions," J. Neurochem., 23, 833-837.

Yurisova, M.N. (1970). "Changes in the hypothalamo-hypophyseal neurosecretory system during winter hibernation of the ground

squirrel *Citellus erythrogenus*," Zh. Evolytus. Biokhim. i.
Fiziol., 6, 516–522.

Zadunaisky, J.A., Wald, F., and De Robertis, E.D. (1965). "Osmotic
behaviour and glial changes in isolated frog brains," Progr.
Brain Res., 15, 196–218.

Zajicek, J., and Zeuthen, E. (1956). "Quantitative determination
of cholinesterase activity in individual cells," Exp. Cell Res.,
11, 568–579.

Zakhar'evskii, A.N., and Kuznetsova, A.F. (1961a). "Interference
microscopes in biology," Tsitologiya, 3, 213–224.

Zakhar'evskii, A.N., and Kuznetsova, A.F. (1961b). "Use of the
interference microscope in biology," Tsitologiya, 3, 245–253.

Zakhar'in, Yu.L. (1968). "Changes in glucose-6-phosphatedehydro-
genase and 6-phosphogluconatedehydrogenase activity in liver
and brain of rat under the influence of various physiological
factors," Vopr. Med. Khim., 14, 348–355.

Zaks, S.M. (1958). "Methods of measurement in thin biological
sections," Sb. Statei Len. Inst. Tochn. Mekhan. i Optiki, 27,
140–147.

Zanetta, J.P., Benda, P., Gombos, G., and Morgan, I.G. (1972).
"The presence of 2', 3'-cyclic AMP 3'-phosphohydrolase in glial
cells in tissue culture," J. Neurochem., 19, 881–883.

Zeuthen, E. (1961). "The Cartesian diver balance," Gen. Cytochem.
Methods, 2, 61–91.

Zomzely-Neurath, C., York, C., and Moore, B.W. (1973). "*In vitro*
synthesis of two brain-specific proteins (S-100 and 14-3-2)
by polyribosomes from rat brain. I. Site by synthesis and
programming by polysome-derived messenger RNA," Arch. Biochem.
Biophys., 155, 58–69.